A Little Less Arctic

Steven H. Ferguson · Lisa L. Loseto
Mark L. Mallory
Editors

A Little Less Arctic

Top Predators in the World's Largest Northern Inland Sea, Hudson Bay

Editors

Steven H. Ferguson
Fisheries and Oceans Canada
University of Manitoba
Freshwater Institute
501 University Crescent
Winnipeg, Manitoba
R3T 2N6
Canada
steve.ferguson@dfo-mpo.gc.ca

Lisa L. Loseto
Fisheries and Oceans Canada
Freshwater Institute
501 University Crescent
Winnipeg, Manitoba
R3T 2N6
Canada
lisa.loseto@dfo-mpo.gc.ca

Mark L. Mallory
Environment Canada
Canadian Wildlife Service
PO Box 1714
Iqaluit, Nunavut
X0A 0H0
Canada
mark.mallory@ec.gc.ca

This book is published as part of the International Polar Year 2007–2008, which is sponsored by the International Council for Sciences (ICSU) and the World Meteorological Organization (WMO).

ISBN 978-90-481-9120-8   e-ISBN 978-90-481-9121-5
DOI 10.1007/978-90-481-9121-5
Springer Dordrecht Heidelberg London New York

Library of Congress Control Number: 2010926248

Printed on acid-free paper

Springer is part of Springer Science+Business Media (www.springer.com)

This book is a tribute to the scientists, students, and northerners who have worked for decades trying to learn more about top predators in Hudson Bay. Learning more for their conservation and their sustainable harvest, and more recently to understand changes associated with global warming. We hope this book will not be an historical record of the way things were but rather a call to increase efforts to monitor, learn, and adapt.

Foreword

Coats Island, in the centre of northern Hudson Bay, lies at the heart of the region dealt with by this book. For my sins or for my virtues, I have found myself there for weeks or months at a time over many summers since 1981. Jo Nakoolak, who has worked with us in our camp near Cape Pembroke every year since 1993 was a member of the last family to overwinter on the island, in the 1960s. The remains of his family's sod-banked cabin still emerge from the tundra, as do the outlines of older house-pits and grave mounds, evidence of the *Sadlermiut* who inhabited northern Hudson Bay before the arrival of Europeans. I think of Jo as the last true Coats Islander.

Now, visits by local people from the nearby community at Coral Harbour are rare and usually occur only to conduct tourists or to hunt walrus at the several large haul-outs along the coast. We see them no more than twice a season. For weeks at a time our only reminder of the world outside Coats Island is the daily flight from London-Calgary which passes over us about mid-day. In the past the faint radio voice emanating from the Nunavut Research Institute and its predecessor, the DIAND Iqaluit Laboratory was a daily link to the outside, but since we got a satellite phone it has been our choice to call out, rather than a daily routine. On a crowded planet, northern Hudson Bay stands out as an enduringly uncrowded place.

In 1992, I sailed across the northeast corner of Hudson Bay, from Coral Harbour to Ivujivik, via Coats and Mansel islands, in an elderly Peterhead, the *Terregluk*, crewed by members of the Nakoolak and Alogut families. The weather as we left Coats was calm and clear, with the early morning sun lighting up the steam rising over a great herd of walrus on the Cape Pembroke haul-out. We lingered to photograph the haul-out, but within an hour of heading eastwards the wind rose out of the northeast and the sea kicked up to Beaufort 5. The non-Inuit members began to look a little green. It was at this point that I realised our only navigation aid was a rather battered looking binnacle. There was no chart. At moments such as that you become acutely conscious of how empty a place northern Hudson Bay is.

Given the size of the sea that rose quickly the captain could not steer a straight course but had to continually adjust to the waves, now running with them, now taking advantage of some minor amelioration to broadside them. As I watched the needle of the compass swing backwards and forwards over 180° I wondered where on earth we were going to land up and how we would figure out where we were when we did. Ten hours later, after a day of grey skies and lumpen seas, we sighted a nondescript line of shingle and rock on the horizon. As we coasted agonisingly slowly eastwards I finally made out a navigation beacon near the shore and realised that our landfall was precisely as planned, at the northern tip of Mansel Island. Somehow, through that heaving grey waste, with only a gyrating compass needle as guide, Jimmy, our captain, had kept us on course: a reminder that the sea and atmosphere have their signs for those who can read them.

Reading the signs is what this book is about. The Hudson Bay region is experiencing unprecedented changes in climate and consequently in ice conditions. These changes are setting in motion a torrent of biological changes that seem set to transform marine ecosystems from Arctic to sub-Arctic and then perhaps to something akin to Boreal. The truly unprecedented events of 2007, when Arctic sea ice reduced to 23% below the previous record minimum, have put off all bets on the rate of Arctic de-icing. The prospect of an ice-free summer for the Arctic Ocean, still distant as recently as the mid-1990s, now seems less than a decade away.

Predicting temperature change and trends in ice conditions is one thing, but predicting the biological consequences is quite another. In terrestrial ecosystems, the occurrence of permafrost is a dominant ecological factor. Likewise, sea-ice has a huge impact in the marine environment. As Hudson Bay normally clears of sea-ice every summer, the global warming trend will affect mainly the duration of the ice-free season and the size and persistence of polynyas and flaw-leads. For species which carry out important aspects of their life history in association with ice, its reduction will surely bring about changes in populations and distributions, but which less ice-tolerant organisms will move in the take their place? There are many candidates occupying the Sub-arctic waters of the western Atlantic, but which of them have the behavioural equipment to take advantage of the potential niches opening up in the huge inland sea of Hudson Bay?

What we are witnessing with climate change is a vast, uncontrolled, ecological experiment – planetary in scale, but having its most immediate effects in the Arctic. The authors of this book catalogue many changes underway and make many educated guesses about the future of marine ecosystems in the Bay, especially their vertebrate constituents. Despite the solid research and scholarship that has gone into creating this landmark publication, I am confident that there will be many future surprises that will, as we continue to study them, increase profoundly our understanding of how marine ecosystems in Hudson Bay function. I very much hope that I am still around in 20 years to see how the scenarios envisaged in this book will play out.

Environment Canada   Tony Gaston
National Wildlife Research Centre
Ottawa K1A 0H3

Acknowledgments

It is not possible to put together a book of this type, with diverse information from a multitude of scientists, without the help of many people and organizations.

Funding for the preparation of this book, as well as financial and logistic support for much of research contained herein, came from a variety of sources which the authors gratefully acknowledge: ArcticNet Network of Centre of Excellence, Canada Foundation for Innovation, Canada Research Chairs program, Churchill Northern Study Center, Environment Canada (Canadian Wildlife Service, Science and Technology Branch), Fisheries and Oceans Canada, Indian and Northern Affairs Canada (Northern Contaminants Program, Northern Scientific Training Program), International Polar Year program, Makivik Incorporated, Manitoba Hydro, Natural Sciences and Engineering Research Council, Natural Resources Canada (Polar Continental Shelf Program), Nunavut Wildlife Management Board (Nunavut Wildlife Research Trust), Rhodes Trust, and a number of universities (University of Manitoba, University of Alberta, University of British Columbia, Carleton University, University of Ottawa).

Research was undertaken by hundreds of students, field assistants and members of communities around the Hudson Bay Region, all of whom deserve a huge "thanks" for their efforts, often under very challenging conditions. In particular, we thank the Hunters' and Trappers' Organizations and communities of Arviat, Cape Dorset, Chesterfield Inlet, Churchill, Coral Harbour, Hall Beach, Igloolik, Ivujivik, Kimmirut, Moosonee, Rankin Inlet, Repulse Bay, and Sanikiluaq.

Data, maps and other helpful bits of information were provided by: the Arctic Monitoring and Assessment Program, Canadian Arctic Resources Committee, S. Cosens and R. Reeves, R. Dietz, M. P. Heide-Jørgensen and K. Laidre, W. G. Ross and the Eastern Arctic Bowhead Whale Recovery Team, the Giovanni online data system (NASA GES DISC), and MODIS mission scientists and associated NASA personnel.

The Canadian Arctic scientific community pulled together to help get chapters reviewed and edited, and we thank the following people for their efforts: M. Bailey, G. Boila, T. Bortoluzzi, L. Dahlke, J. DeLaronde, B. Dunn, S. Ferguson, K. Fisher, J. Garlich-Miller, C. Garroway, J. Higdon, J. Justus, M. Keast, T. Kelley, N. Koper, L. Loseto, S. Luque, A. MacHutchon, M. Mallory, O. Nielson, J. Orr, S. Petersen, D. Pike, C. Piroddi, T. Pitcher, P. Richard, E. Richardson, R. Riewe, T. Stephenson,

B. Stewart, R. Stewart, D. Varkey, R. Vickery, M. Dowsley, M. Hipfner and D. Preikshot. Our apologies to those we've forgotten here!

Thanks to Tony Gaston for writing a revealing foreword to the book. Finally, the editors extend a sincere thank you to their families and friends who supported them during the time needed to put this book together.

Contents

Contributors

David G. Barber
Centre for Earth Observation Science (CEOS), Faculty of Environment, Earth, and Resources, University of Manitoba Winnipeg, Manitoba, R3T 2N2, Canada

Birgit M. Braune
Environment Canada, Science and Technology Branch, National Wildlife Research Centre, Carleton University, Raven Road, Ottawa, ON K1A 0H3, Canada

Magaly Chambellant
Fisheries and Oceans Canada, 501 University Crescent, Winnipeg, MB R3T 2N6, Canada

Elly G. Chmelnitsky
Department of Biological Sciences, University of Manitoba, Winnipeg, MB, Canada

Andrew E. Derocher
Department of Biological Sciences, Biological Sciences Centre, University of Alberta, Edmonton, AB, Canada

Steven H. Ferguson
Fisheries and Oceans Canada, Central and Arctic Region, Winnipeg, MB, Canada; Department of Environment and Geography, University of Manitoba, Winnipeg, MB, Canada; Department of Biological Scienes, University of Manitoba, Winnipeg, MB, Canada

Ashley Gaden
Department of Environment and Geography, University of Manitoba, 440 Wallace Building, Winnipeg, MB R3T 2N2, Canada

Anthony J. Gaston
Environment Canada, Science and Technology Branch, National Wildlife Research Centre, Carleton University, Raven Road, Ottawa, ON K1A 0H3, Canada

H. Grant Gilchrist
Wildlife Research Division, Science and Technology Branch, Environment
Canada, National Wildlife Research Centre, 1125 Colonel By Drive, Raven Road,
Carleton University, Ottawa, Ontario, K1A 0H3, Canada

Meagan Hainstock
Ducks Unlimited Canada, Oak Hammock March Conservation Centre, Stonewall,
Manitoba, R0C 2Z0, Canada

Dominique Henri
School of Geography and the Environment, Oxford University Centre for the
Environment, South Parks Road, Oxford, 0X1 3QY, UK

Jeff W. Higdon
Department of Environment and Geography, University of Manitoba and Fisheries
and Oceans Canada, Central and Arctic Region, Winnipeg, MB, Canada

Klaus Hochheim
Centre for Earth Observation Science (CEOS), Department of Environment
and Geography, University of Manitoba, MB, Canada

Carie Hoover
Fisheries Centre, University of British Columbia, 2202 Main Mall, Vancouver,
BC V6T1Z4, Canada

Trish C. Kelley
Department of Environment and Geography, University of Manitoba, Winnipeg,
MB R3T 2N2, Canada

Lisa L. Loseto
Fisheries and Oceans Canada, Freshwater Institute, 501 University Crescent,
Winnipeg, MB R3T 2N6, Canada

Jennifer V. Lukovich
Centre for Earth Observation Science (CEOS), Department of Environment
and Geography, University of Manitoba, MB, Canada

Nick J. Lunn
Wildlife Research Division, Wildlife and Landscape Science Directorate,
Science & Technology Branch, Environment Canada, Edmonton, AB, Canada

Mark L. Mallory
Environment Canada, Canadian Wildlife Service, Box 1714, Iqaluit,
NU X0A 0H0, Canada

Martyn E. Obbard
Wildlife Research and Deveopment Section, Ontario Ministry of Natural Resources, DNA Building, Trent University, 2140 East Bank Dr., Peterborough, ON K9J 7B8, Canada

Jack R. Orr
Fisheries and Oceans Canada, 501 University Crescent, Winnipeg, MB R3T 2N6, Canada

Elizabeth Peacock
Wildlife Research Section, Department of Environment, Government of Nunavut, Igloolik, NU, Canada
and
US Geological Survey, Alaska Science Center, Anchorage, AK 99516, USA

Stephen D. Petersen
Fisheries and Oceans Canada, 501 University Crescent, Winnipeg, MB R3T 2N6, Canada

Pierre R. Richard
Fisheries and Oceans Canada, 501 University Crescent, Winnipeg, MB R3T 2N6, Canada

Gregory J. Robertson
Environment Canada, Science and Technology Branch, 6 Bruce Street, Mount Pearl, NF A1N 4T3, Canada

Gary A. Stern
Department of Environment and Geography, 440 Wallace Building, University of Manitoba, Winnipeg, Manitoba, Winnipeg, MB R3T 2N2, Canada; Fisheries and Oceans Canada, 501 University Crescent, Winnipeg, MB R3T 2N6, Canada

D. Bruce Stewart
Arctic Biological Consultants, 95 Turnbull Dr., Winnipeg, MB R3V 1X2, Canada

Robert E.A. Stewart
Department of Fisheries and Oceans, 501 University Crescent, Winnipeg, MB R3T 2N6, Canada

Kristin H. Westdal
Department of Environment and Geography, University of Manitoba, 440 Wallace Building, Winnipeg, MB R3T 2N2, Canada

Paul J. Wilson
Natural Resources DNA Profiling and Forensic Centre, Biology Department &
Forensic Science Program, Trent University, Peterborough, Ontario, K9J 7B8,
Canada

Michael Yurkowski
Department of Fisheries and Oceans, 501 University Crescent, Winnipeg,
MB R3T 2N6, Canada

The Ocean-Sea Ice-Atmosphere System of the Hudson Bay Complex

D.B. Stewart and D.G. Barber

Abstract Arctic marine water, vast inputs of fresh water, nearly complete seasonal ice cover, and dynamic coastal morphology make the Hudson Bay (HB) complex remarkable among the world's large marine areas (LMEs). Each of these features is influenced by the climate. Together they enable Foxe Basin, Hudson Strait, Ungava Bay, Hudson Bay and James Bay which comprise the HB complex, to support a well developed Arctic marine food web far south of its normal range. The life history characteristics that enable biota to thrive in the region also make them sensitive and vulnerable to changes in the seasonal ice cover, freshwater inputs, and/or water levels that can stem from changes in climate.

   This introductory chapter summarizes current understanding of the physical, chemical, and biological conditions in the HB complex. It also provides examples of how some of the key climate-sensitive oceanographic parameters influence species ecology. The level of research effort in this region has been low relative to the Atlantic and Pacific coasts of Canada. This limits understanding and modeling of many key processes, particularly with respect to seasonal and inter-annual change. It also limits the differentiation of natural cyclical changes from those that are anthropogenically-driven and systematic. Over the coming decades, long-term, stable research will be required to better predict and mitigate the impacts of climate change, a growing human population, hydroelectric and non-renewable resource developments, shipping, and the long range transport of contaminants on this fascinating and potentially vulnerable ecosystem.

Keywords Arctic Canada • Oceanography • Flora and fauna • Coastal zone • Environmental sensitivity

D.B. Stewart (✉)
Arctic Biological Consultants, 95 Turnbull Dr., Winnipeg, MB R3V 1X2, Canada
e-mail: stewart4@mts.net

S.H. Ferguson et al. (eds.), *A Little Less Arctic: Top Predators in the World's Largest Northern Inland Sea, Hudson Bay*, DOI 10.1007/978-90-481-9121-5_1,
© Springer Science+Business Media B.V. 2010

Introduction

Arctic marine water, vast inputs of fresh water, nearly complete seasonal ice cover, and dynamic coastal morphology make the Hudson Bay (HB) complex remarkable among the world's large marine areas (LMEs) (Fig. 1). Each of these features is

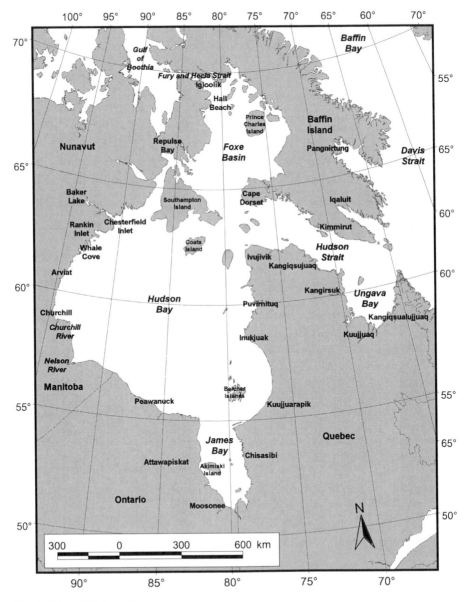

Fig. 1 Map of Hudson Bay complex

influenced by the climate. Together they enable Foxe Basin, Hudson Strait, Ungava Bay, Hudson Bay and James Bay which comprise the HB complex, to support a well developed Arctic marine food web far south of its normal range. Polar bears (*Ursus maritimus*), for example, live and breed in James Bay at the same latitude as Amsterdam in The Netherlands! These features also create favourable seasonal habitats for internationally important migratory populations of seabirds, shorebirds, waterfowl and marine mammals. Millions of geese forage and breed along its coasts, and the largest population of beluga (*Delphinapterus leucas*) in the world summers in the Nelson River estuary. In the south, the coastal waters are so dilute in early summer that they attract fishes typical of fresh water to feed. The life history characteristics that enable biota to thrive in the HB complex also make them sensitive and vulnerable to changes in the seasonal ice cover, freshwater inputs, and/or water levels that can stem from changes in climate.

This introductory chapter provides background on the physical, chemical, and biological conditions in the Hudson Bay complex. Because this region is large, has only limited commercial and sovereignty interests associated with it, and is remote and seasonally ice covered, the level of research effort has been low relative to the Atlantic and Pacific coasts of Canada. This limits our current understanding and modeling of many key processes, particularly with respect to seasonal and inter-annual change. It also limits our ability to differentiate natural cyclical changes from those that are anthropogenically-driven and systematic.

Geographical Setting

The HB complex is bounded by Baffin Island and the Canadian mainland, with a connection to the Arctic Ocean via Fury and Hecla Strait and larger one to the Labrador Sea (North Atlantic) via Hudson Strait. It encompasses 1,242,000 km² of sea surface and 83,000 km² of island surface (Stewart 2007), and receives freshwater runoff from 4,041,400 km² of the North American mainland and islands – a larger catchment than the St Lawrence and Mackenzie rivers combined (NAC 1986) (Fig. 2). The basin is shallow for such a large marine area, except in Hudson Strait, and there is often little coastal development or bottom relief to promote mechanical mixing or upwelling that might increase the availability of chemical nutrients in the surface waters (Fig. 3). Hudson Bay and James Bay, in the south, are abnormally cold relative to other marine areas at the same latitude and they can affect both the generation of local climate (e.g., cyclogenesis) and advection of storms over southern Canada.

Coastline and Seafloor

The basin of the HB complex is composed of crystalline rock of the Canadian Precambrian Shield, much of it overlain by thick layers of younger Paleozoic sedimentary rock of the Hudson or Arctic Platform that are covered by unconsolidated materials.

Fig. 2 Hudson Bay watershed (From Stewart and Lockhart 2005)

Coasts and seafloor underlain closely by Shield rock tend to be irregular, often rugged, with more exposed bedrock and greater relief and/or coastal development (Fig. 3) (Dionne 1980; EAG 1984; Martini 1986; Norris 1986; Stewart and Lockhart 2005). In contrast, those underlain closely by platformal rock are characterized by very gradual slopes, low relief, shallow nearshore waters, and extensive tidal flats that give way inland to low-lying, marshy coastal plains. The seafloor is generally covered by glacial till or fine-grained glaciomarine deposits that typically grade from coarse gravels nearshore to fine silt and clay offshore (Meagher et al. 1976 in Dionne 1980; Henderson 1989; Josenhans and Zevenhuizen 1990; NAC 2007).

Low-lying, complex, and cliff and headland coastlines have particular biological significance (Fig. 4) (Dionne 1980; EAG 1984; Martini 1986; Stewart and Lockhart 2005; Stewart and Howland 2009). Low-lying coasts underlain by platformal rock occur from southern James Bay to Arviat in Nunavut, on the larger islands of James Bay, along the southern shores of the islands in northern Hudson Bay, and along the east coast and on the islands of Foxe Basin. The low-lying coastal sections underlain by platformal rock provide vital habitat for many migratory waterfowl and shorebird species. Large rivers such as the Nelson, Churchill, Albany, Moose, Nottaway, and

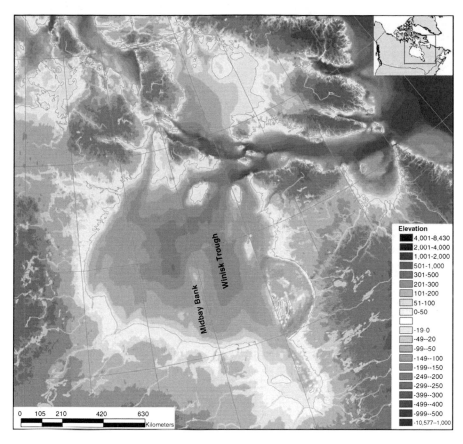

Fig. 3 Topography and bathymetry (m) of the Hudson Bay complex (Adapted from Arctic Monitoring and Assessment Programme)

Nettilling that dissect these coasts carry much of the runoff into the HB complex, and provide important estuarine habitats for fish and marine mammals. Complex coastal sections underlain by Shield rock and consisting of small headlands and bays offer the greatest variety of landforms and biological habitats. Eastern James Bay, western Hudson Bay, and northwestern Hudson Strait have well-developed sections of complex coastline. Their tidal flats, intertidal salt marshes, and subtidal eelgrass meadows (*Zostera marina* L.) have particular ecological importance for migratory shorebirds and waterfowl (see also Curtis 1975; Martini et al. 1980; Dignard et al. 1991). Local relief is generally <30 m and the shores are a mixture of exposed bedrock and unconsolidated materials. Major rivers that dissect the complex coasts include the Eastmain, La Grande and Povungnituk in Québec, and the Thelon and Kazan in Nunavut. Cliff and headland sections that are underlain by Shield rock support important cliff-nesting bird species that are not common elsewhere in the HB complex. Akpatok Island, shores facing Foxe Channel and

Fig. 4 Raised beaches (*top left*) where coastal materials have been pushed by sea ice and then stranded by isostaic rebound. Flights of raised beaches on the north coast of Southampton Island (*top right*). Low-lying coast at the Nelson River estuary (*bottom left*). Cliff and headland coast near Rankin Inlet (*bottom right*) (Photo credits: D.B. Stewart)

Hudson Strait, Wager Bay, southeastern and northwestern Hudson Bay, and the Belcher Islands have well-developed sections of cliff and headland coast. Local relief in these sections can exceed 100 m and their topography ranges from gently rolling hills to steep cliffs that rise abruptly from the sea to 500 m above sea level. Exposed bedrock is common in these areas, and tidal flats are lacking. Grande rivière de la Baleine is the only major river that dissects these coastal sections.

Fluvial dynamics have created a rich history of sediment deposition in the estuaries of the rivers entering into the HB complex. Paleoclimatic studies of these sediments provide a detailed record of freshwater-marine coupling in these areas and can be used as a means of examining hydrologic changes in the watersheds which can be linked either to hydroelectric development or to climate variability and change.

The submarine geology and physiography of the HB complex tend to be extensions of its coastal formations and features (Pelletier 1986; Josenhans and Zevenhuizen 1990). The eastern half of Foxe Basin, wide bands along the south coasts of Hudson Bay and Ungava Bay, and most of James Bay are less than 50 m deep (Fig. 3). Foxe Basin is seldom deeper than 100 m, except in the southwest (Sadler 1982; Prinsenberg 1986c). A deep channel extends from southwestern Foxe Basin through Hudson Strait, bisecting the complex. The channel depth increases from over 400 m in the west to about 1,000 m north of Ungava Bay, and then rises at the east end of Hudson Strait to a sill <400 m deep that limits exchange of deep water with the Labrador Sea (Drinkwater 1986).

In the main basin of Hudson Bay the bottom extends well offshore as a broad coastal shelf <80 m deep and then slopes gradually to a smooth sea floor with an average depth of 250 m (Josenhans and Zevenhuizen 1990; Stewart and Lockhart 2005). This pattern is interrupted by the Midbay Bank, a ridge-like feature that extends from the south shore and rises to a depth of <40 m, and by the Winisk Trough, a deep trench-like feature that extends offshore the Winisk River estuary towards Coats Island. In the north this trough is about 1.6 km wide with steep walls that drop from the seafloor to a depth of 370 m. In southeastern Hudson Bay, where the depth seldom exceeds 120 m, enclosed bathymetric deeps (>200 m depth) parallel and resemble the adjacent cuesta coastline (Josenhans et al. 1991). Surface exchange with Foxe Basin occurs through Roes Welcome Sound, which has a sill at 60 m depth (Defossez et al. 2005). Deepwater exchange with Hudson Strait is limited by a sill at 185 m depth across the northeast exit to Hudson Bay (Prinsenberg 1986b). James Bay, which extends south from Hudson Bay, is seldom deeper than 50 m and extremely shallow for such a large marine area.

The relatively shallow offshore depths in James Bay, Hudson Bay, and much of Foxe Basin likely exclude many of the deepwater fishes that occur in Hudson Strait, including commercially valuable species such as the Greenland halibut (turbot) (*Reinhardtius hippoglossoides*) and redfish (*Sebastes* spp.) (Morin and Dodson 1986; Stewart and Lockhart 2005). The whales, most seals, and perhaps walruses (*Odobenus rosmarus*) can dive to the bottom to feed throughout James Bay, and most, if not all, of Foxe Basin and Hudson Bay.

Successive glaciations have shaped the coasts and seafloor of the HB complex, and caused profound changes in its oceanography. Each glaciation obscured evidence of the last, so the glacial history is a matter of debate except for the Laurentide Ice Sheet, which covered the region most recently (Shilts 1986). It reworked the surface and its enormous weight depressed the earth's flexible crust up to 300 m below its present level (Andrews 1974). During deglaciation this permitted marine waters to flood large areas inland from the present-day coasts, despite a sea level that was 100 m lower. Subarctic oceanographic conditions, which persisted between about 6,500 and 4,000 yBP (Bilodeau et al. 1990), were likely responsible for the relict Subarctic species that inhabit James Bay and southeastern Hudson Bay.

The earth's crust has been rebounding since the ice load was removed, an order of magnitude faster initially but still continuing at a vertical rate of up to 1.3 m per century (Webber et al. 1970; Hillaire-Marcel 1976). Most coastlines exhibit a variety of emergent glacial deposits (Fig. 4). Where there are very gradual mud flats, such as in Rupert Bay, isostatic rebound can create up to 15 m (horizontally) of new land each year (d'Anglejan 1980). This emergence of coastal sediments also appears to play an important role in resuspension processes due to wave interactions with the coastal environments. The sediments become quite mobile and may also play a role in nutrient supply to coastal ecosystem processes. The future shape of the HB complex will be very different than it is today (Barr 1979).

Climate Controls

Climate controls such as the radiation regime, nature of immediate and adjacent surfaces, physical geography, and upper-air circulation also influence oceanographic conditions in the HB complex. Because of its distance from the Equator, large seasonal variations occur in the incoming solar radiation that provides heat and powers photosynthesis. These variations are greatest in the north. At Igloolik, for example, the sun did not set from 21 May to 22 July 2009 or rise from 1 December 2009 to 10 January 2010 (www.timeanddate.com). This means that northern Foxe Basin receives the highest incoming solar radiation in the HB complex in June, but none at the height of winter (Maxwell 1986).

A large percentage of the incoming solar radiation is reflected (albedo) by snow, ice, wet surfaces, fog and cloud (Maxwell 1986; Bergmann et al. 1991; Ehn et al. 2006). At Churchill, Manitoba, the mean daily net radiation is positive from mid-March through October, but negative during the remainder of the year when incoming solar radiation is more than offset by long wave and turbulent heat losses (Maxwell 1986). To the north this period is shorter and vice versa. The mean daily net radiation increases from <5 kcal·cm^{-2} in northern Foxe Basin to >30 kcal·cm^{-2} in southern James Bay (Hay 1978).

The wide seasonal variation in incoming solar radiation and significant reduction of light penetration into the water by snow and ice cover severely constrain when, where, and how much primary production can occur. Photosynthesis by ice algae, for example, begins in spring as light intensity increases (Roff and Legendre 1986). Early in the season it is limited by the low irradiance, so that algal patches tend to develop in lighted areas under thin snow-ice cover (Gosselin et al. 1986). Later in the season the maximum growth occurs in areas that offer protection from photoinhibition. These algae are associated with nutrient-rich brine channels in the sea ice (Demers et al. 1989), and provide important food for planktonic copepods (Runge and Ingram 1991; Tourangeau and Runge 1991). They are a particularly vulnerable link in the HB complex food web to changes in the snow and ice cover.

In winter, ice cover causes much of the marine surface to experience the same climate as the adjacent snow-covered tundra (Maxwell 1986; Stewart and Lockhart 2005) (Fig. 5). Landfast ice and consolidated pack ice also limit wind mixing of the surface waters. In summer, the cold Arctic water cools air flowing over adjacent coastal areas, and contributes to the presence of permafrost well past its normal southern latitude.

While it is exposed to cold Arctic air masses year-round and to occasional intrusions of warm air from the south in summer, the HB complex is buffered from the direct entry of air masses from the Pacific and Atlantic oceans by mountains and distance (Maxwell 1986). There are few topographical features within the region to modify local climates, but some modification occurs in the lee of elevated cliff and headland coasts.

Fig. 5 Winter ice cover limits wind mixing and scours and depopulates the nearshore benthic zone (Photo credit *top*: D.B. Stewart). Wind forcing and wave action can be significant during the open water season (Photo credit *bottom*: ArcticNet, University of Manitoba)

The upper air circulation controlling movement of weather systems over the water is mainly related to the persistent, counterclockwise air flow around a low pressure vortex or trough (polar vortex) that is situated over Baffin Island in winter but weakens and retreats northward in summer (Maxwell 1986). In winter there is a general flow of cold Arctic air from the northwest or west over the western half of the HB complex and from the west or southwest over the east. In summer there is a general but less intense flow of air from the west through to the north (EAG 1984).

The position and intensity of the trough strongly influence year-to-year variability of the region's climate and affect timing of ice breakup and freezeup (Cohen et al. 1994; Wang et al. 1994c; Mysak et al. 1996; Mysak and Venegas 1998). A more intense trough or an eastward shift in its position brings cold northerly air into the region more frequently, and vice versa. These shifts can occur in a very short time, forced primarily by natural variation in the atmospheric pressure field associated with the Arctic Oscillation (Macdonald et al. 2003). These atmospheric pressure systems can also influence water circulation through wind stress or by tilting the sea surface (Larouche and Dubois 1988, 1990).

Due to its size the HB complex is also susceptible to the influence of large scale atmospheric teleconnection patterns (Hochheim and Barber 2009). Hemispheric teleconnections are climate indices, which relate the surface pressure patterns of the northern annular mode to the local surface pressure within the geographic expanse of the HB complex. Various climate indices have been identified as potentially significant in relation to HB surface air temperatures (SATs), including the North Atlantic Oscillation (NAO), Arctic Oscillation (AO), Southern Oscillation Index (SOI), East Pacific/North Pacific Oscillation (EP/NP) and the Pacific Decadal Oscillation (PDO). Details of how each index functions are well presented in the literature and will not be repeated here. Each index has an associated seasonal pressure and SAT pattern. These patterns drive the local cyclogenesis and advection of cyclones within the HB complex. The patterns also explain differing amounts of variability in the sea ice concentration anomalies seasonally (Hochheim and Barber 2009).

Marine Environment

Circulation and water mass characteristics within the HB complex are strongly influenced by the freshwater dynamics, which vary seasonally and geographically in response to ice formation and melt, freshwater runoff from the land, and inputs of marine surface water (Harvey et al. 2006). The large freshwater input and the extensive formation and melting of sea ice lead to pronounced vertical layering of the water masses. This stratification is particularly strong in Hudson Bay and James Bay, but is moderated in Hudson Strait and shallow eastern areas of Foxe Basin by tidal mixing. Summer surface salinity values over most of this region are low relative to the Atlantic and Pacific, where they typically exceed 34 psu (Antonov et al. 2006). In Foxe Basin, northern Hudson Strait and northwestern Hudson Bay this is predominately due to sea ice meltwater (Tan and Strain 1996). While in James Bay, northeastern Hudson Bay, southern Hudson Strait, and Ungava Bay it is predominately from runoff and rainfall.

Seawater

Seawater in the HB complex originates from the Pacific Ocean, is cooled and diluted by passage through the Arctic Archipelago, and enters via Fury and Hecla Strait and Hudson Strait (Jones et al. 2003). The water passing through Fury and Hecla Strait is well-mixed by the tides before it flows into Foxe Basin (Fig. 6). Very high currents, up to 3 m·s^{-1}, have been observed in the Strait (Collin and Dunbar 1964; Sadler 1982).

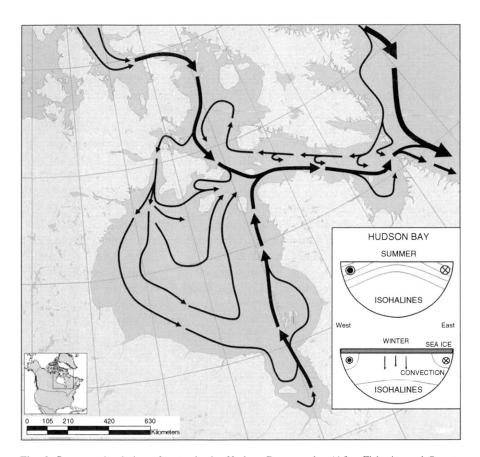

Fig. 6 Summer circulation of water in the Hudson Bay complex (After Fisheries and Oceans Canada poster 8: Eastern Canadian Arctic). Seasonal differences in the salinity profile (isohalines) are shown schematically for an east-west cross section of Hudson Bay (inset after Ingram and Prinsenberg 1998). Flow directions are similar but surface currents are stronger in summer than winter. Summer stratification is characterized by a warmer and fresher shallow upper water layer

In winter, this circulation may be temporarily disturbed or reversed during the passage of cyclones (Godin and Candela 1987). The inflow to Foxe Basin is rather small, 0.04 Svd in the winter (Sadler 1982; 1 Svd = 10^6 m^3·s^{-1}) and 0.1 Svd in the summer (Barber 1965), but has a substantial influence on the circulation, filling the northern half in 1 year and driving the counterclockwise (cyclonic) gyre (Prinsenberg 1986c). This lighter, less saline Arctic water flows south on the surface along the coast of Melville Peninsula. When it reaches southern Foxe Basin, some of the water flows west via Frozen Strait and Roes Welcome Sound into Hudson Bay; some flows northeast to join water entering Foxe Basin from Hudson Strait; and some continues through Foxe Channel and into Hudson Strait (Prinsenberg 1986c; Tan and Strain 1996).

The surface water from Foxe Basin circulates cyclonically around Hudson Bay, creating Arctic oceanographic conditions much farther south than elsewhere along the North American continent (Prinsenberg 1986b; Wang et al. 1994; Tan and Strain 1996). Some of it enters James Bay while the rest is deflected north to exit northeast into Hudson Strait. In summer, the warm, dilute water from James Bay is recognizable as it flows northward along the east coast of Hudson Bay and into Hudson Strait. A westward, wind-driven return flow across the top of Hudson Bay has been predicted and there is a small, perhaps intermittent, intrusion of water from Hudson Strait at the northeast corner of Hudson Bay. This circulation is maintained by inflow/outflow forcing that likely occurs year-round, and reinforced during the open water season by wind and buoyancy forcing. Estimates of the average residence time of water in Hudson Bay range from 1 or 2 to 6.6 years (Prinsenberg 1984; Drinkwater 1988; Jones and Anderson 1994; Ingram and Prinsenberg 1998).

Most water exchange with other marine regions is via Hudson Strait. Cold, saline (32.5 < S < 33.5) water from the Baffin Island Current flows northwest into the HB complex along the north side of Hudson Strait (Drinkwater 1988; Straneo and Saucier 2008a, b). Some of this inflow penetrates into Foxe Basin, where it contributes to a small counterclockwise loop before joining the southward outflow of Arctic water on the western side of the basin (Prinsenberg 1986c). The rest crosses the Strait to join warmer, less saline water from Hudson Bay (<30) and Foxe Basin (<32) that flows southeast out of Hudson Strait, some of it via Ungava Bay. Some warmer, high salinity water from the Labrador Sea penetrates into the eastern half of the Strait below about 200 m (Drinkwater 1988).

Tides

Powerful tides that originate in the North Atlantic Ocean surge into the Hudson Bay complex twice daily (semidiurnal) via Hudson Strait (Dohler 1968; Drinkwater 1988). They progress as a Kelvin wave counterclockwise around Foxe Basin and Hudson/James Bay, and overshadow local tides and any tidal influence from the Arctic Ocean. In Leaf Bay, at the head of Ungava Bay, a maximum spring rise of 16.7 m has been recorded, the highest in the world (Dohler 1968; Canadian

Hydrographic Service 1988; NIMA 2002; Kuzyk et al. 2008). In Hudson Strait the spring tide range is 7.9 m at the eastern entrance, increasing to 12.5 m along the north shore at Kimmirut, and then decreasing to 4.9 m in the west at Nottingham Island. In Foxe Basin, the spring range is 7 m in Foxe Channel in the south, and declines to between 3 and 3.4 m in the north at Rowley Island. In Hudson Bay, the range increases from 3 m in the northeast to 4 m along the west coast at Churchill Harbour and then decreases to about 2 m along the east side of James Bay and in the Belcher Islands and to 0.5 m at Inukjuak (Dohler 1968; Godin 1974). The breadth of the intertidal zone is an important determinant of the nearshore benthic flora and fauna, as it influences their exposure to freezing (Dale et al. 1989) and to ice scour (Bursa 1968; Dadswell 1974; Conlan et al. 1998; Conlan and Kvitek 2005).

The tides set up strong currents at the eastern entrance to Hudson Strait (up to 2 $m \cdot s^{-1}$; Drinkwater 1986), in Foxe Basin (0.3 $m \cdot s^{-1}$; Prinsenberg 1986c), at the entrance to Hudson Bay (0.9 $m \cdot s^{-1}$), and within the Bay (0.3 $m \cdot s^{-1}$) (Prinsenberg and Freeman 1986). Chesterfield Inlet is an estuary with stronger tidal currents, higher tidal amplitudes, and a greater degree of mixing than elsewhere in Hudson Bay (Dohler 1968; Budgell 1976, 1982). There are also strong tidal currents at the narrow entrance to Wager Bay, and to Lac Guillaume Delisle (Zoltai et al. 1987; NIMA 2002).

Tidal currents contribute to biological productivity in the HB complex by replenishing nutrients in the surface waters. In concert with surface currents, they can also create oceanographic conditions that concentrate invertebrate and fish biomass (Hudon et al. 1993). These biological "hotspots" are a vital food resource for the large cliff nesting seabird colonies (Cairns and Schneider 1990). Where the currents exceed 1.5 $m \cdot s^{-1}$ the sea surface may remain open year-round (Prinsenberg 1986c), creating polynyas wherein marine mammals and birds overwinter (Mallory and Fontaine 2004).

Freshwater Runoff

The HB complex, particularly James Bay and southern Hudson Bay, receives a very large input of freshwater runoff, about 940 $km^3 \cdot year^{-1}$ or roughly one fifth the river discharge into the Arctic Ocean (Déry et al. 2005; Straneo and Saucier 2008a). Most of this freshwater is added from May through October (Prinsenberg 1986c). It has a strong influence on the timing and pattern of ice breakup, surface circulation, water column stability, species distributions, and biological productivity (Prinsenberg 1988; Stewart and Lockhart 2005). There are extensive freshwater plumes off the larger river mouths in Hudson Bay and James Bay year-round. These plumes spread farther and deeper under the ice than during ice-free conditions. In James Bay, where precipitation is much greater than evaporation and runoff is high there is an annual net gain of 473 cm of fresh water over the entire surface (Prinsenberg 1984, 1986a). This is much greater than the average for Hudson/James Bay (64 cm). Hudson Bay loses more fresh water through evaporation than it gains from precipitation.

In southern and western James Bay, which are shallow and receive a great deal of sediment laden runoff, the water clarity is low relative to other parts of the HB complex, except possibly in shallow eastern Foxe Basin where wave action stirs up bottom sediments (Barber 1972; Prinsenberg 1986c). The clear water in Hudson Bay enables macroalgae to grow on the seafloor at a depth of at least 75 m (Barber 1983).

Runoff contributes to the dynamics of the marine circulation by diluting the salt water and creating density-driven currents. These currents are restricted to coastal waters (up to 50 km wide) (Prinsenberg 1982; Granskog et al. 2009). They are greatest in early summer when runoff is high, and weakest between February and May, when there is little runoff and thick ice cover limits wind stress and insulates the surface water. Indeed, from summer to winter the magnitude of the density-driven currents at the inflow to James Bay decreases from 3.5 to 1.3 cm s^{-1} and at the outflow from 15 to 5 cm s^{-1} (Prinsenberg 1982). Because the magnitude of density-driven currents is proportional to the runoff rate, hydroelectric developments that increase winter runoff will also increase winter circulation (Prinsenberg 1982, 1991). This is now the case with large rivers impounded by hydroelectric development such as the Nelson and La Grande, where spring runoff is stored in large reservoirs for release later in the year. This has flattened the seasonal hydrograph, reducing the spring freshet and increasing flow under the sea ice in winter (e.g., Hernández-Henríquez et al. 2009). The environmental impacts of shifting the seasonal runoff regime are not well understood.

The presence of freshwater runoff under the sea ice also affects primary production of sea ice algae. Freshwater plumes, especially from the large volume rivers (e.g., La Grande and Nelson), can travel significant distances as a buoyant layer under the landfast ice and mobile pack ice (Messier et al. 1986; Ingram and Prinsenberg 1998). This advection of the freshwater layer is facilitated by the ice cover, which limits vertical mixing by tides and wind. The salinity of surface waters underlying the ice is a major determinant of the standing crop and taxonomic composition of the ice algae, both of which are lower near freshwater outflows such as the Grand rivière de la Baleine (Poulin et al. 1983). The freshwater layer appears to reduce productivity of the underice habitat, likely due to a combination of factors, including nutrient limitation (Ingram and Prinsenberg 1998) and changes in the osmotic pressure of the ice/water interface. The overall effect of these freshwater plumes on the productivity of ice algae in Hudson Bay is significant yet still poorly understood (e.g., Gosselin et al. 1986; Legendre et al. 1992).

Remote sensing of chromomorphic dissolved organic matter (CDOM) offers an important new tool for distinguishing freshwater components (runoff and sea ice melt) of the surface circulation in Hudson Bay and Hudson Strait (Granskog et al. 2009). The CDOM fingerprint of terrestrial runoff into the coastal waters is apparent in the ocean colour. Offshore, where the main freshwater input is from sea ice melt the CDOM levels are lower than they are inshore, and similar to the central Arctic Ocean and Beaufort Sea (Granskog et al. 2009). These differences make it possible to follow the dispersal and dynamics of river runoff during the ice-free season. Ocean colour can also be used indirectly to detect potential source and sink regions of atmospheric CO_2 (Granskog et al. 2009).

The delivery of coloured dissolved organic matter (CDOM) in runoff and its circulation result in significant inshore-offshore gradients in, for example, UV light attenuation, photic depth and radiant heating in Hudson Bay, which themselves influence other photoprocesses such as the potential photochemical production of nutrients (Granskog et al. 2009). These CDOM results highlight the need to investigate the composition and role of organic matter as a potential additional source of biologically labile substrates in the otherwise low-nutrient Hudson Bay, and perhaps a key factor in the higher biomass observed in Hudson Bay coastal waters. They also suggest that CDOM may play an important role in the penetration of UV radiation, vertical distribution of biomass and the radiant heating of the surface ocean.

Seasonal Circulation

In eastern and central Foxe Basin, where the water is relatively shallow, the water mass is homogeneous in summer due to strong tidal and wind mixing; whereas to the west and south the water is stratified (Campbell 1964; Griffiths et al. 1981; Prinsenberg 1986c). Seasonal ranges in surface temperature are very small, from −1.7°C to about 3.1°C. In the spring, surface salinities as low as four have been observed offshore, reflecting dilution by meltwater and river runoff. By late summer and early autumn, surface salinities have returned to 29.0–32.0. The density of the surface water increases during the winter due to cooling and brine rejection by ice formation. These density effects are greatest in eastern Foxe Basin where the water is shallow and thick ice forms. The cold, saline water (t = −1.97°C, S = 34.07) flows downhill to become the bottom water in deeper areas of Foxe Basin and Foxe Channel (Campbell 1964; Prinsenberg 1986c). Intermittent flow of this dense water over the sill that separates Foxe Channel from Hudson Bay likely maintains the homogeneous bottom layer in Hudson Bay.

In spring and summer, the water entering northwestern Hudson Bay is diluted by meltwater and runoff from the land, warmed by the sun, and mixed by the wind as it circulates cyclonically around Hudson Bay and James Bay. This produces strong vertical stratification of the water column that is characteristic of these areas in summer, particularly offshore (Fig. 7). Summer surface temperatures typically range from 1 to 9°C and salinities from 21 to 32 but can be lower close to shore and in bays receiving discharge from rivers (Fig. 8) (Barber 1967; Anderson and Roff 1980a; Prinsenberg 1986a). Under open water conditions, the rates of longshore bedload and suspended sediment transport by waves and currents are higher when water temperatures are near freezing than under warmer summer conditions (Héquette and Tremblay 2009). A strong pycnocline (density gradient) develops at about 20 m depth that slows vertical mixing, thereby limiting nutrient additions to surface waters and hence biological productivity. Inshore areas generally have lower water temperatures, salinities, and clarities and higher chlorophyll *a*, ATP (adenosine triphosphate), and pelagic biomass than the offshore areas (Stewart and Lockhart 2005).

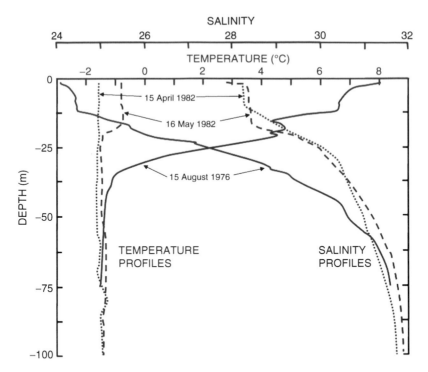

Fig. 7 Representative vertical profiles of temperature and salinity in southeastern Hudson Bay at various times of the year (different years) (Redrawn from Ingram and Prinsenberg 1998, p. 851)

These differences may be attributable to mixing processes which bring colder, deeper, relatively nutrient-rich water to the surface, and to dilution and nutrient addition by freshwater runoff. In winter, lower runoff, salt rejection from the growing ice cover, and surface cooling weaken the vertical stratification and permit very slow vertical mixing (Roff and Legendre 1986). Temperature and salinity are relatively stable below a depth of 50 m, but small changes related to the seasonal disappearance of the pycnocline have been observed to 65 m in James Bay and 100 m in Hudson Bay. The water becomes progressively colder and more saline with depth, approaching the same deep water type (T = −1.4°C, S > 33) at about 100 m. The bottom water in James Bay is subject to considerable seasonal and interannual variation in temperature and salinity, due in part to the relative shallowness of the bay.

Temperature and salinity conditions in the large estuaries can vary widely over the year. In the Churchill River estuary, for example, salinity in the upper few meters can range from near zero during the spring freshet to 17 or 18 in winter and pre-melt (Kuzyk et al. 2008). Below the halocline at 5–6 m salinity can range from 29 in the fall to 33.5 in winter and pre-melt. In 2005, water temperatures during the winter and pre-melt period were consistently near freezing.

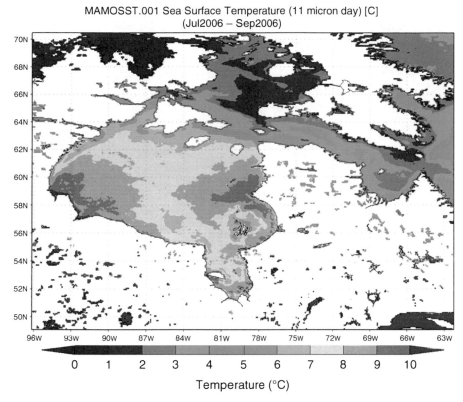

Fig. 8 Summer (July–September) sea surface temperature in 2006 (Prepared using the Giovanni online data system, developed and maintained by the NASA Goddard Earth Sciences (GES) Data and Information Services Center (DISC)

The westward inflow of Arctic surface water along the south shore of Baffin Island and eastward outflows from Foxe Basin and Hudson Bay along the Quebec coast create a marked cross-channel gradient in the surface temperature and salinity of Hudson Strait (Drinkwater 1986, 1988; Straneo and Saucier 2008b). Recent river runoff from Hudson Bay and James Bay is concentrated in a narrow wedge along the south side of the Strait, raising the water temperature, lowering the salinity, and strengthening stratification of the water column (Tan and Strain 1996). Midway along the Strait the bulk of the fresh water passes within 20 km of shore (Straneo and Saucier 2008b). The average summer current along the south shore is intense, reaching a maximum of 0.3 m·s^{-1} in the Cape Hopes Advance area (Drinkwater 1988). This freshwater forcing is seasonal (June to March) and accelerates as the freshest waters pass, peaking in November and December (Straneo and Saucier 2008b).

The warm, dilute surface outflow from Hudson Bay is mixed with colder surface water from Foxe Basin and colder, higher-salinity deep water as it passes through Hudson Strait (Straneo and Saucier 2008b). Vertical mixing is intense in eastern

Hudson Strait (Collin and Dunbar 1964). Arctic properties (low temperature and salinity) of the combined flow are fed into the Labrador Current at the north end of the Labrador Shelf and carried southward to influence coastal ocean climates of Atlantic Canada (Stewart and Lockhart 2005). The freshwater flux from the HB complex exiting Hudson Strait is about 30 mSv, which is significant and on the order of one third the freshwater flux from Davis Strait (Straneo and Saucier 2008a).

Sea Ice

The HB complex is essentially ice-covered in winter and ice-free in summer. Each spring the melting of this ice contributes significantly to the freshwater budget. While most ice forms locally, a small amount of multi-year ice can enter via Fury and Hecla Strait (Markham 1986) although this is a very rare event given the recent reduction in sea ice in the northern hemisphere (Smith and Barber 2007). Multi-year ice is rare in Hudson Bay and absent from James Bay. Icebergs are common in Hudson Strait and Ungava Bay but extremely rare elsewhere in the region.

Ice formation progresses from northwest to southeast, beginning in early October in northern Foxe Basin, spreading into northern Hudson Bay and Hudson Strait by mid-October, and finally into southern James Bay and eastern Hudson Strait by early December (Fig. 9). The extent and thickness of the ice increase rapidly dur-

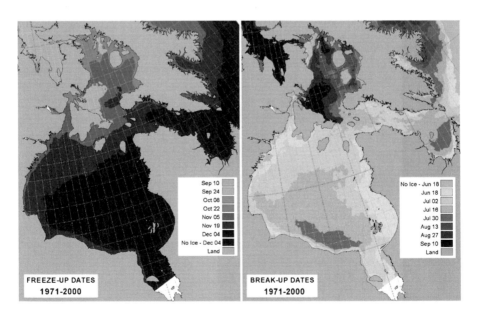

Fig. 9 Ice formation and breakup pattern in the Hudson Bay Complex over the 21 year period 1971–2000 (From http://www.ice.ec.gc.ca/IA_NWCA_SM/ar_freezeup.gif; http://www.ice.ec.gc.ca/IA_NWCA_SM/ar_breakup.gif). © Her Majesty The Queen in Right of Canada, 2010. Reproduced with the permission of the Minister of Public Works and Government Services Canada

ing November and December, with maximum coverage in late April or early May. Maximum ice thickness varies but generally increases from south to north, ranging from about 1 m in southern James Bay (Moosonee 71–130 cm) to 2 m in northern Foxe Basin (Hall Beach 156–293) (Prinsenberg 1988; Loucks and Smith 1989). Freshwater budgets of Hudson Bay and Foxe Basin indicate that up to 90% more ice is produced than is indicated by the thickness data, some of it as ice pressure ridges that can reach heights of 3–3.5 m (Prinsenberg 1988).

Landfast ice forms along most of the coastline, extending farther offshore from protected or shallow coastlines than from exposed or cliff and headland coasts (http://www.natice.noaa.gov/). During severe winters it covers much of southeastern Hudson Bay and can form a bridge between the Belcher Islands and mainland. During such years, Wager Bay, Frozen Strait, and northern Foxe Basin may also be covered by landfast ice.

In winter and early spring the ice floes that cover most of the HB complex are kept in constant motion by the wind (Markham 1986; Stewart and Lockhart 2005) (Fig. 10). Ice scouring of nearshore ecosystems to a depth of 5–10 m below the low tide mark is common in Arctic coastal waters (e.g., Heine 1989; Conlan et al. 1998), and the scour zone is very wide in the areas of Hudson Strait and Ungava Bay with extreme tidal ranges (Fig. 5). The floes in Foxe Basin have a rugged appearance from being formed under stormy conditions and moved constantly by strong tidal currents and winds (Markham 1986). In shallow areas where bottom sediments are suspended by autumn storms this ice is a distinctive brown colour (Prinsenberg 1986c).

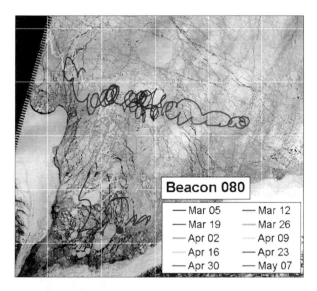

Fig. 10 Ice beacon drift track in the Nelson Estuary from 5 March to 7 May 2009. Data are colour coded weekly, with each colour beginning at the date shown in the legend. The underlying Landsat TM image was recorded on 27 March. The *red dot* marks the beacon position at the time of the overpass, 10:50 CST 27 March (Prepared by ArcticNet, University of Manitoba)

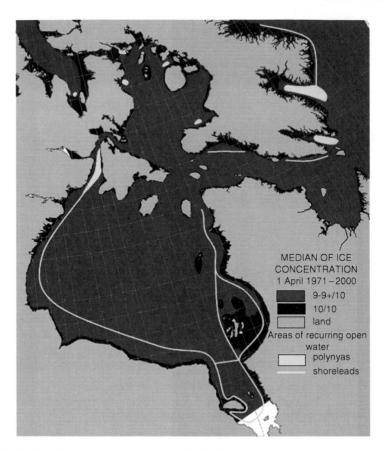

Fig. 11 Median of sea ice concentration in the Hudson Bay complex over the 21 year period 1971–2000 (From http://www.ice.ec.gc.ca/IA_NWCA_MCSI/ar_ctmed0401.gif), overlain by a schematic depicting areas of recurring open water (Stirling and Cleator 1981). © Her Majesty The Queen in Right of Canada, 2010. Reproduced with the permission of the Minister of Public Works and Government Services Canada

The pack ice in Hudson Strait and Ungava Bay tends to be less concentrated than elsewhere in the region, enabling more overwintering by whales and walrus.

Openings in the sea ice, particularly flaw leads and polynyas, are vitally important to overwintering species and early spring migrants (Stirling and Cleator 1981; Barber and Massom 2007). A flaw lead complex develops along the shore and at fast ice edges when the wind blows offshore (Fig. 11). These leads are most common along the west shore of Hudson Bay from Churchill to Coral Harbour, and in northern Hudson Strait from Cape Dorset to Big Island (62°43' N, 70°43' W) (Markham 1986). Recurring polynyas form in the northwest and southeast of Foxe Basin; in Hudson Bay near islands along the southeast coast, in the Belchers, and in Roes Welcome Sound; and in James Bay off the southwest tip of Akimiski Island (Martini and Protz 1981; Stirling and Cleator 1981; Nakashima 1988; Gilchrist and Robertson 2000; Barber and Massom 2007). The latter polynya is one of the most southerly in Canadian seas. A large polynya in the northwest corner of Hudson Bay also plays an important role in bottom water formation throughout the winter, and as an 'ice factory' producing first-year ice which is then exported into the anticlockwise circulation of sea ice in Hudson Bay (Hochheim and Barber 2009). Ice edges at the polynyas and along the well developed flaw lead system, where the coastal fast ice and mobile pack ice meet, are areas of upwelling and associated stimulation of phytoplankton blooms. The flaw lead polynya system around the bay can also play an important role in both biogeochemical processes and associated ecosystem function, particularly when a large freshwater input occurs spatially with the flaw lead polynya (e.g., Kuzyk et al. 2008).

Breakup begins in late May or early June along the coasts downstream from large estuaries and marine inflows (Fig. 3). James Bay is normally ice free by early July; Hudson Bay, Hudson Strait and Ungava Bay by late July or early August; and Foxe Basin by mid-September. Winds and currents carry the melting ice south causing a massive transfer of fresh water (Saucier et al. 2004). In Hudson Bay, the last ice to melt is usually concentrated along the Ontario coast, and in Hudson Strait/Ungava Bay it is in eastern Ungava Bay. Foxe Basin has a longer ice season and sometimes has remnant ice amidst the northern islands and along the north coast of Southampton Island in September (Prinsenberg 1986c). Depending upon weather conditions, the timing of freeze-up or breakup may be retarded or advanced by up to a month, but the basic pattern of ice formation is similar from year-to-year (Cohen et al. 1994; Stewart and Lockhart 2005). Further details of the sea ice extent and concentration anomalies and their association with atmospheric and oceanic forcing are found elsewhere in this book (Hochheim et al. this volume).

Biota and Habitat Use

Plants and animals that inhabit the HB complex are well adapted to the existing marine conditions and their seasonal changes. The following sections provide examples of how some of the key climate-sensitive oceanographic parameters in the HB complex, in particular the fresh water inputs, nearly complete seasonal ice cover and dynamic coastal morphology, influence species ecology.

Influence of Freshwater Inputs

In the spring and early summer the surface of the HB complex receives enormous volumes of freshwater, mostly as runoff and meltwater from the sea ice but some as precipitation. The volume, form, and timing of these inputs strongly influence water column stability, biological productivity, the timing and pattern of ice breakup, species distributions, and surface circulation (Prinsenberg 1988; Stewart and Lockhart 2005). This freshwater also affects sea ice thermodynamic processes directly (growth and ablation) and these, in turn, are affected by large scale atmospheric circulation patterns related to the northern annual mode (Hochheim and Barber 2009). A significant reduction in freshwater inputs to the surface of the HB complex could have profound impacts on aspects of the ecology such as the primary production by phytoplankton under ice and in the ocean surface mixed layer, use of coastal waters by fish and beluga, fish recruitment, and overwintering by Hudson Bay eider (*Somateria mollissima sedentaria*).

Water Column Stability and Biological Productivity

The seasonal dilution of surface waters creates a strong density gradient (pycnocline) that stabilizes the water column, limiting vertical movement of nutrients into the surface waters (euphotic zone) and thereby primary production. In areas such as northern Hudson Bay, where tides are weak and there is little coastal development or bottom relief to promote mixing, phytoplankton can become nutrient limited (Drinkwater and Jones 1987). Nitrogen depletion can limit primary production between the sea ice and the pycnocline in late spring (Maestrini et al 1986; Demers et al. 1989; Gosselin et al. 1990; Bergmann et al. 1991; Welch et al. 1991) and above the pycnocline after breakup (Anderson and Roff 1980a, b; Roff and Legendre 1986). In consequence, the primary production of inshore areas of Hudson Bay is comparable to seasonally open-water areas of Canada's Arctic Archipelago (Roff and Legendre 1986; Subba Rao and Platt 1984). Such nutrient limitation is less likely in Hudson Strait where there is greater mixing.

During the summer, chlorophyll-*a* concentrations at the surface tend to be highest nearshore, particularly in bays and estuaries where there is more mixing and nutrient input from the land, and near islands where there is periodic entrainment or upwelling of deeper, nutrient-rich water (Fig. 12). In offshore waters of Hudson Bay and in western Hudson Strait primary production maxima occur below the pycnocline, where nutrient concentrations are higher and clear water allows sufficient light penetration for photosynthesis (0.1–1% of surface light) (Anderson 1979; Anderson and Roff 1980b; Roff and Legendre 1986; Harvey et al. 1997). In southeastern Hudson Bay, James Bay, and coastal waters of Hudson Bay the midsummer chlorophyll *a* maximum generally occurs in the upper 20–25 m (Grainger 1982), at or above the pycnocline (Anderson and Roff 1980b).

Use of Coastal Waters by Fish

Runoff dilutes coastal waters around the mouths and downstream of the large rivers that flow into southern Hudson Bay and James Bay, creating an extensive brackish zone (Granskog et al. 2009). The ability to exploit this zone, particularly the large estuaries, is an important ecological adaptation for both the freshwater and Arctic marine fish species in the HB complex (Morin and Dodson 1986; Ponton et al. 1993; Stewart and Lockhart 2005). The estuaries provide important seasonal

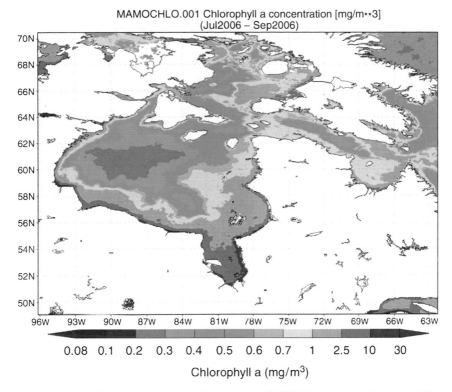

Fig. 12 Chlorophyll-*a* concentration in the summer of 2006 (July through September) from MODIS-Aqua remote sensing data. (Prepared using the Giovanni online data system, developed and maintained by the NASA Goddard Earth Sciences (GES) Data and Information Services Center (DISC)

foraging habitat and nursery habitat for many freshwater, anadromous and marine species; spawning habitat for marine species such as Arctic cod (*Boreogadus saida*), shannies (Stichaeidae) and sculpins (Cottidae); and year-round habitat for fourhorn sculpin (*Myoxocephalus quadricornis*). To take advantage of these habitats many of the species undertake complex seasonal movements that are influenced by variations in temperature and salinity (Morin et al. 1981). These movements can be extensive. Anadromous Arctic char (*Salvelinus alpinus*) in Hudson Bay, for example, can make marine sojourns of at least 800 km (G.W. Carder, personal communication 1991). They feed heavily in the marine environment and shed their freshwater parasites, enabling them to grow faster and larger than Arctic char that remain in fresh water.

Estuarine habitats in the south support more freshwater and anadromous species and fewer Arctic and deepwater species than those in the north (Morin et al. 1980; Morin and Dodson 1986; Stewart and Lockhart 2005). Warm, shallow, dilute estuaries along the Quebec coast, from the Eastmain River north to Lac Guillaume Delisle (Richmond Gulf), attract typically freshwater species such as lake trout

(*S. namaycush*), lake whitefish (*Coregonus clupeaformis*), lake cisco (*C. artedi*), and burbot (*Lota lota*). Even species that seldom enter brackish water, such as the white sucker (*Catostomous commersoni*) and walleye (*Sander vitreum*) visit the estuaries of James Bay. Farther north and moving offshore, where the salinity is higher and water colder, there are fewer freshwater species and more Arctic species such as Arctic char, Greenland cod (*Gadus ogac*) and shorthorn sculpin (*M. scorpius*).

Use of Estuaries by Belugas

The largest summering concentration of belugas in the world occurs in the Nelson River estuary area, and there are smaller concentrations at the estuaries of the Seal, Churchill, Winisk, Severn, and Nastapoka rivers (Richard 1991, 2005). Use of these estuaries by belugas may be related to moulting and/or neonate survival (Stewart and Lockhart 2005). It may also be a strategy for avoiding predation by killer whales when ice is absent (P. Richard, DFO Winnipeg, personal communication 2010). Their persistence at these locations, in the face of intense harvesting pressure, suggests that these estuaries serve an extremely important habitat function for the species. The sensitivity of this function to changes in the volume or seasonality of the runoff is unknown. But, belugas in the Nelson River estuary occupy different habitat during low and high runoff years, suggesting that runoff plays an important role in the species' habitat use (Smith 2007).

Fish Recruitment

The timing and extent of the spring freshet influences the estuarine distributions of marine fish larvae, and determines when anadromous and freshwater fish larvae enter the estuaries (Ponton et al. 1993). In the spring and summer of 1988–1990, Arctic cod and sandlance (*Ammodytes* spp.) were the most abundant larvae in and around the plume of Grande rivière de la Baleine. The larval densities of these species, and of slender eelblenny (*Lumpenus fabricii*) and gelatinous seasnail (*Liparis fabricii*), were greatest in salinities >25 psu; Arctic shanny (*Stichaeus punctatus*), sculpins, and capelin (*Mallotus villosus*) larvae were more abundant at salinities of 1–25 psu; and burbot and coregonine larvae were associated with fresh or brackish water, even in Hudson Bay.

In the spring, first-feeding larval Arctic cod and sandlance accumulated under the ice at the pycnocline below the river's brackish plume (Ponton and Fortier 1992). This habitat offered them salinities they could tolerate and the optimal combination of prey density and visibility. When the plume became too thick and the depth of the 25 ppt [psu] isohaline exceeded 9 m the larvae stopped feeding (Fortier et al. 1996). Outside the plume area, where the surface meltwater layer was clearer than the river plume, the under-ice distribution of larval Arctic cod and sandlance was less stratified and appeared to be less constrained by light penetration and salinity. By affecting both prey density and light, plume thickness is an important determinant of feeding success by larval fishes (Fortier et al. 1995, 1996).

Feeding success and growth of these fish larvae also depend upon how well the timing of their hatch matches the reproductive cycle of the copepods they prey upon (Fortier et al. 1995). While the fish species hatch at about the same time each year, the copepod lifecycle is not always in synchrony so the nauplii they eat are not always available at the right time. Larval survival is closely coupled with the reproductive cycle of the copepods they prey upon (Fortier et al. 1995, 1996), so environmental changes that disrupt these cycles could also reduce recruitment.

Surface Circulation

The Hudson Bay subspecies of the common eider, *Somateria mollissima sedentaria*, winters in southeastern Hudson Bay where open water and shallow depth coincide (Abraham and Finney 1986; Nakashima 1988). Belcher Islands Inuit report their presence, sometimes in quantity, at almost every ice edge that is accessible from Sanikiluaq in winter and in a number of polynyas (Nakashima 1988). The eiders rely on winds and currents to maintain these areas of open water throughout the winter. In the winter of 1991–1992, many eiders were found frozen into areas where the water usually remains open in winter. Inuit attributed the closure of these openings to calm wind conditions and unusually weak winter currents (McDonald et al. 1997; Robertson and Gilchrist 1998; Mallory et al. this volume).

Importance of Seasonal Ice Cover

The quality, quantity, and duration of the seasonal sea ice cover strongly influence the ecology of the nearshore and ice biota, pelagic systems under the ice and at ice edges, and use of the ice surface as a platform for travel and reproduction (Stewart and Lockhart 2005). As interfaces between air, ice, and water the ice edge habitats are important sites of energy transfer, where mixing occurs and biota are attracted to feed. The presence of nearly complete ice cover, with extensive areas of landfast ice and nearshore scouring, is an important determinant of which species can survive in the region (Stewart and Howland 2009). Significant changes to ice cover in the HB complex could have profound impacts on aspects of its ecology such as the well-developed epontic (under-ice) community, nearshore benthic flora and fauna, food webs, and presence of ice-adapted marine mammals.

Seasonal ice cover also influences human activities within the HB complex and the timing and routing of vessel traffic to and from the region (Stewart and Howland 2009). The reliance of aboriginal peoples (Inuit and coastal Cree) on sea ice for travelling and hunting is reflected in their detailed knowledge of its processes, characteristics, and annual cycles (McDonald et al. 1997).

Fig. 13 The underside of the sea ice is an important site of primary production by ice algae (*top*) and the upper surface provides a platform for marine mammals (*bottom*) (Photo credits: top: Jeremy Stewart, bottom: ArcticNet, University of Manitoba)

Epontic Community

The bottom surface of the sea ice supports a diverse community of ice adapted flora and fauna (Stewart and Lockhart 2005) (Fig. 13). The species composition and growth of the sea ice microalgae change seasonally in response to temperature, light, salinity and nutrients (Demers et al. 1986; Roff and Legendre 1986; Bergmann et al. 1991; Welch et al. 1991; Legendre et al. 1992, 1996; Monti et al. 1996). Their under-ice distribution is patchy (Gosselin et al. 1986; Bergmann et al. 1991). On a large scale (30 km) it is directly related to salinity, which affects the ice surface available for colonization; on a smaller scale (0.3–500 m) it is controlled by variations in the thickness of snow-ice cover which affects shortwave extinction. During the spring blooms some of the ice algal production is exported to the benthos, as sinking cells or fecal pellets of herbivores, but most may be retained in the pelagic environment (Tremblay et al. 1989).

Planktonic nematodes, rotifers, ciliates, and copepods also live in the lower 3 cm of the sea ice (Hsiao et al. 1984; Grainger 1988). Their abundance has been positively related to salinity, and to the presence of sea ice microflora (Grainger 1988; Tourangeau 1989). The ice fauna is generally denser but less diverse than the zooplankton living under the ice. During and after the bloom ice algae are an important source of food for the marine planktonic copepods *Calanus glacialis* and *Pseudocalanus minutus* (Runge and Ingram 1991; Tourangeau and Runge 1991), which are themselves important links in the marine food web. Contributions of the ice flora and fauna to the food web of the HB complex are vulnerable to changes in salinity and illumination, and to the continued availability of ice during periods when the illumination is sufficient for photosynthesis.

Nearshore Benthic Flora and Fauna

Grounding ice modifies the seabed topography, reworks the sediment, and ploughs and crushes the seabed biota in Arctic marine environments (Conlan et al. 1998; Conlan and Kvitek 2005). In the HB complex this scouring action prevents the establishment of a rich bottom flora in shallows and nearshore (Bursa 1968). Few species of benthic invertebrates inhabit the intertidal zone on a permanent basis (Dadswell 1974; Stewart and Lockhart 2005). Exposure to freezing in the winter also limits the permanent establishment of flora and fauna in the intertidal zone (Dale et al. 1989), which in some areas can be very extensive.

Food Webs

The presence of seasonal ice cover is an important determinant of which species can live in the HB complex year-round, which are seasonal visitors, and which cannot survive. The timing of ice formation and breakup determine when habitats and food resources are available. If the timing is not suitable the affected biota must be capable of adapting or moving if they are to survive.

The plant, invertebrate and fish faunas of the HB complex consist of a mixture of freshwater, estuarine, and marine forms (Roff and Legendre 1986, Stewart and Lockhart 2005). Many of the species are widely distributed outside this region, generally in Arctic waters. Species assemblages in James Bay and southeastern Hudson Bay reflect the massive seasonal freshwater inputs and estuarine character of the circulation (Grainger and McSween 1976), while those in Hudson Strait include more Atlantic and deepwater species (Stewart and Lockhart 2005). With the exception of anadromous fishes that winter in fresh water and undertake marine sojourns in the summer, these biota are resident in the HB complex year-round.

Seasonal movements of marine invertebrates and fishes into and out of the region have not been described.

The mammal fauna consists largely of migratory species that are either unable to access the surface when ice is present, or require ice to gain offshore access. Harp (*Pagophilus groenlandicus*) and hooded (*Cystophora cristata*) seals and at least six species of whales including beluga, bowhead (*Balaena mysticetus*), narwhal (*Monodon monoceros*), killer (*Orcinus orca*), minke (*Balaenoptera acutorostrata*), and humpback (*Megaptera novaeanglia*; S. Ferguson, 2007, DFO Winnipeg, personal communication) whales are typically seasonal visitors to the region, although the first three species do overwinter in Hudson Strait and sometimes in leads and polynyas elsewhere (Stewart and Lockhart 2005). The timing of their seasonal movements can vary by a month or so from year to year depending upon ice conditions, and late departure can lead to ice entrapment mortalities. Polar bears and Arctic foxes (*Vulpes lagopus*) frequent coastal areas in summer and ice habitats during other seasons. Typically, only the Atlantic walrus and ringed (*Pusa hispida*), bearded (*Erignathus barbatus*), and harbour (*Phoca vitulina*) seals inhabit the waters of the HB complex year-round.

While the HB complex provides resources of critical national and international importance to migratory seabirds, waterfowl and shorebirds, few species other than the Hudson Bay eider are resident year-round (Morrison and Gaston 1986; Stewart and Lockhart 2005). These eiders are uniquely equipped to access benthic food resources in polynyas, where strong currents keep the surface open year-round and create rich feeding habitats (Heath et al. 2006). Other species winter farther south or east where they have easier access to marine resources, and the air and water temperatures are warmer. Migrants are sensitive to changes in ice conditions. Persistent spring ice cover, for example, will delay breeding by thick-billed murre (*Uria lomvia*) and alters their diet (Gaston and Hipfner 1998).

Presence of Ice-Adapted Marine Mammals

Ice cover provides a vital platform for ringed seals that live year-round in the HB complex, and for the polar bears that prey upon them. In winter, adult ringed seals generally occupy stable landfast ice where they maintain breathing holes through the ice and build subnivean lairs in which to haul out and/or pup (McLaren 1958; Frost and Lowry 1981; Smith 1987; Smith et al. 1991). In spring, the highest densities of breeding adults occur on stable landfast ice in areas with good snow cover, whereas non-breeders occur at the floe edge or in the moving pack ice (Smith 1975; Lunn et al. 1997; Holst et al. 1999; Chambellant this volume). Ringed seals can often be seen in spring when they come out onto the ice beside their breathing holes to bask in the sun and moult their hair coat. Ringed seal pups are born in late March or early April, nursed on the ice, and abandoned at ice breakup (McLaren 1958) (Fig. 13). Recruitment may suffer if the landfast ice breakup occurs earlier in the spring or there is less than 32 cm of snow cover (Ferguson et al. 2005).

Polar bears rely on the ice to provide a travel platform for hunting seals, and prefer areas where wind, water currents or tides cause the ice to crack and re-freeze (Fig. 13) (Urquhart and Schweinsburg 1984; Stirling and Ramsay 1986; Kolenosky et al. 1992; Stirling et al. 2004; Lunn et al. 2005). They move onto the ice when it forms in the fall and travel widely until melting forces them ashore to fast until freeze-up. While on the ice, they hunt ringed and bearded seals to build up the fat stores they require to survive their summer fast (see also Iverson et al. 2006). These bears are vulnerable to any change in the ice regime that lengthens their fast or reduces the availability of their prey.

Importance of Coastal Emergence

Tidal salt marshes are diverse and highly productive ecosystems that exist within a small elevation range relative to sea level, and rely on a variety of processes for their continued existence (Stewart and Lockhart 2005). These habitats often have seaward slopes of <2 m·km^{-1}, and show species zonations that are clearly linked to salinity and elevation, water content, and soil texture (Ringius 1980; Glooschenko and Clarke 1982; Protz 1982; Riley 1982; Martini and Glooschenko 1984; Ewing and Kershaw 1986; Earle and Kershaw 1989). Postglacial emergence is continuously exposing new land along these shallow coastlines for colonization by a succession of vegetation (Hik et al. 1992; Stewart and Lockhart 2005). Provided the rate of emergence remains constant, new marsh created on the seaward side offsets the encroachment of inland vegetation, maintaining the amount and composition of the marsh. If glacial meltwaters were to offset isostatic rebound, the rate of coastal emergence would be effectively reduced, possibly for centuries, and these marshes would no longer support large numbers of migratory birds and other species.

Concluding Remarks

In the coming decades it will be important to gain a better understanding of Canada's vast inland sea. It is through this understanding that the impacts of climate change, a growing human population, hydroelectric and non-renewable resource developments, shipping, and the long range transport of contaminants on this fascinating and potentially vulnerable ecosystem are to be predicted and mitigated. Long-term, stable research is required to achieve this goal. Scientific studies and traditional knowledge have important, complementary roles to play in assessing long-term environmental trends in the HB complex, and whether they are driven by natural cycles, human activities or, more likely, a combination of both.

References

Abraham, K.F., and Finney, G.H. 1986. Eiders of the eastern Canadian Arctic. Environment
    Canada, Canadian Wildlife Service, Report Series Number 47: 55–73.
Anderson, J.T. 1979. Seston and phytoplankton data from Hudson Bay, 1975. Fisheries and
    Environment Canada, Canada Centre for Inland Waters, Technical Note Series No. 80–2:
    xii + 76 p.
Anderson, J.T., and Roff, J.C. 1980a. Seston ecology of the surface waters of Hudson Bay. Can.
    J. Fish. Aquat. Sci. 37: 2242–2253.
Anderson, J.T., and Roff, J.C. 1980b. Subsurface chlorophyll a maximum in Hudson Bay. Nat.
    Can. (Que.) 107: 207–213.
Andrews, J.T. 1974. Post glacial rebound, pp. 35–36. In National atlas of Canada. Department of
    Energy, Mines, and Resources, Ottawa.
Antonov, J.I., Locarnini, R.A., Boyer, T.P., Mishonov, A.V., and Garcia, H.E. 2006. World Ocean
    Atlas 2005, Volume 2: Salinity. In S. Levitus (ed) NOAA Atlas NESDIS 62, U.S. Government
    Printing Office, Washington, DC, p. 182 [Available online at: http://www.nodc.noaa.gov/OC5/
    indprod.html].
Barber, D.G., and Massom, R.A. 2007. The role of sea ice in Arctic and Antarctic Polynyas,
    pp. 1–54. In W.O. Smith and D.G. Barber (eds) Polynyas: Windows to the World. Elsevier
    Oceanography Series 74.
Barber, F.G. 1965. Current observations in Fury and Hecla Strait. J. Fish. Res. Board Can. 22:
    225–229.
Barber, F.G. 1967. A contribution to the oceanography of Hudson Bay. Department of Energy,
    Mines and Resources, Ottawa, Marine Sciences Branch, Manuscript Report Series No. 4: 69.
Barber, F.G. 1972. On the oceanography of James Bay. Canada Department of the Environment,
    Marine Sciences Branch, Manuscr. Rep. Ser. 24: 3–96.
Barber, F.G. 1983. Macroalgae at 75 m in Hudson Bay. Can. Tech. Rep. Fish. Aquat. Sci. 1179:
    iv + 17 p.
Barr, W. 1979. Hudson Bay: the shape of things to come. Musk-ox 11: 64.
Bergmann, M.A., Welch, H.E., Butler-Walker, J.E., and Siferd, T.D. 1991. Ice algal photosynthesis
    at Resolute and Saqvaqjuac in the Canadian Arctic. J. Mar. Syst. 2: 43–52.
Bilodeau, J., de Vernal, A., Hillaire-Marcel, C., and Josenhans, H. 1990. Postglacial paleoceanog-
    raphy of Hudson Bay: stratigraphic, microfaunal, and palynological evidence. Can. J. Earth
    Sci. 27: 946–963.
Budgell, W.P. 1976. Tidal propogation in Chesterfield Inlet, N.W.T. Fisheries and Environment
    Canada, Canada Centre for Inland Waters, Manuscript Report Series No. 3: xiv + 99 p.
Budgell, W.P. 1982. Spring-neap variation in the vertical stratification of Chesterfield Inlet,
    Hudson Bay. Nat. Can. (Que.) 109: 709–718.
Bursa, A.S. 1968. Marine plants, pp. 343–351. In C.S. Beals and D.A. Shenstone (ed) Science,
    history and Hudson Bay, Vol. 1. Department of Energy, Mines and Resources, Ottawa.
Cairns, D.K., and Schneider, D.C. 1990. Hot spots in cold water: Feeding habitat selection by
    thick-billed murres, pp. 52–60. In S.G. Sealy (ed) Auks at sea. Studies in Avian Biology 14.
Campbell, N.J. 1964. The origin of cold high-salinity water in Foxe Basin. J. Fish. Res. Board
    Can. 21: 45–55.
Canadian Hydrographic Service. 1998. Tides and current tables 1999, Vol. 4, p. 98. Arctic and
    Hudson Bay. Department of Fisheries and Oceans, Ottawa, ON.
Cohen, S.J., Agnew, T.A., Headley, A., Louie, P.Y.T., Reycraft, J., and Skinner, W. 1994. Climate
    variability, climatic change, and implications for the future of the Hudson Bay bioregion. An
    unpublished report prepared for the Hudson Bay Programme. Canadian Arctic Resources
    Council (CARC), Ottawa, ON. viii + 113 p.
Collin, A.E., and Dunbar, M.J. 1964. Physical oceanography in Arctic Canada. Oceanogr. Mar.
    Biol. Annu. Rev. 2: 45–75.

Conlan, K.E., and Kvitek, R.G. 2005. Recolonization of soft-sediment ice scours on an exposed Arctic coast. Mar. Ecol. Prog. Ser. 286: 21–42.

Conlan, K.E., Lenihan, H.S., Kvitek, R.G., and Oliver, J.S. 1998. Ice scour disturbance to benthic communities in the Canadian High Arctic. Mar. Ecol. Prog. Ser. 166: 1–16.

Curtis, S.C. 1975. Distribution of eelgrass: east coast, James Bay. Map at a scale of 1:125,000. Canadian Wildlife Service, Ottawa, ON.

d'Anglejan, B. 1980. Effects of seasonal changes on the sedimentary regime of a subarctic estuary, Rupert Bay (Canada). Sediment. Geol. 26: 51–68.

Dadswell, M.J. 1974. A physical and biological survey of La Grande River estuary, James Bay, Quebec. Can. Field-Nat. 88: 477–480.

Dale, J.E., Aitken, A.A., Gilbert, R., and Risk, M.J. 1989. Macrofauna of Canadian Arctic fjords. Mar. Geol. 85: 331–358.

Defossez, M., Saucier, F.J., Myers, P.G., and Caya, D. 2005. Observation and simulation of deep water renewal in Foxe Basin, Canada. EOS Trans. AGU 86: Fall Meet. Suppl., Abstract OS41B-0598.

Demers, S., Legendre, L., Therriault, J.C., and Ingram, R.G. 1986. Biological production at the ice-water ergocline, pp. 31–54. *In* J.C.J. Nihoul (ed) Marine interfaces ecohydrodynamics. Elsevier, Amsterdam.

Demers, S., Legendre, L., Maestrini, S.Y., Rochet, M., and Ingram, R.G. 1989. Nitrogenous nutrition of sea-ice microalgae. Polar Biol. 9: 377–383.

Déry, S.J., Stieglitz, M., McKenna, E.C., and Wood, E.F. 2005. Characteristics and trends of river discharge into Hudson, James, and Ungava Bays, 1964-2000. J. Climate 18: 2540–2557.

Dignard, N., Lalumiere, R., Reed, A., and Julien, M. 1991. Habitats of the northeastern coast of James Bay. Environment Canada, Canadian Wildlife Service, Occasional Paper Number 70: 27 p. + map.

Dionne, J.-C. 1980. An outline of the eastern James Bay coastal environments. Geol. Surv. Can. Pap. 80-10: 311–338.

Dohler, G.C. 1968. Tides and currents, pp. 824–837. *In* C.S. Beals and D.A. Shenstone (ed) Science, history and Hudson Bay, Vol. 2. Department of Energy, Mines and Resources, Ottawa.

Drinkwater, K.F. 1986. Physical oceanography of Hudson Strait and Ungava Bay, pp. 237–264. *In* I.P. Martini (ed) Canadian inland seas. Elsevier Oceanography Series 44. Elsevier Science, Amsterdam.

Drinkwater, K.F. 1988. On the mean and tidal currents in Hudson Strait. Atmosphere-Ocean 26: 252–266.

Drinkwater, K.F., and Jones, E.P. 1987. Density stratification, nutrient and chlorophyll distributions in the Hudson Strait region during summer in relation to tidal mixing. Cont. Shelf Res. 7: 599–607.

EAG (Environmental Applications Group Limited). 1984. Environmental literature review Hudson Bay offshore petroleum exploration. Prepared for Canadian Occidental Petroleum Limited, Toronto, ON. viii + 179 p (Draft Final Report revised March 1984).

Earle, J.C., and Kershaw, K.A. 1989. Vegetation patterns in James Bay coastal marshes. III. Salinity and elevation as factors influencing plant zonations. Can. J. Bot. 67: 2967–2974.

Ehn, J.K., Granskog, M.A., Papakyriakou, T., Galley, R., and Barber, D.G. 2006. Surface albedo observations of Hudson Bay (Canada) landfast sea ice during the spring melt. Ann. Glaciol. 44: 23–29.

Ewing, K., and Kershaw, K.A. 1986. Vegetation patterns in James Bay coastal marshes. I. Environmental factors on the south coast. Can. J. Bot. 64: 217–226.

Ferguson, S.H., Stirling, I., and McLoughlin, P. 2005. Climate change and ringed seal (*Phoca hispida*) recruitment in western Hudson Bay. Mar. Mamm. Sci. 21: 121–135.

Fortier, L., Gilbert, M., Ponton, D., Ingram, R.G., Robineau, B., and Legendre, L. 1996. Impact of fresh waters on a subarctic coastal ecosystem under seasonal sea ice (southeastern Hudson Bay, Canada). III. Feeding success of marine fish larvae. J. Mar. Syst. 7: 251–265.

Fortier, L., Ponton, D., and Gilbert, M. 1995. The match/mismatch hypothesis and the feeding success of fish larvae in ice-covered southeastern Hudson Bay. Mar. Ecol. Prog. Ser. 120: 11–27.

Frost, K.J., and Lowry, L.F. 1981. Ringed, Baikal and Caspian seals: *Phoca hispida* Schreber, 1775; *Phoca sibirica* Gmelin, 1788 and *Phoca caspica* Gmelin, 1788, pp. 29–53. *In* S.H. Ridgway and R.J. Richardson (ed) Handbook of marine mammals, Vol. 2 Seals. Academic Press, London.

Gaston, A.J., and Hipfner, M. 1998. The effect of ice conditions in northern Hudson Bay on breeding by Thick-billed Murres (*Uria lomvia*). Can. J. Zool. 76: 480–492.

Gilchrist, H.G., and Robertson, G.J. 2000. Observations of marine birds and mammals wintering at polynyas and ice edges in the Belcher Islands, Nunavut, Canada. Arctic 53: 61–68.

Glooschenko, W.A. and Clarke, K. 1982. The salinity cycle of a subarctic salt marsh. Nat. Can. (Que.) 109: 483–490.

Godin, G. 1974. The tide in eastern and western James Bay. Arctic 27: 104–110.

Godin, G., and Candela, J. 1987. Tides and currents in Fury and Hecla Strait. Estuarine Coastal Shelf Sci. 24: 513–525.

Gosselin, M., Legendre, L., Therriault, J.-C., Demers, S., and Rochet, M. 1986. Physical control of the horizontal patchiness of sea-ice microalgae. Mar. Ecol. Prog. Ser. 29: 289–298.

Gosselin, M., Legendre, L., Therriault, J.-C., and Demers, S. 1990. Light and nutrient limitation of sea-ice microalgae (Hudson Bay, Canadian Arctic). J. Phycol. 26: 220–232.

Grainger, E.H. 1982. Factors affecting phytoplankton stocks and primary productivity at the Belcher Islands, Hudson Bay. Nat. Can. (Que.) 109: 787–791.

Grainger, E.H. 1988. The influence of a river plume on the sea-ice meiofauna in south-eastern Hudson Bay. Estuarine Coastal Shelf Sci. 27: 131–141.

Grainger, E.H., and McSween, S. 1976. Marine zooplankton and some physical-chemical features of James Bay related to La Grande hydro-electric development. Can. Fish. Mar. Serv. Tech. Rep. 650: 94.

Granskog, M.A., Macdonald, R.W., Kuzyk, Z.Z.A., Senneville, S., Mundy, C.-J., Barber, D.G., Stern, G.A., and Saucier, F. 2009. Coastal conduit in southwestern Hudson Bay (Canada) in summer: rapid transit of freshwater and significant loss of colored dissolved organic matter. J. Geophys. Res. 114: C08012, doi:10.1029/2009JC005270.

Griffiths, D.K., Pingree, R.D., and Sinclair, M. 1981. Summer tidal fronts in the near-arctic regions of Foxe Basin and Hudson Bay. Deep-Sea Res. 28A: 865–873.

Harvey, M., Therriault, J.-C., and Simard, N. 1997. Late-summer distribution of phytoplankton in relation to water mass characteristics in Hudson Bay and Hudson Strait (Canada). Can. J. Fish. Aquat. Sci. 54: 1937–1952.

Harvey, M., Starr, M., Therriault, J.-C., Saucier, F., and Gosselin, M. 2006. MERICA-Nord Program: monitoring and research in the Hudson Bay complex. AZMP Bulletin 5: 27–32.

Hay, J.E. 1978. Hydrological atlas of Canada, Plate 12. Net radiation. Environment Canada, Ottawa, ON.

Heath, J.P., Gilchrist, H.G., and Ydenberg, R.C. 2006. Regulation of stroke pattern and swim speed across a range of current velocities: diving by common eiders wintering in polynyas in the Canadian Arctic. J. Exp. Biol. 209: 3974–3983.

Heine, J.N. 1989. Effects of ice scour on the structure of sublittoral marine algal assemblages of St. Lawrence and St. Matthew islands, Alaska. Mar. Ecol. Prog. Ser. 52: 253–260.

Henderson, P.J. 1989. Provenance and depositional facies of surficial sediments in Hudson Bay, a glaciated epeiric sea. Ph.D. thesis, Department of Geology, University of Ottawa, Ottawa, ON. xvii + 287 p.

Hik, D.S., Jeffries, R.L., and Sinclair, A.R.E. 1992. Foraging by geese, isostatic uplift and asymmetry in the development of salt marsh plant communities. J. Ecol. 80: 395–406.

Hillaire-Marcel, C. 1976. La déglaciation et le relèvement isostatique sur la côte est de la baie d'Hudson. Cahiers de Géographie du Québec 20: 185–220.

Hochheim, K., and Barber, D.G. 2009. Atmospheric forcing of sea ice in Hudson Bay during the fall period, 1980–2005. J. Geophys. Res. – Oceans (in press).

Holst, M., Stirling, I., and Calvert, W. 1999. Age structure and reproductive rates of ringed seals (*Phoca hispida*) on the northwestern coast of Hudson Bay in 1991 and 1992. Mar. Mamm. Sci. 15: 1357–1364.

Hsiao, S.I.C., Pinkewycz, N., Mohammed, A.A., and Grainger, E.H. 1984. Sea ice biota and under ice plankton from southeastern Hudson Bay in 1983. Can. Data Rep. Fish. Aquat. Sci. 494: 49.

Hudon, C., Crawford, R.E., and Ingram, R.G. 1993. Influence of physical forcing on the spatial distribution of marine fauna near Resolution Island (eastern Hudson Strait). Mar. Ecol. Prog. Ser. 92: 1–14.

Héquette, A., and Tremblay, P. 2009. Effects of low water temperature on longshore sediment transport on a subarctic beach, Hudson Bay, Canada. J. Coastal Res. 25: 171–180.

Hernández-Henríquez, M.A., Mlynowski, T.J., and Déry, S.J. 2009. Reconstructing the natural streamflow of La Grande Rivère, Québec, Canada. I3P/WC2N Joint Annual Workshop, Lake Louse Inn, Lake Louise, Alberta. Poster. [Available online at: http://www.usask.ca/ip3/download/ws4/posters/hernandez_poster.pdf, Accessed 17 March 2010].

Ingram, R.G., and Prinsenberg, S. 1998. Chapter 29. Coastal oceanography of Hudson Bay and surrounding eastern Canadian Arctic waters, coastal segment (26, P), pp. 835–861. *In* A.R. Robinson and K.H. Brink (ed) The Sea, Volume 11. Wiley, New York.

Iverson, S.J., Stirling, I., and Lang, S.L.C. 2006. Spatial and temporal variation in the diets of polar bears across the Canadian Arctic: indicators of changes in prey populations and environment, pp. 98–117. *In* I.L. Boyd, S. Wanless, and C.J. Camphuysen (ed) Top Predators in Marine Ecosystems. Cambridge University Press, Cambridge.

Jones, E.P., and Anderson, L.G. 1994. Northern Hudson Bay and Foxe basin: water masses, circulation and productivity. Atmosphere-Ocean 32: 361–374.

Jones, E.P., Swift, J.H., Anderson, L.G., Lipizer, M., Civitarese, G., Falkner, K.K., Kattner, G., and McLaughlin, F. 2003. Tracing Pacific water in the North Atlantic Ocean. J. Geophys. Res. 108: 3116, doi: 10.1029/2001JC001141.

Josenhans, H.W., and Zevenhuizen, J. 1990. Dynamics of the Laurentide Ice Sheet in Hudson Bay, Canada. Mar. Geol. 92: 1–26.

Josenhans, H., Zevenhuizen, J., and Veillette, J. 1991. Baseline marine geological studies off Grande Rivère de la Baleine and Petite Rivère de la Baleine southeastern Hudson Bay. Geol. Surv. Can. Pap. 91-1E: 347–354.

Kolenosky, G.B., Abraham, K.F., and Greenwood, C.J. 1992. Polar bears of southern Hudson Bay. Polar Bear Project, 1984–1988, Final Report. Ontario Ministry of Natural Resources, Maple, ON. vii + v + 107 p.

Kuzyk, Z.A., Macdonald, R.W., Granskog, M.W., Scharien, R.K., Galley, R.J., Michel, C., Barber, D., and Stern, G. 2008. Sea ice, hydrological, and biological processed in the Churchill River estuary region, Hudson Bay. Estuarine Coastal Shelf Sci. 77: 369–384.

Larouche, P., and Dubois, J.-M. 1988. Évaluation dynamique de la circulation de surface utilisant la dérive de blocs de glace (baie d'Hudson, Canada). Photo-Interprétation 1988-6: 19–28.

Larouche, P., and Dubois, J.-M. 1990. Dynamical evaluation of the surface circulation using remote sensing of drifting ice floes. J. Geophys. Res. 95: 9755–9764.

Legendre, L., Martineau, M.-J., Therriault, J.-C., and Demers, S. 1992. Chlorophyll a biomass and growth of sea-ice microalgae along a salinity gradient (southeastern Hudson Bay, Canadian Arctic). Polar Biol. 12: 445–453.

Legendre, L., Robineau, B., Gosselin, M., Michel, C., Ingram, R.G., Fortier, L., Therriault, J.-C., Demers, S., and Monti, D. 1996. Impact of freshwater on a subarctic coastal ecosystem under seasonal sea ice (southeastern Hudson Bay, Canada). II. Production and export of ice algae. J. Mar. Syst. 7: 233–250.

Loucks, R.H., and Smith, R.E. 1989. Hudson Bay and Ungava Bay ice-melt cycles for the period 1963–1983. Can. Contract. Rep. Hydrogr. Ocean Sci. 34: vi + 48 p.

Lunn, N.J., Branigan, M., Carpenter, L., Chaulk, K., Doidge, B., Galipeau, J., Hedman, D., Huot, M., Maraj, R., Obbard, M., Otto, R., Stirling, I., Taylor, M., and Woodley, S. 2005. Polar bear management in Canada 2001–2004, pp. 101–116. *In* J. Aars, N.J. Lunn, and A.E. Derocher (ed) Polar Bears. Proceedings of the 14th Working Meeting of the IUCN/SSC Polar Bear specialist Group 20–24 June 2005, Seattle, Washington, DC. Occas. Pap. IUCN Species Surviv. Comm. 32.

Lunn, N.J., Stirling, I., and Nowicki, S.N. 1997. Distribution and abundance of ringed (*Phoca hispida*) and bearded seals (*Erignathus barbatus*) in western Hudson Bay. Can. J. Fish. Aquat. Sci. 54: 914–921.

Macdonald, R.W., Harner, T., and Fyfe, J. 2003. Interaction of climate change with contaminant pathways to and within the Arctic, pp. 224–279. *In* J. Van Oostdam, S. Donaldson, M. Feeley, and N. Tremblay (ed) Sources, Occurrence, Trends and Pathways in the Physical Environment – Canadian Arctic Contaminants Assessment Report II (CACAR II). Canada Department of Indian Affairs and Northern Development, Ottawa, ON.

Maestrini, S., Rochet, M., Legendre, L., and Demers, S. 1986. Nutrient limitation of the bottom-ice microalgal biomass (southeastern Hudson Bay, Canadian Arctic). Limnol. Oceanogr. 31: 969–982.

Mallory, M.L., and Fontaine, A.J. 2004. Key marine habitat sites for migratory birds in Nunavut and the Northwest Territories. Canadian Wildlife Service Occasional Paper No. 109.

Mansfield, A.W. 1959. The walrus in the Canadian arctic. Fish. Res. Board Can. Arctic Unit Circ. 2: 1–13.

Mansfield, A.W. 1967. Seals of arctic and eastern Canada. Fish. Res. Board Can. Bull. 137: 35.

Markham, W.E. 1986. The ice cover, pp. 101–116. *In* I.P. Martini (ed) Canadian inland seas. Elsevier Oceanography Series 44. Elsevier Science, Amsterdam.

Martini, I.P. 1986. Coastal features of Canadian inland waters, pp. 117–142. *In* I.P. Martini (ed) Canadian inland seas. Elsevier Oceanography Series 44. Elsevier Science, Amsterdam.

Martini, I.P., and Glooschenko, W.A. 1984. Emergent coasts of Akimiski Island, James Bay, Northwest Territories, Canada: geology, geomorphology, and vegetation. Sediment. Geol. 37: 229–250.

Martini, I.P., and Protz, R. 1981. Coastal geomorphology, sedimentology and pedology of Hudson Bay, Ontario, Canada. University of Guelph, Dept. Land Resource Science, Tech. Memo 81-1: 322.

Martini, I.P., Morrison, R.I.G., Glooschenko, W.A., and Protz, R. 1980. Coastal studies in James Bay, Ontario. Geoscience Can. 7: 11–21.

Maxwell, J.B. 1986. A climate overview of the Canadian inland seas, pp. 79–99. *In* I.P. Martini (ed) Canadian inland seas. Elsevier Oceanography Series 44. Elsevier Science, Amsterdam.

McDonald, M., Arragutainaq, L., and Novalinga, Z. 1997. Voices from the Bay: traditional ecological knowledge of Inuit and Cree in the Hudson Bay bioregion. Canadian Arctic Resources Committee; Environmental Committee of Municipality of Sanikiluaq, Ottawa, ON. xiii + 98 p.

McLaren, I.A. 1958. The biology of the ringed seal (*Phoca hispida* Schreber) in the eastern Canadian Arctic. Fish. Res. Board Can. Bull. 118: vii + 93 p.

Meagher, L.J., Ruffman, A., and Stewart, J.M. 1976. Marine geological data synthesis. James Bay. Geol. Surv. Can. Open File 497: 561.

Messier, D., Ingram, R.G., and Roy, D. 1986. Physical and biological modifications in response to La Grande hydroelectric complex, pp. 403–424. *In* I.P. Martini (ed) Canadian inland seas. Elsevier Oceanography Series 44. Elsevier Science Publishers, Amsterdam.

Monti, D., Legendre, L., Therriault, J.-C., and Demers, S. 1996. Horizontal distribution of seaice microalgae: environmental control and spatial processes (southeastern Hudson Bay, Canada). Mar. Ecol. Prog. Ser. 133: 229–240.

Morin, R., and Dodson, J.J. 1986. The ecology of fishes in James Bay, Hudson Bay and Hudson Strait, pp. 293–325. *In* I.P. Martini (ed) Canadian inland seas. Elsevier Oceanography Series 44. Elsevier Science, Amsterdam.

Morin, R., Dodson, J.J., and Power, G. 1980. Estuarine fish communities of the eastern James-Hudson Bay coast. Environ. Biol. Fishes 5: 135–141.

Morin, R., Dodson, J.J., and Power, G. 1981. The migrations of anadromous cisco (*Coregonus artedii*) and lake whitefish (*C. clupeaformis*) in estuaries of eastern James Bay. Can. J. Zool. 59: 1600–1607.

Morrison, R.I.G., and Gaston, A.J. 1986. Marine and coastal birds of James Bay, Hudson Bay and Foxe Basin, pp. 355–386. *In* I.P. Martini (ed) Canadian inland seas. Elsevier Oceanography Series 44. Elsevier Science, Amsterdam.

Mysak, L.A., and Venegas, S.A. 1998. Decadal climate oscillations in the Arctic: a new feedback loop for atmosphere-ice-ocean interactions. Geophys. Res. Lett. 25: 3607–3610.

Mysak, L.A., Ingram, R.G., Wang, J., and Van Der Baaren, A. 1996. The anomalous sea-ice extent in Hudson Bay, Baffin Bay and the Labrador Sea during three simultaneous NAO and ENSO episodes. Atmosphere-Ocean 34: 313–343.

Nakashima, D.J. 1988. The common eider (*Somateria mollissima sedentaria*) of eastern Hudson Bay: a survey of nest colonies and Inuit ecological knowledge, Part II: Eider ecology from Inuit hunters. Environmental Studies Revolving Funds Report No. 102: 112–174.

NAC (National Atlas of Canada 5th edn.). 1986. Drainage basins, 1985. Natural Resources Canada, Ottawa, ON. [http://atlas.gc.ca/site/english/maps/archives/5thedition/environment/water/mcr4055].

NAC (National Atlas of Canada 6th edn.), 2007. Surficial materials. Natural Resources Canada, Ottawa, ON. [http://atlas.nrcan.gc.ca/site/english/maps/environment/land/surficialmaterials; Accessed 4 November 2007].

NIMA (National Imagery and Mapping Agency). 2002. Newfoundland, Labrador, and Hudson Bay. Sailing directions (enroute). Pub. 146 (8th edn.). National Imagery and Mapping Agency, Maritime Safety Information Division, Bethesda, Maryland. xi + 376 p.

Norris, A.W. 1986. Review of Hudson Platform Paleozoic stratigraphy and biostratigraphy, pp. 17–42. *In* I.P. Martini (ed) Canadian inland seas. Elsevier Oceanography Series 44. Elsevier Science, Amsterdam.

Pelletier, B.R. 1986. Seafloor morphology and sediments, pp. 143–162. *In* I.P. Martini (ed) Canadian inland seas. Elsevier Oceanography Series 44. Elsevier Science, Amsterdam.

Ponton, D., and Fortier, L. 1992. Vertical distribution and foraging of marine fish larvae under the ice cover of southeastern Hudson Bay. Mar. Ecol. Prog. Ser. 81: 215–227.

Ponton, D., Gagne, J.S., and Fortier, L. 1993. Production and dispersion of freshwater, anadromous, and marine fish larvae in and around a river plume in subarctic Hudson Bay, Canada. Polar Biol. 13: 321–331.

Poulin, M., Cardinal, A., and Legendre, L. 1983. Réponse d'une communauté de diatomées de glace à un gradient de salinité (baie d'Hudson). Mar. Biol. 76: 191–202.

Prinsenberg, S.J. 1982. Time variability of physical oceanographic parameters in Hudson Bay. Nat. Can. (Que.) 109: 685–700.

Prinsenberg, S.J. 1984. Freshwater contents and heat budgets of James Bay and Hudson Bay. Cont. Shelf Res. 3: 191–200.

Prinsenberg, S.J. 1986a. Salinity and temperature distributions of Hudson Bay and James Bay, pp. 163–186. *In* I.P. Martini (ed) Canadian inland seas. Elsevier Oceanography Series 44. Elsevier Science, Amsterdam.

Prinsenberg, S.J. 1986b. The circulation pattern and current structure of Hudson Bay, pp. 187–204. *In* I.P. Martini (ed) Canadian inland seas. Elsevier Oceanography Series 44. Elsevier Science, Amsterdam.

Prinsenberg, S.J. 1986c. On the physical oceanography of Foxe Basin, pp. 217–236. *In* I.P. Martini (ed) Canadian inland seas. Elsevier Science, Amsterdam.

Prinsenberg, S.J. 1988. Ice-cover and ice-ridge contributions to the freshwater contents of Hudson Bay and Foxe Basin. Arctic 41: 6–11.

Prinsenberg, S.J. 1991. Effects of hydroelectric projects on Hudson Bay's marine and ice environments. James Bay Publication Series, Potential Environmental Impacts – Paper No. 2. North Wind Information Services Inc., Montréal, QC, H2W 2M9. 8 p.

Prinsenberg, S.J., and Freeman, N.G. 1986. Tidal heights and currents in Hudson Bay and James Bay, pp. 205–216. *In* I.P. Martini (ed) Canadian inland seas. Elsevier Oceanography Series 44. Elsevier Science, Amsterdam.

Protz, R. 1982. Development of gleysolic soils in the Hudson Bay and James Bay coastal zone, Ontario. Nat. Can. (Que.) 109: 491–500.

Richard, P.R. 1991. Status of the belugas, *Delphinapterus leucas*, of southeast Baffin Island, Northwest Territories. Can. Field-Nat. 105: 206–214.

Richard, P.R. 2005. An estimate of the Western Hudson Bay beluga population size in 2004. Candian Stock Assessment Secretariat Research Document 2005/017: 32. [Available at: http://www.dfo-mpo.gc.ca/csas/csas/publications/resdocs-docrech/2005/2005_017_e.htm].

Riley, J.L. 1982. Hudson Bay Lowland floristic inventory, wetlands catalogue and conservation strategy. Nat. Can. (Que.) 109: 543–555.

Ringius, G.S. 1980. Vegetation survey of a James Bay coastal marsh. Can. Field-Nat. 94: 110–120.

Robertson, G.J., and Gilchrist, H.G. 1998. Evidence of population declines among common eiders breeding in the Belcher Islands, Northwest Territories. Arctic 51: 378–385.

Roff, J.C., and Legendre, L. 1986. Physio-chemical and biological oceanography of Hudson Bay, pp. 265–291. *In* I.P. Martini (ed) Canadian inland seas. Elsevier Oceanography Series 44. Elsevier Science, Amsterdam.

Runge, J.A., and Ingram, R.G. 1991. Under-ice feeding and diel migration by the planktonic copepods *Calanus glacialis* and *Pseudocalanus minutus* in relation to the ice algal production cycle in southeastern Hudson Bay, Canada. Mar. Biol. 108: 217–225.

Sadler, H.E. 1982. Water flow into Foxe Basin through Fury and Hecla Strait. Nat. Can. (Que.) 109: 701–707.

Saucier, F.J., Senneville, S., Prinsenberg, S., Roy, F., Smith, G., Gachon, P., Caya, D., and Laprise, R. 2004. Modelling the sea ice-ocean seasonal cycle in Hudson Bay, Foxe Basin and Hudson Strait, Canada. Clim. Dyn. 23: 303–326.

Shilts, W.W. 1986. Glaciation of the Hudson Bay region, pp. 55–78. *In* I.P. Martini (ed.). Canadian inland seas. Elsevier Oceanography Series 44. Elsevier Science, Amsterdam.

Smith, T.G. 1975. Ringed seals in James Bay and Hudson Bay: population estimates and catch statistics. Arctic 28: 170–182.

Smith, A.J. 2007. Beluga whale (*Delphinapterus leucas*) use of the Nelson River estuary, Hudson Bay. M.Sc. thesis, University of Manitoba, Winnipeg, MB.163 p.

Smith, T.G. 1981. Notes on the bearded seal, *Erignathus barbatus*, in the Canadian arctic. Can. Tech. Rep. Fish. Aquat. Sci. 1042: v + 49 p.

Smith, T.G. 1987. The ringed seal, *Phoca hispida*, of the Canadian western arctic. Can. Bull. Fish. Aquat. Sci. 216: 81 p.

Smith, T.G., Hammill, M.O., and Taugbol, G. 1991. A review of the developmental, behavioural and physiological adaptations of the ringed seal, *Phoca hispida*, to life in the Arctic winter. Arctic 44: 124–131.

Smith, W.O., and Barber, D.G. 2007. Polynyas and climate change. A view to the future, pp. 411–419. *In* W.O. Smith and D. G. Barber (eds) Polynyas: Windows to the World. Elsevier Oceanography Series 74.

Stewart, D.B. 2007. Hudson Bay Large Marine Ecosystem (LME): overview and assessment of vulnerability to hydrocarbon development. Canadian contribution to Chapter 6 of the Arctic Monitoring and Assessment Programme's (AMAP) scientific report entitled "Assessment of Oil and Gas Activities in the Arctic". Prepared by Arctic Biological Consultants, Winnipeg, MB for Indian and Northern Affairs Canada, Ottawa, ON. 60 pp.

Stewart, D.B., and Howland, K.L. 2009. An ecological and oceanographical assessment of the alternate ballast water exchange zone in the Hudson Strait Region. DFO Can. Sci. Advis. Sec. Res. Doc. 2009/008: vi +92 p.

Stewart, D.B., and Lockhart, W.L. 2005. An overview of the Hudson Bay marine ecosystem. Can. Tech. Rep. Fish. Aquat. Sci. 2586: vi + 487 p.

Stirling, I., and Cleator, H. (ed). 1981. Polynyas in the Canadian Arctic. Environment Canada, Canadian Wildlife Service, Occasional Paper Number 45: 73.

Stirling, I., and Ramsay, M.A. 1986. Polar bears in Hudson Bay and Foxe Basin: present knowledge and research opportunities, pp. 341–354. *In* I.P. Martini (ed) Canadian inland seas. Elsevier Oceanography Series 44. Elsevier Science, Amsterdam.

Stirling, I., Lunn, N.J., Iacozza, J., Elliott, C., and Obbard, M. 2004. Polar bear distribution and abundance on the southwestern Hudson Bay coast during open water season, in relation to populaiton trends and annual ice patterns. Arctic 57: 15–26.

Straneo, F., and Saucier, F. 2008a. Chapter 10. The Arctic - Subarctic exchange through Hudson Strait, pp. 249–261. In R.R. Dickson, J. Meincke, and P. Rhines (ed) Arctic-Subarctic Ocean Fluxes. Springer, Netherlands.

Straneo, F., and Saucier, F. 2008b. The outflow from Hudson Strait and its contribution to the Labrador Current. Deep-Sea Res. 55: 926–946.

Subba Rao, D.V., and Platt, T. 1984. Primary production of Arctic waters. Polar Biol. 3: 191–201.

Tan, F.C., and Strain, P.M. 1996. Sea ice and oxygen isotopes in Foxe Basin, Hudson Bay, and Hudson Strait, Canada. J. Geophys. Res. 101: 20,869–920,876.

Tourangeau, S. 1989. Reproduction of Calanus glacialis in relation to ice microalgal productivity in southeastern Hudson Bay. Rapp. P-V. Reun. Cons. Int. Explor. Mer 188: 175.

Tourangeau, S., and Runge, J.A. 1991. Reproduction of Calanus glacialis under ice in spring in southeastern Hudson Bay, Canada. Mar. Biol. 108: 227–233.

Tremblay, C., Runge, J.A., and Legendre, L. 1989. Grazing and sedimentation of ice algae during and immediately after a bloom at the ice-water interface. Mar. Ecol. Prog. Ser. 56: 291–300.

Urquhart, D.R., and Schweinsburg, R.E. 1984. Polar bear: life history and known distribution of polar bear in the Northwest Territories up to 1981. Northwest Territories Renewable Resources, Yellowknife, N.W.T. v + 70 p.

Wang, J., Mysak, L.A., and Ingram, R.G. 1994. A three-dimensional numerical simulation of Hudson Bay summer ocean circulation: Topographic gyres, separations, and coastal jets. J. Phys. Oceanogr. 24: 2496–2514.

Webber, P.J., Richardson, J.W., and Andrews, J.T. 1970. Post-glacial uplift and substrate age at Cape Henrietta Maria, southeastern Hudson Bay, Canada. Can. J. Earth Sci. 7: 317–325.

Welch, H.E., Bergmann, M.A., Siferd, T.D., and Amarualik, P.S. 1991. Seasonal development of ice algae near Chesterfield Inlet, N.W.T., Canada. Can. J. Fish. Aquat. Sci. 48: 2395–2402.

Zoltai, S.C., Holroyd, G.L., and Scotter, G.W. 1987. A natural resource survey of Wager Bay, Northwest Territories. Environment Canada, Canadian Wildlife Service, Technical Report Series 25: ix + 129 p.

Changing Sea Ice Conditions in Hudson Bay, 1980–2005

K. Hochheim, D.G. Barber, and J.V. Lukovich

Abstract We present an overview of changes in Hudson Bay sea ice in the context of thermodynamic forcing due to increased surface air temperatures and dynamic wind and current forcing mechanisms. Examined in particular is the correspondence between sea ice extent, surface air temperatures, and atmospheric indices during spring and fall from 1980 to 2005. Changes in the timing of freeze-up and break-up over the last several decades were significant. In the spring, temperature trends were consistently positive with temperature increases of 0.23°C/decade from 1950 to 2005. With increasing temperatures in the Hudson Bay region, sea ice concentrations and sea ice extents have decreased significantly as well. Warmer surface air temperatures have also shifted the mean freeze-up and break-up dates by 0.8–1.6 weeks in each of the seasons. Dynamic forcing of sea ice is further explored using the concept of relative vorticity, or the tendency for sea ice to rotate clockwise (or counterclockwise) within Hudson Bay in response to changes in atmospheric circulation. Surface air temperatures and hence ice extent showed cyclical patterns over the time period studied and appear to be driven by large scale atmospheric circulation patterns. This cyclical behaviour has been previously associated with various hemispheric indices including the North Atlantic Oscillation. The implications of changing ice conditions in Hudson Bay for marine mammal habitat are discussed.

Keywords Atmosphere • Circulation • Freeze-up • Break-up • Forcing • Rotation • Ice phenology • Hemispheric indices • Surface air temperature

K. Hochheim (✉)
Centre for Earth Observation Science (CEOS), Department of Environment and Geography, University of Manitoba, MB Canada
e-mail: hochheim@cc.umanitoba.ca

S.H. Ferguson et al. (eds.), *A Little Less Arctic: Top Predators in the World's Largest Northern Inland Sea, Hudson Bay*, DOI 10.1007/978-90-481-9121-5_2, © Springer Science+Business Media B.V. 2010

Introduction

Sea ice forms an integral part of the marine ecosystem at high latitudes. Marine mammals and sea birds are affected directly through change in sea ice-related habitats or indirectly through changing marine productivity processes, food web energetics and caloric transfer (Hoover this volume). In general, changes in sea ice growth, decay, and the spatial and temporal variability of these processes all affect the marine cryosphere and associated physical–biological coupling, ultimately impacting higher trophic level biota. Changes in dynamic processes (i.e., ice motion) affect habitat suitability through the formation of open water in winter, sea ice ridging, rubbling, and movement of ice over preferred feeding areas (e.g., walrus). Changes in thermodynamic processes can affect access to preferred habitats, destruction of habitats in the early spring (e.g., rainfall effect on snow lairs) and the timing of critical habitat formation (e.g., ice formation suitable for polar bears in the fall).

Changes in sea ice and its associated snow cover (hereinafter referred to as the marine cryosphere) have impacts on all aspects of the Arctic marine system due to the control which sea ice has on transmission of light, heat and momentum across the ocean-sea ice-atmosphere (OSA) interface (Carmack et al. 2006). Changes in the light environment result principally from changes in the snow cover on sea ice, due to the higher extinction of photosynthetically active radiation (PAR) in snow relative to sea ice. These changes have a direct connection to primary production both beneath the sea ice in spring (Mundy et al. 2009) and in the summer open water (Arrigo et al. 2008). Control of heat fluxes between the atmosphere and ocean are dominated by the presence and timing of sea ice formation (and are a consequence of these fluxes). The rate of sea ice formation is important to marine mammals and birds due to the relationship between fall storms and sea ice dynamic processes. For example, sea ice which tends to form later in the season tends to rubble and ridge more easily. Momentum exchanges across the OSA interface influence the presence and distribution of snow on sea ice, which both feedback into radiative transfer and heat exchanges. These processes can also have a strong impact on habitats for marine mammals, particularly in the formation of ringed seal lairs (Furgal et al. 1996; Chambellant this volume). Momentum exchange can also control upwelling of nutrients into the euphotic zone thereby feeding back directly into ecosystem productivity (Kuzyk et al. in press). This is true both of surface melt water stabilizing the ocean surface mixed layer and for upwelling at ice edges (Mundy et al. 2009).

In this chapter we review the general nature of sea ice from the perspective of a climatic analysis spanning the last 30 years. We examine whether the trends in sea ice extent and concentration are linear or periodic (cyclic) over this period. This will then shed light on how marine mammals and sea birds may evolve in the changing marine cryosphere of Hudson Bay.

Background

Our best knowledge of sea ice conditions is based on satellite data that has been acquired on a regular basis since 1979. These data show that trends in sea ice concentration during the 1979–1996 period were relatively small throughout the Arctic, −2.2% and −3.0%/decade, in contrast to the 1997–2007 period which showed that declines in sea ice concentration accelerated to −10.1% and −10.7%/decade (Comiso et al. 2008). We also note that trends in sea ice concentration in ice marginal zones within the polar seas during early part of the satellite record (1979–1993) varied geographically (Deser and Teng 2008). In particular during the winter, Eastern Canada (Labrador Sea) and Bering Sea had large positive trends in sea ice concentration (indicating more ice), while the Greenland and Barents seas and the Sea of Okhotsk had large negative trends (less ice). In 1993–2007 sea ice concentration trends were consistently negative throughout all of the Arctic and sub-arctic seas indicating a more general global warming trend. Geographic variations in sea ice concentration within the Arctic are also evident for the summer melt period during the early part of the satellite record while more recent trends in sea ice concentration are dominated by negative trends throughout the Arctic.

In Hudson Bay, Parkinson et al. (1999) noted that only very slight negative trends in sea ice extent existed within the Hudson Bay Region (HBR) (including Foxe Basin) from 1979 to 1996, and that these trends were statistically non-significant. Based on data from Environment Canada, Gagnon and Gough (2006) showed that temperature trends from 1960 to 1990 were predominately negative (cooling) and ice thickness trends were positive (thickening) during the fall and winter periods thus supporting the satellite observations showing increased sea ice concentrations in Eastern Canada from 1979 to 1993 (Deser and Teng 2008). Freeze-up and break-up of sea ice in Hudson Bay showed statistically significant trends with freeze-up dates occurring later in northern and northeastern HBR of 0.32–0.55 days/year, 1971–2003 (Gagnon and Gough 2005). During the spring melt period, Gagnon and Gough (2005) noted that break-up was occurring earlier, with magnitude ranging from −0.49 to −1.25 days/year. These trends were typically occurring in James Bay, along the southern shore of Hudson Bay and in the western half of Hudson Bay. Parkinson and Cavalieri (2008) also noted significant negative trends in sea ice extent (less ice) in the HBR using satellite passive microwave data for years 1979–2006. Changes in sea ice extent during the fall were estimated at -8.5×10^3 km^2/year ± 1.9. Spring trends in sea ice extent were estimated at -3.4×10^3 km^2/year ± 0.8 and these were all significant at 99% probability. Also using satellite passive microwave data, Markus et al. (2009) examined trends in the time of melt onset and freeze-up for ten different Arctic regions. In all regions except one (Sea of Okhotsk) trends in melt onset were negative (toward earlier melt) ranging from −1 to −7.3 days/decade, while the trend for the HBR was −5.3 days/decade and significant at the 99% level. Trends in freeze onset were consistently positive throughout the regions, ranging from 1.0 to 7.0 days/decade (indicating later freeze-up), with Hudson Bay having one of the larger trends at 5.4 days/decade. Hochheim and Barber (2010) show sea ice concentration and their spatial distribution within the HBR, linking both sea ice concentrations and sea ice extent to interannual variations in surface air temperatures and hemispheric indices; some of these results are highlighted below.

In the sections below, we use various data sources to better appreciate recent changes in the sea ice regime in the HBR by: (1) showing mean surface air temperature trends surrounding Hudson Bay (1950–2005); (2) producing maps showing trends in sea ice concentration to gain a better understanding of the spatial distribution and extent of changes in sea ice; and (3) linking the observed ice trends to surface air temperature and dynamic forcing.

Methods

Surface air temperature anomaly trends for the HBR were computed using CANGRID data developed for climate change studies by the Climate Research Division of Environment Canada. The bounds used to compute the mean HBR regional temperature anomalies (per month per year) were 50–65°N and 72.5–100°W. The use of temperature anomalies in gridding data has the advantage of removing location, physiographic, and elevation effects. Monthly temperature anomalies were computed for each month per year relative to the 1980–2005 mean to match the normals computed for sea ice data. A three month running mean was applied to the monthly surface air temperature anomaly data ending on (including) the month of interest; the intent here was to incorporate lead-up surface air temperatures to obtain a (moving) seasonal temperature index (anomaly) value.

The sea ice concentration trends and sea ice extent data were generated based on Canadian Ice Service digital ice charts (Environment Canada, Canadian Ice Service). Although the Canadian Ice Service data goes back to 1970, the charts produced since the early 1980s are of more consistent quality due to improvements in earth observation technology. To determine trends in sea ice concentration anomalies, a least squares linear regression was calculated for each grid point over the 26 year period where the slope of the regression indicated the trend per year, following an approach established by Parkinson et al. (1999), Parkinson and Cavalieri (2008), and Galley et al. (2008).

In order to monitor sea ice dynamics, ice relative vorticity, or tendency for ice to move clockwise (or counterclockwise) within Hudson Bay was computed using weekly sea ice motion vectors from the NSIDC data set for the Hudson Bay region, and spatially averaged over the region extending from 55°N to 64°N, and from 77°W to 95°W. The analysis was computing using vorticity was computed using a numerical finite-differencing scheme on the zonal and meridional ice motion components, as in Lukovich and Barber (2006). This analysis was used to explain in part the spatial distribution of positive and negative sea ice concentration anomalies in Hudson Bay.

Below we present a short overview of changes in freeze-up and break-up dates in Hudson Bay. The threshold used to determine freeze onset and break-up was 50%

sea ice concentration. Instead of computing linear trends over the 26 year period we compared mean differences in dates between two periods, 1980–1995 versus 1996–2005; the former being representative of the cooler temperature regime, the latter being representative of a significantly warmer temperature regime.

Air Temperature Trends

Based on gridded, historically-adjusted temperature data from Environment Canada using 3 month running means ending on the month of interest, temperature trends surrounding Hudson Bay during the fall were positive indicating a warming of 0.2–1.8°C/decade depending on month and location (Hochheim and Barber 2010). In general the largest increases were on the eastern half of Hudson Bay and the lowest were along the southwestern coast of Hudson Bay between the Nelson River Estuary and James Bay. During spring, temperature trends were positive (warming) but generally weaker, ranging from 0.2°C to 1.0°C/decade with the greatest warming occurring in the northeast and northern portions of Hudson Bay (June), and northwestern and eastern Hudson Bay during June–August. The lowest (but positive) temperatures trends in the spring were also observed along the southwestern shore of the Bay.

Overall mean regional surface air temperature trends surrounding Hudson Bay are shown in Fig. 1 for both spring and fall. The graphs suggest that: (1) surface air temperature anomalies for a given month vary interannually; (2) the temperature fluctuations have a cyclical nature (smoothing spline fit $\lambda = 0.0957$ (minimal smoothing)); and (3) temperatures in the past have been relatively cooler. During the fall period the 1950s to the early 1990s were relatively cooler: temperature trends during this period were negative (cooling) from −0.12°C to −0.28°C/decade (for November and December, respectively), although none of these trends were statistically significant. Since 1989 surface air temperatures have increased dramatically from 1.8°C and 2.3°C/decade for November and December, respectively (Hochheim and Barber 2010). Interestingly, warming trends from 1950 to 2005 for both November and December were statistically non-significant, owing to large interannual variations in temperature and the more semi-curvilinear temperature trend.

Comparing semi-decadal mean surface air temperature anomalies between 1980–1995 and 1996–2005 for October and November, mean surface air temperatures were significantly higher for the latter period (0.99°C and 1.44°C respectively). In December, 1996–2005 was identified as statistically different from the two preceding periods; 1.94°C warmer than 1970–1979 and 1.85°C warmer than 1980–1995.

During the spring and summer break-up period (June–August) negative surface air temperature anomalies occurred in the mid 1950s to early 1970 in contrast to the fall period where negative anomalies were observed from the 1970s to early 1990s. Mean 3 month surface air temperatures ending in June showed warming trends on the order of 0.23°C/decade ($p = 0.0436$) from 1950 to 2005 (Hochheim et al. 2010). Similar trends are observed for July and August (Fig. 1).

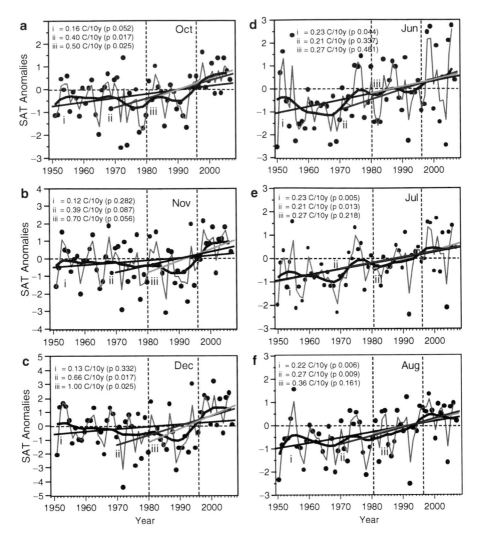

Fig. 1 Mean surface air temperature (SAT) anomalies based on seasonal 3 month running means ending for the freeze-up period from October to November (**a–c**) and for the break-up period from June to August (**d–f**). Two spline fits are added to show the higher frequency cyclical patterns surface air temperature anomalies ($\lambda = 0.0957$) and the longer pattern ($\lambda = 100$) of SAT anomalies. Three linear trends are shown, (i) 1950–2005, (ii) 1970–2005, (iii) 1980–2005

Sea Ice Freeze-up and Break-up Patterns

Freeze-up starts in the northern portion of the Hudson Bay around the shores of Southampton Island and along the northwestern coast of Hudson Bay. The ice then extends southward over Hudson Bay and along the coast to the Nelson Estuary and in a narrow band along the southwestern coast towards James Bay. The nearshore

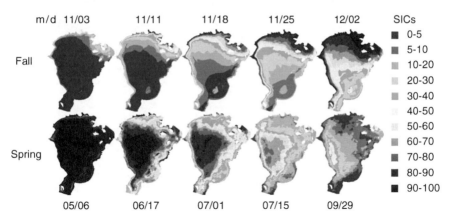

Fig. 2 Fall freeze-up and spring break-up sequences for Hudson Bay based on Canadian Ice Service data using mean sea ice concentrations computed over 1980–2005

waters are typically shallow (≤40 m) and less saline due to riverine inputs and therefore subject to early freezing. By 2 December the coastal areas along the southwestern portion of Hudson Bay are consolidated. The southeastern portion of the Bay including James Bay is typically the last to freeze (Fig. 2).

During the winter months the sea ice is consolidated ≥90% and typically consists of very large floes. The ice tends to slowly circulate in a counterclockwise fashion in response to currents and prevailing winds and thus has a positive vorticity.

During the spring, break-up occurs along the northern and northwestern portions of the Hudson Bay and along the east coast. Currents and winds tend to keep sea ice concentration highest in the central and southwestern portions. The last remnants of ice typically occur along the southwestern coast and the Bay is typically ice free following the second week of August.

The mean sea ice concentration map ending 6 May shows typical late season latent heat polynyas in Hudson Bay. Latent heat polynas are areas of open water (or low sea ice concentration) which are created by prevailing winds. These areas are biologically productive and thus are loci for organisms at higher trophic levels. The largest polynya is located along the northwestern coast of Hudson Bay including Southampton Island, while other polynyas are located southwest of Cape Churchill towards the Nelson River Estuary, along the east and west shores of James Bay, the Belcher islands, and Coates and Mansel islands and off the west coast of Québec (Barber and Massom 2007).

Sea Ice Concentration Trends, 1980–2005

Trends in sea ice concentration anomalies were computed for weeks ending 3 October to 2 December for the fall period, and weeks ending 24 June to 24 July using the CIS data (Fig. 3). The sea ice concentration anomaly trends were

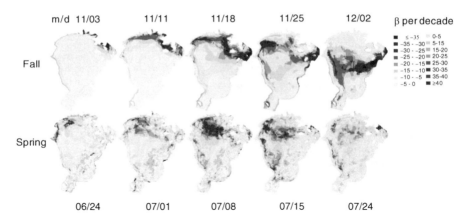

Fig. 3 Linear trends (β) in sea ice concentration anomalies using Canadian Ice Service data (1980–2005) for the fall and spring period

predominantly negative indicating reduction in sea ice concentration (note: near-shore positive anomalies are attributed to improved detection capabilities in sea ice due to RADARSAT-1 introduced in 1996). During the fall period, significant trends were found suggesting reductions in sea ice concentration of −23.3% to −26.9%/decade. During the week ending 2 December the most significant trends in ice reduction occurred off the southwestern and northeastern coasts of Hudson Bay.

During spring the largest negative trends in sea ice concentration were observed in western and northwestern Hudson Bay (Fig. 3). The statistically significant trends in sea ice concentration ranged from −15.1% to 20.4%/decade. Weak positive sea ice concentration trends (i.e., more ice) occurred in the eastern half of Hudson Bay, although these trends were statistically non-significant. The positive anomalies are, in part, explained by dynamic forcing resulting from changes in atmospheric circulation in the HBR (Hochheim et al. 2010).

Air Temperatures and Teleconnections

Hudson Bay is a relatively shallow sea (≤150 m) and more enclosed relative to other Arctic seas, in that it is more isolated from the effects of open–ocean circulation, which affects warm water intrusions and sea ice export (Wang et al. 1994). Thus unlike some other Arctic seas, variations in sea ice concentration and sea ice extent in Hudson Bay are more a function of atmospheric forcing, specifically changes in air temperature and wind circulation.

The cyclical nature of interannual variations in surface air temperature in Hudson Bay have largely been attributed to a number of standardized hemispheric indices (oscillations) which are associated with characteristic wind, temperature and precipitation patterns. These oscillations operate at various scales ranging from 2 to 7 years to decadal scales of 20–50 years. In Hudson Bay, several indices have been linked to air temperature and include the North Atlantic Oscillation (NAO) (Prinsenberg et al. 1997; Kinnard et al. 2006; Qian et al. 2008) and the Southern Oscillation index (SOI) (Wang et al. 1994; Mysak et al. 1996). Positive NAOs predict cooler temperatures over Hudson Bay. A strong NAO is associated with more northerly winds and the negative NAO phase is associated with more southerly winds and warmer temperatures. When strong positive NAOs occur together with strong negative SOIs, Hudson Bay regional temperatures are exceptionally cold (Wang et al. 1994; Mysak et al. 1996).

More recently, evidence shows that the East Pacific/North Pacific (EP/NP) predicts fall (September–November) surface air temperatures surrounding Hudson Bay. About 62% of the variance in surface air temperature surrounding Hudson Bay was explained by the EP/NP index over 1951–2005, and 79% explained over 1980–2005 (Hochheim and Barber 2010). Although most atmospheric indices are generally poorly correlated with spring surface air temperature anomalies, recent results suggest that both the EP/NP index West Pacific (WP) and Arctic Oscillation (AO) together can explain up to 68% of the variation in interannual surface air temperatures for July and August in the HBR (Hochheim et al. 2010).

Air Temperature and Sea Ice Extent

During the fall period the seasonal surface air temperature anomalies surrounding Hudson Bay explain up to 62% ($p < 0.001$) of the interannual variation in sea ice extent for consolidated ice (sea ice concentration ≥80%) computed with CIS data. Results for the week ending 2 December (1980–2005), a period of maximum inter-annual variation for sea ice extent, show that for every 1°C increase in surface air temperature, the area of consolidated ice decreased by 116,600 km². Similar results were obtained in spring. Using sea ice extent ≥60% in the spring for the week ending 15 July ($R^2 = 0.62$; $p < 0.0001$), for every 1°C increase in temperature there was a reduction in sea ice extent of 118,000 km². Using multiple regression and integrating both spring and fall surface air temperatures (ending November) consistently improved the capability of predicting weekly spring sea ice extents; coefficients of determination (R^2) were as high as 74% (Hochheim et al. 2010).

Relative Vorticity

The strong negative trends in sea ice concentration in the western portion of Hudson Bay and the positive trends in the eastern portion of Hudson Bay during the spring (Fig. 3) were partially explained by large-scale dynamic forcing due to changes in atmospheric circulation patterns that affect the distribution of late season (mobile) ice within Hudson Bay (Hochheim et al. 2010). Since 1990 the sea ice has had a predominantly positive vorticity (a counterclockwise rotation) causing ice to circulate eastward as a result of prevailing winds, thus enhancing negative sea ice concentration trends in western Hudson Bay, while creating posi- tive trends in eastern Hudson Bay (Fig. 4). During 1983–1989, the vorticity was strongly negative (sea ice circulating clockwise), thus contributing to heavier ice conditions in the west during the early part of the satellite record and relatively lower sea ice concentrations in the eastern portion of Hudson Bay. Wind compos- ites for negative and positive sea ice relative vorticity regimes (clockwise and counterclockwise circulation, respectively) highlight correspondence between late-season sea ice and atmospheric dynamic variability, with implications for marine ecosystems.

When including 3 month average relative vorticity ending late June with fall (November) and spring surface air temperature, as much as 84% of the variation in sea ice extent for the spring period was explained during the break-up period (Hochheim et al. 2010).

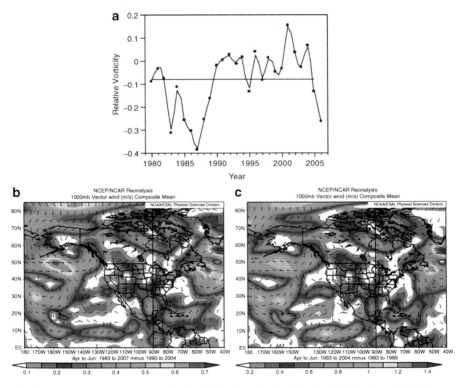

Fig. 4 (**a**) Sea ice relative vorticity for the Hudson Bay Region. Negative (positive) vorticity indicates a counterclockwise (clockwise) circulation of sea ice within Hudson Bay; the relative vorticity is averaged over a 3 month period ending June, (**b**) mean wind vectors over Hudson Bay associated with negative vorticity, (**c**) mean wind vectors over Hudson Bay associated with positive vorticity

Recent Changes in Freeze-up and Break-up Dates

Increasing regional surface air temperatures have led to later freeze-up dates (defined as sea ice concentrations ≥50%). Temporal shifts in mean freeze-up dates for the weeks ending 28 October–2 December (week of year 43–48) over 1980–1995 and 1996–2005 are depicted in Fig. 5a, b, with the change shown in Fig. 5c. During the fall period, freeze-up is 0.4–0.8 weeks earlier over 16% of Hudson Bay (or 1.35×10^5 km^2), 0.8–1.2 weeks earlier over 25% of Hudson Bay (or 1.97×10^5 km^2), and 1.2–1.6 weeks earlier over 38% of Hudson Bay (or 3.02×10^5 km^2), and about 1.6–2.0 weeks earlier over about 10% of the Bay.

Similar shifts were observed for the spring period suggesting earlier break-up of sea ice. During the spring, 18% of the Hudson Bay area exhibited a 04–0.8 week earlier break-up (or 1.48×10^5 km^2), 30% of the Bay had break-up 0.8–1.2 weeks earlier (2.42×10^5 km^2), 28% of the Bay had break-up occur 1.2–1.6 weeks (2.27×10^5 km^2), and 10% of Hudson Bay had a 1.6–2.0 week earlier break-up in the spring (or 8.11×10^4 km^2).

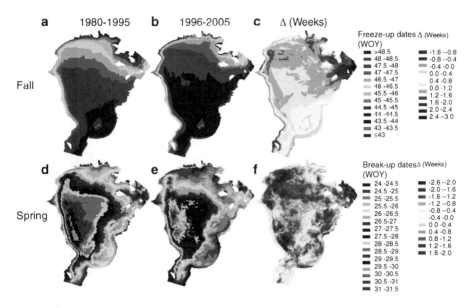

Fig. 5 Mean break-up dates based of 50% sea ice concentration for the fall period, 28 October–2 December (week of year 43–48) and the spring period 17 June–30 July (week of year 24–30). The shifts in freeze-up and break-up dates are computed for 1980–1995 which is representative of a cooler regime (**a, d**) and for 1996–2005 which is representative of the warmer regime (**b, e**). Difference maps show shifts in freeze-up (**c**) where positive values indicate shifts to later freeze-up dates (in weeks) and (**f**), where negative values indicate shifts to earlier break-up dates (in weeks)

Conclusions

Based on the CANGRID data, the HBR has recently undergone significant warming during the fall period, particularily since 1989. The largest differences in mean temperature before and after 1995 were in the month of December (1.85–1.94°C). In the spring, temperature trends were consistently positive estimating temperature increases of 0.23°C/decade from 1950 to 2005. With increasing temperatures in the HBR, sea ice concentrations and sea ice extents have decreased significantly as well. Warmer surface air temperatures have also shifted the mean freeze-up and break-up dates by 0.8–1.6 weeks in each of the seasons.

Surface air temperatures and hence ice extent are very cyclical and appear to be driven by large scale atmospheric circulation patterns. This cyclical behaviour has been associated previously with various hemispheric indices including the NAO, SOI, NP/EP and PDO, suggesting that Hudson Bay may become increasingly sensitive to hemispheric changes expected due to a warming planet. There is already ample evidence that the delay in sea ice formation is having an adverse effect on polar bears (*Ursus maritimus*) in Hudson Bay (Stirling et al. 1999) since the bears must extend their fasting period on land. Changes in sea ice phenology are also having a negative effect on thick-billed murres (*Uria lomvia*; Gaston et al. 2009). The late formation of sea ice can also be expected to increase the ridging and rubbling of sea ice which may in turn catch more of the fall snow thereby enhancing ringed seal habitats. We should expect that changes in the movement and phenology of formation and break-up of sea ice in Hudson Bay will have direct and strong impacts on all pagophilic species (Chambellant this volume; Mallory et al. this volume; Peacock et al. this volume). Moreover, with the possibility of increasing precipitation we can expect increasing snow depth on sea ice and this will have commensurate effects on various levels of the marine ecosystem due to control of light and heat across the ocean-sea ice-atmosphere interface (Hoover this volume).

References

Arrigo, K.R., G. van Dijken, and S. Pabi. 2008. Impact of a shrinking Arctic ice cover on marine primary production. Geophys. Res. Lett. 35: L19603. doi:10.1029/2008GL035028.

Barber, D.G., and R.A. Massom. 2007. The role of sea ice in Arctic and Antarctic Polynyas. In W.O. Smith and D. G. Barber (eds). Polynyas: Windows to the World. Elsevier Oceanogr. Ser. 74(1–54).

Carmack, E., D. Barber, J. Christensen, R. Macdonald, B. Rudels, and E. Sakshaug. 2006. Climate variability and physical forcing of the food webs and the carbon budget on panarctic shelves. Prog. Oceanogr. 71: 145–181.

Comiso, J.C., C.L. Parkinson, R. Gersten, and L. Stock. 2008. Accelerated decline in the Arctic sea ice cover. Geophys. Res. Lett. 35: L01703. doi:10.1029/2007GL031972.

Deser, C., and H. Teng. 2008. Evolution of Arctic sea ice concentration trends and the role of atmospheric circulation forcing, 1979–2007. Geophys. Res. Lett. 35: L02504. doi:10.1029/2007 GL032023.

Furgal, C.M., S. Innes, and K. Kovacs. 1996. Characteristics of ringed seal, *Phoca hispida*, subnivean structures and breeding habitat and their effects on predation. Can. J. Zoo. 74: 858–874.

Gagnon, A.S., and W.A. Gough. 2005. Trends in the dates of ice freeze-up and break-up over Hudson Bay, Canada. Arctic 58(4): 370–382.

Gagnon, A.S., and W.A. Gough. 2006. East-West asymmetry in ice thickness trends over Hudson Bay, Canada. Clim. Res. 32: 177–186.

Galley, R.J., E. Key, D.G. Barber, B.J. Hwang, and J.K. Ehn. 2008. Spatial and temporal variability of sea ice in the southern Beaufort Sea and Amundsen Gulf: 1980–2004. J. Geophys. Res. 113: C05S95. doi:10.1029/2007JC004553.

Gaston, A.J., H.G. Gilchrist, M.L. Mallory, and P.A. Smith. 2009. Changes in seasonal events, peak food availability, and consequent breeding adjustment in a marine bird: a case of progressive mismatching. Condor 111: 111–119.

Hochheim, K., and D.G. Barber. 2010. Atmospheric forcing of sea ice in Hudson Bay during the Fall Period, 1980–2005. J. Geophys. Res. – Oceans (in press).

Hochheim, K, J. Lukovich, and D.G. Barber. 2010. Atmospheric forcing of sea ice in Hudson Bay during the Spring period, 1980–2005. J. Marine Syst. (in review).

Kinnard, C., C.M. Zdanowicz, D.A. Fisher Bea Alt, and S. McCourt. 2006. Climatic analysis of sea-ice variability in the Canadian Arctic from operational charts, 1980–2004. Annal. Glaciol. 44: 391–402.

Kuzyk, Z.Z.A., R.W. Macdonald, J.E. Tremblay, and G.A. Stern 2010. Elemental and stable isotopic constraints on river influence and patterns of nitrogen cycling and biological productivity in Hudson Bay. Continental Shelf Research 30: 163–176.

Lukovich, J.V., and D.G. Barber. 2006. Atmospheric controls of sea ice motion in the Southern Beaufort Sea. J. Geophys. Res. 111: D18103. doi:10.1029/2005JD006408.

Markus, T., J.C. Stroeve, and J. Miller. 2009. Recent changes in Arctic sea ice melt onset, freeze-up, and melt season length. J. Geophys. Res. 114: C12024. doi:10.1029/2009JC005436.

Mysak, L.A., R.G. Ingram, J. Wang, and A. Van Der Baaren. 1996. The anomalous sea-ice extent in Hudson Bay, Baffin Bay and the Labrador Sea during three simultaneous NAO and ENSO episodes. Atmos. Ocean 34: 313–343.

Mundy, C.J., M. Gosselin, J.K. Ehn, Y. Gratton, A. Rossnagel, D.G. Barber, J. Martin, J.-É. Tremblay, M. Palmer, K. Arrigo, G. Darnis, L. Fortier, B. Else, and T. Papakyriakou. 2009. Contribution of under-ice primary production to an ice-edge upwelling phytoplankton bloom in the Canadian Beaufort Sea. Geophys. Res. Lett. 36: L17601. doi:10.1029/2009GL038837.

Parkinson, C.L., and D.J. Cavalieri. 2008. Arctic sea ice variability and trends, 1979–2006. J. Geophys. Res. 113: C07003. doi:10.1029/2007JC004558.

Parkinson, C.L., D.J. Cavalieri, P. Gloersen, H.J. Zwally, and J.C. Comiso. 1999. Arctic sea ice extents, areas and trends, 1978–1996. J. Geophys. Res. 104C: 20837–20856.

Prinsenberg, S.J., I.K. Peterson, S. Narayanan, and J.U. Umoh. 1997. Interaction between atmosphere, ice cover, and ocean off Labrador and Newfoundland from 1962 to 1992. Can. J. Fish. Aquat. Sci. 54(Suppl. 1): 30–39.

Qian, M., C. Jones, R. Laprise, and D. Caya. 2008. The Influences of NAO and the Hudson Bay sea-ice on the climate of eastern Canada. Clim. Dyn. 31: 169–182.

Stirling, I., N. Lunn, and J. Iacozza. 1999. Long-term trends in the population ecology of polar bears in western Hudson Bay in Relation to climatic change. Arctic 52: 294–306.

Wang, J., L.A. Mysak, and R.G. Ingram, 1994. Interannual variability of sea-ice cover in Hudson Bay, Baffin Bay and the Labrador Sea. Atmosphere-Ocean 32: 421–447.

Importance of Eating Capelin: Unique Dietary Habits of Hudson Bay Beluga

T.C. Kelley, L.L. Loseto, R.E.A. Stewart, M. Yurkowski, and S.H. Ferguson

Abstract Beluga whales (*Delphinapterus leucas*) fill an important ecological and economic role in Hudson Bay. However, little is known about their diet and a better understanding of beluga populations is required. Though Arctic cod (*Boreogadus saida*) are important forage fish species for many circumpolar marine predators, beluga are opportunistic feeders and may feed on a variety of prey items. Here, we compare the fatty acid profile of two key forage fish, Arctic cod and capelin (*Mallotus villosus*), to determine the relative importance of each species to the diets of beluga during the 1980s in three Canadian Eastern Arctic beluga populations: Western Hudson Bay, Cumberland Sound, and the High Arctic. First, we compared the two prey species using a Principle Component Analysis (PCA) to determine the fatty acids that best described each species. Five fatty acids dominated the Arctic cod profile (the 20 and 22 carbon length monounsaturates 20:1n7, 20:1n9, 22:1n9, 22:1n11, 22:1n7), and five fatty acids were representative of the capelin profile (18:2n6, 16, 22:6n3, 22:5n6, and 20:4n6). The levels of these ten fatty acids were significantly different between the two fish species. A discriminant function analysis followed by univariate tests, were performed on beluga fatty acid profiles to determine if populations could be differentiated. Results demonstrated significant differences among the three beluga populations. Finally, to examine the qualitative dietary importance of Arctic cod and capelin among the three beluga populations all fatty acid profiles were evaluated together with a PCA. We found the fatty acid profiles that segregated the Hudson Bay beluga population from others appeared to be associated with a capelin diet relative to the other beluga populations that appeared to feed more heavily on Arctic cod. The difference in fatty acid profiles and diet between the northern populations and the Hudson Bay population is discussed relative to possible environmental explanations.

T.C. Kelley (✉)
Department of Environment and Geography, University of Manitoba,
Winnipeg, MB R3T 2N2, Canada
e-mail: umkelle0@cc.umanitoba.ca

S.H. Ferguson et al. (eds.), *A Little Less Arctic: Top Predators in the World's Largest Northern Inland Sea, Hudson Bay*, DOI 10.1007/978-90-481-9121-5_3,
© Springer Science+Business Media B.V. 2010

Keywords Feeding ecology • Fatty acids • Food web • Blubber • Hudson Bay • High Arctic • Cumberland Sound • *Delphinapterus leucas* • *Mallotus villosus* • *Boreogadus saida*

Introduction

Beluga whales (*Delphinapterus leucas*) are an integral part of the eastern Canadian Arctic marine ecosystem, an important species in the diet of local Inuit, and provide economic resources to the community of Churchill, Manitoba through whale-watching tourism. Little is known about their diet in Hudson Bay, although Arctic cod (*Boreogaidus saida*) is an important prey species for many beluga populations (Seaman et al. 1982; Welch et al. 1992; Dahl et al. 2000; Loseto et al. 2009). However, beluga are opportunistic feeders and prey on a variety of items including redfish (*Sebastes marinus*), halibut (*Reinhardtius hippoglossoides*), and shrimp (*Pandalus borealis*) in Greenland (Heide-Jorgensen and Teilmann 1994). Pacific salmon (*Oncorhynchus* spp.) were dominant prey items to the Alaskan beluga populations (Frost and Lowry 1981). Finally, beluga have been observed feeding on capelin (*Mallotus villosus*) in the Churchill River estuary in summer (Watts and Draper 1986).

Capelin is a temperate cold-water fish species (Carscadden and Vilhjálmsson 2002). They are adapted to living at the edge of Arctic waters (Vilhjálmsson 2002) and move into warmer and coastal waters to spawn (Vesin et al. 1981; Carscadden et al. 1989). Capelin respond quickly to changes in climate and a simplified prediction of temperature increases of 2–4°C would result in capelin shifting distribution 4–18° latitude north (Rose 2005). Increased ice cover and decreased water temperatures associated with prolonged cold periods has been associated with the movement of capelin to non-traditional, warmer waters (Carscadden et al. 2001). Thus, increased temperatures in Hudson Bay may lead to changes in capelin habitat use and prevalence, resulting in unknown impacts on beluga.

Conversely, Arctic cod are typically associated with sea ice, especially along ice pressure ridges (Moskalenko 1964) and ice cracks (Lønne and Gulliksen 1989). These habitats are thought to provide refuge from predators such as ringed seals (*Phoca hispida*) as well as provide prey to feed on (Gradinger and Bluhm 2004). Evidently, Arctic cod is better adapted than other fish species to the environmental conditions in the far north (Tynan and DeMaster 1997) as they are adapted to feed and live under the ice (Dunbar 1981; Gradinger and Bluhm 2004).

Sea ice cover in Hudson Bay is seasonal, with a near complete ice cover for most of the year (November–June) and ice free conditions in the summer (Saucier et al. 2004). The effects of climate change are greatest at higher latitudes and Hudson Bay is expected to exhibit one of the highest rates of warming (Parkinson et al. 2008; Gagnon and Gough 2005; Parkinson et al. 1999). Ice-associated species, such as beluga, Arctic cod, and the food web they depend on will be affected by this shift in climate (Stirling et al. 2004; Ferguson et al. 2005; Stirling and Parkinson 2006). Changes in Arctic cod distribution may lead to changes in whale migration, distribution, and abundance patterns (Tynan and DeMaster 1997; Laidre et al. 2008).

Beluga diets are poorly understood because whales harvested in summer typically have empty stomachs and faeces are difficult to collect due to their deliquescence. Direct observation of feeding by marine mammals has been accomplished (Watts and Draper 1986; Harwood and Smith 2002), though it is often difficult. Recently, techniques using fatty acid signatures have been used to obtain information on diet, foraging locations, and population structure of many animals, including marine mammals, because fatty acids are predictably incorporated into consumer fat tissue, integrating diet over weeks to months (Kirsch et al. 2000; Iverson et al. 2004; Thiemann et al. 2007). This technique assumes that a prey species' fatty acid signatures are reflected in the tissues of its predator, making fatty acids an ideal diet biomarker.

In the 1980s, as part of investigations into beluga stock identity (Stewart 1994), blubber and prey samples were collected and processed to examine stock differences, but were never published. The dataset is unique as it provides a glimpse into beluga feeding habits during the 1980s that is otherwise unavailable.

Here we resurrect those data to: (1) determine if Arctic cod and capelin can be distinguished from each other based on their fatty acid profiles; (2) determine if Hudson Bay beluga have different diets than the other two populations; (3) evaluate the relative importance of these species in beluga whales harvested in Hudson Bay, Cumberland Sound and Grise Fjord; (4) assess whether we can statistically discriminate among the three populations of beluga using fatty acid markers; and (5) provide retrospective data by which modern analysis can assess climate change impacts.

Methods

Sample Collection

Samples used for this study were collected in the Eastern Arctic, from Arviat, Pangnirtung and Grise Fjord (Fig. 1). The whale samples represent three different beluga whale populations; beluga from Arviat are whales belonging to the western Hudson Bay population; beluga from Grise Fjord are from the High Arctic population; and samples from Pangnirtung belong to the Cumberland Sound population (Stewart 1994). The samples ($n = 97$) were collected between 1983 and 1987 from whales harvested by Inuit during annual subsistence hunts. Sixty-five samples were collected in Arviat, 17 in Grise Fjord, and 15 in Pangnirtung. Eight capelin samples were collected in Arviat in 1987 and 28 Arctic cod samples were collected in Resolute Bay in 1988.

Fig. 1 Map showing beluga harvest locations that took place at Grise Fiord, Pangnirtung and Arviat. Beluga from Grise Fiord are representative of the High Arctic population, whales from Pangnirtung are part of the Cumberland Sound population and whales from Arviat are from the western Hudson Bay population

Lipid and Fatty Acid Extraction

Whole fish were homogenized for lipid extraction, whereas beluga blubber samples were first trimmed to remove oxidized surfaces, and sub-sampled for analysis. Lipids were extracted using 40 ml of 1:1 chloroform-methanol. The lipid phase was collected, washed and filtered (using fiber-glass paper). This lipid phase was then used to prepare fatty acid methyl esters. The lipids were transesterfied using 5% methanolic HCl and the samples flushed with nitrogen gas then heated at 100°C for 90 min. The fatty acid methyl esters were extracted twice with distilled hexane and purified with toluene on Silica Gel G thin-layer plates. Methyl esters in hexane were analyzed using a Varian 2100 gas chromatograph converted to use Supelcowax 10 capillary glass columns (60 m × 0.75 ID). The peak areas were measured and converted to area percentage by a computing integrator (Shimadzu Chromatopak model C-R3A, Kyoto, Japan). Fatty acids were identified by relative retention times based on comparisons to authenticated fatty acid retention times, by use of log relative retention time versus carbon number plots, and by comparison of equivalent chain lengths as described by Jamieson (1975). Each fatty acid was described using the shorthand nomenclature of A:Bn-X, where A represents the number of carbon atoms, B the number of double bonds, and X the position of the double bond closest to the terminal methyl group.

Data Analysis

Fatty acids were recorded as a percent of the total number of fatty acids present. Of the 107 fatty acids recorded, for subsequent analyses we used 29 fatty acids that are not manufactured by higher level predators, but are incorporated into the blubber as a result of consuming specific, identifiable prey items (Iverson et al. 2004). Fatty acid percents were log-transformed prior to the multivariate analyses (Kenkel 2006; Aitchison 1986).

Iverson et al. (2004) developed the Quantitative Fatty Acid Signature Analysis (QFASA) method of quantifying the proportion of individual prey species consumed by predators by feeding captive seals a known diet. QFASA has not been calibrated for beluga or cetaceans, so we used principle components analysis (PCA) to perform a qualitative investigation of dietary preferences. PCA has been used to perform qualitative analyses of predator diets (e.g. Loseto et al. 2009; Dahl et al. 2000; Iverson et al. 1997), and though it cannot be used to make quantitative statements about prey consumption, it can provide some insight into the diet of predators. Fish species that plot close to beluga populations on a PCA score plot have relatively similar fatty acid profiles suggest they are important diet items to the particular beluga population (Dahl et al. 2000; Loseto et al. 2009).

Prey Species Identification

First we compared fatty acid signatures of capelin (*n* = 8) and Arctic cod (*n* = 28) using a PCA with a covariance matrix (SYN-TAX© Ordination 2000). The method gives equal weight to abundant and rare fatty acids which was necessary as many of the fatty acids accounted for less than 5% of the total fatty acid composition of the lipid. The PCA is a qualitative test that indicated five fatty acids could characterize Arctic cod and five characterized capelin. We compared these signatures quantitatively, to determine if the prey species had identifiable differences by comparing the mean proportion of the ten fatty acids found to be associated with both prey species (*t*-test).

Beluga Study Site Discrimination

A discriminant function analysis (DFA) (SYSTAT 11®) was performed on beluga samples using the 29 fatty acids known to be incorporated into the blubber from prey with little biotransformation (Iverson et al. 2004). The purpose of this test was to determine if there was a significant difference in the diet among the beluga sampled from the three communities. The DFA showed significant differences between the three populations, so we tested the influence of Arctic cod and capelin consumption in the three populations by comparing the prevalence of ten representative prey fatty acids previously identified (five for each species) using an ANOVA. Significant differences in fatty acids profiles would suggest differences in the feeding behaviour of whales hunted near each community.

We determined that there were significant differences between the fatty acids associated with Arctic cod and capelin in each of the communities so a PCA with a covariance matrix (SYN-TAX© Ordination 2000) was performed to determine which of the prey species most associated with the three beluga populations. This information was analysed qualitatively, by determining the graphical proximity of the prey species to the populations of beluga. Prey species situated in close proximity to a beluga population on the PCA was interpreted as being important in the diet of that population, while a prey species situated far from a population was interpreted as having low importance in the diet of that population.

Results

Prey Species Identification

The first two axes of the prey PCA explained 45.2% and 12.3% of the variance respectively with no overlap along the first PCA axis (PC 1) (Fig. 2). To typify each species we selected fatty acids with the largest absolute values on PC 1. The five fatty acids representative of Arctic cod included the 20 and 22 carbon length mono-unsaturates (20:1n7, 20:1n9, 22:1n9, 22:1n11, 22:1n7). The five fatty acids that represented capelin included several omega 6s (18:2n6, 16:0, 22:6n3, 22:5n6, and 20:4n6) (Fig. 2). T-tests comparing the ten fatty acids associated with Arctic cod

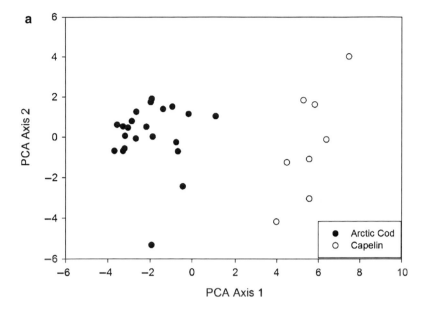

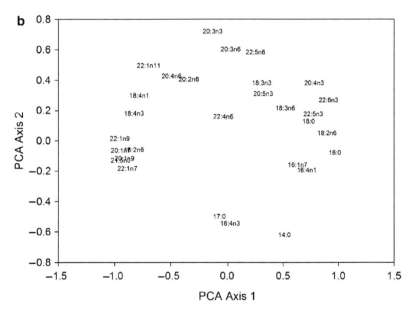

Fig. 2 (a) PCA plot of Arctic cod *(Boreogaidus saida)* and capelin *(Mallotus villosus)*, and (b) the fatty acids contributing to the segregation. Fatty acids associated with Arctic cod are 22:1n9, 22:1n11, 22:1n7, 20:1n7 and 20:1n9. Fatty acids associated with capelin are 18:2n6, 16:0, 22:6n3, 22:5n6 and 20:4n6

and capelin showed the proportion of fatty acids associated with the two prey species were significantly different between capelin and Arctic cod (Fig. 3, all *t*-tests $P \leq 0.001$).

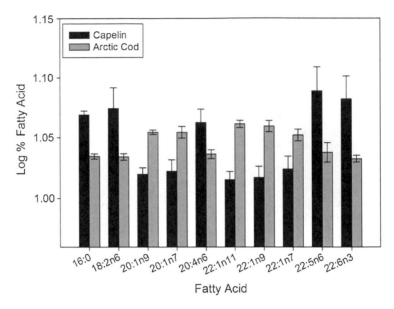

Fig. 3 Select fatty acids representing capelin and Arctic cod. T-tests comparing the mean concentrations of these fatty acids showed they were all significantly different at $P \leq 0.001$. All values have been log adjusted. Error bars represent 95% confidence intervals

Beluga Study Site Discrimination and Diet

The discriminant function analysis showed the fatty acid profiles were significantly different among the three populations of beluga (Wilk's Lambda = 0.056, $F_{58, 132} =$ 7.304, $P \leq 0.001$; Fig. 4). The fatty acid most strongly driving Factor 1 was 22:6n3. The fatty acid most strongly driving Factor 2 was 22:1n11. These fatty acids were representative of capelin and Arctic cod, respectively (Figs. 2 and 3).

Eight of the ten fatty acids were significantly different; 20:4n6 and 20:1n7 were not significantly different between populations (Table 1). Post-hoc comparisons between populations showed that whales from Arviat had significantly higher concentrations of fatty acids associated with capelin ($P = 0.05$) than either Grise Fjord or Pangnirtung, and significantly lower concentrations of fatty acids associated with Arctic cod than the other two communities ($P = 0.05$, Fig. 5). They also showed significant differences between Arviat and Grise Fjord and Arviat and Pangnirtung for fatty acids 16:0, 22:6n3, 22:5n6, 22:1n9, 22:1n11, and 22:1n7, with no significant differences between Grise Fjord and Pangnirtung for these fatty acids. Fatty acids 18:2n6 and 20:1n9 were significantly different between all communities.

In the combined prey and beluga PCA, 82.2% of the variance was explained by the first two PCA axes (PC 1: 54.9%; PC 2: 27.3%, Fig. 6). The beluga populations separated along the X axis with Arviat and Pangnirtung separating with little overlap, and Grise Fjord overlapping both populations. While the populations of beluga were

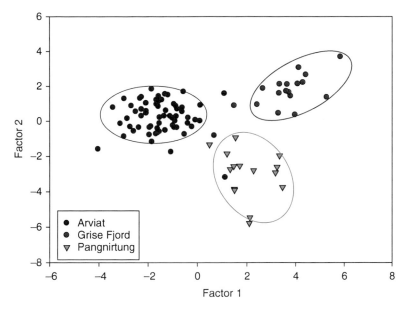

Fig. 4 Discriminant function analysis of beluga fatty acid profiles from western Hudson Bay (Arviat), High Arctic (Grise Fjord), and Cumberland Sound (Pangnirtung) populations. Significant differences among populations were indicated (Wilk's Lambda = 0.056, $F_{58, 132}$ = 7.304, $P \leq 0.001$). The factor most strongly influencing the PC 1 was 22:6n3, which is associated with capelin. The factor influencing the PC 2 axis was 22:1n11, a fatty acid associated with Arctic cod. Ellipses represent 95% confidence levels

Table 1 Results of ANOVA comparison of selected fatty acids between beluga populations

Fatty acid	df	F	p
16:0	2, 94	40.513	<0.001[a]
18:2n6	2, 94	34.800	<0.001[a]
20:1n9	2, 94	18.032	<0.001[a]
20:1n7	2, 94	1.846	0.163
20:4n6	2, 94	2.648	0.076
22:1n11	2, 94	19.224	<0.001[a]
22:1n9	2, 94	18.785	<0.001[a]
22:1n7	2, 94	14.554	<0.001[a]
22:5n6	2, 94	7.214	0.001[a]
22:6n3	2, 94	5.764	0.004[b]

[a]Significant differences were found for 16:0, 18:2n6, 20:1n9, 22:1n11, 22:1n9, 22:1n7, and 22:5n6 at $P \leq 0.001$.

[b]Differences between communities for 22:6n3 were significant at $P \leq 0.05$.

Significant differences were not found for 20:1n7 or 20:4n6.

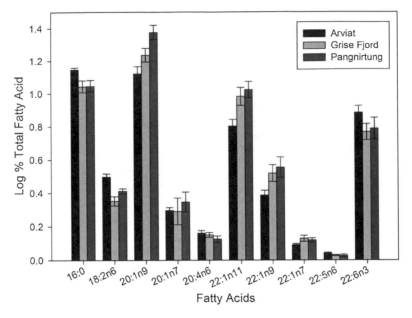

Fig. 5 Comparison of fatty acids representing Arctic cod and capelin among the three communities harvesting beluga representing three populations: western Hudson Bay (Arviat), High Arctic (Grise Fjord), and Cumberland Sound (Pangnirtung). An ANOVA analysis comparing sites found that Arviat had significantly higher concentrations of fatty acids associated with capelin than either Grise Fjord or Pangnirtung, and significantly lower concentrations of fatty acids associated with Arctic cod than the other two communities (except for fatty acids 20:4n6 and 20:1n7, which were not significantly different). All values have been log transformed. Error bars represent 95% confidence intervals

generally separated from one another, there was some overlap among the three groups (Fig. 6a). The prey separated completely on PC 1 with no overlap. The two fish species also separated on the second PCA axis, ranging from −0.2 to 0.1 for Arctic cod and −0.1 to −0.5 for capelin (Fig. 6a). Arctic cod appeared more prevalent in the diet of beluga from Pangnirtung, less prevalent in the diet of beluga from Grise Fjord, and least prevalent in the diet of beluga from Arviat; capelin showed the reverse pattern (Fig. 6). Along the second PCA axis, however, Arctic cod appear equally important to the diets of all three beluga populations, whereas capelin appeared less prevalent.

Discussion

Prey Differentiation

Fatty acid profiles of Arctic cod and capelin were similar within species and different between species. Arctic cod and capelin are both planktivores, consuming copepods, euphasiids and amphipods (Lønne and Gulliksen 1989; Carscadden et al.

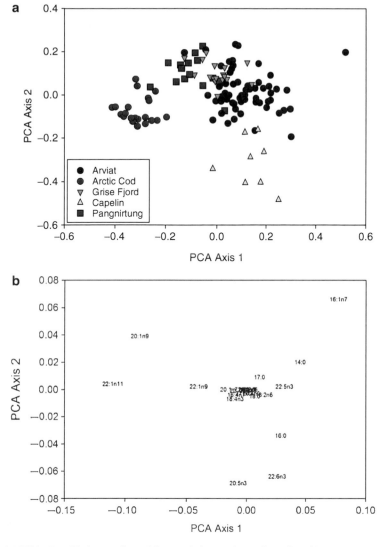

Fig. 6 (a) PCA plot of beluga collected from subsistence hunts from three Nunavut communities representing western Hudson Bay (Arviat), High Arctic (Grise Fjord), and Cumberland Sound (Pangnirtung) beluga populations along with the prey items capelin and Arctic cod. (b) PCA loadings of fatty acids that drive the placement of beluga and prey species. Beluga from Arviat are closely associated with capelin. Beluga from Pangnirtung as well as beluga from Grise Fjord are most closely associated with Arctic cod

2001). Trophic levels for Arctic cod and capelin are similar, at trophic level three (Hobson et al. 2002; Falk-Petersen et al. 2009a). Fatty acids that dominated Arctic cod fatty acid profiles were different than those for capelin; the 20 and 22 monounsaturates were the most prevalent in Arctic cod, while capelin fatty acid profiles were dominated by the omega 3s and 6s. The high proportion of 20 and 22 monounsaturates in Arctic cod has been observed elsewhere (e.g. Dahl et al. 2000; Andersen et al. 2004; Falk-Petersen et al. 2004; Loseto et al. 2009) and reflects a copepod diet. Specifically the fatty acids 20:1n9 and 22:1n11 are known to be copepod biomarkers (Sargent and Falk-Petersen 1988; Fraser et al. 1989; Kattner et al. 1989; Falk-Petersen et al. 2009b). Calanoid copepods biosynthesize those fatty acids de novo from wax esters that are then transferred to consumers (Kattner and Hagen 1995). Thus, lower proportions of the copepod fatty acids in capelin suggests they may not preferentially feed on copepods or that capelin may rapidly metabolize these long fatty acids to shorter chain fatty acids.

Capelin, however, are believed to be reliant on copepods, and differences observed between capelin and Arctic cod may be a function of feeding in different niches with different primary producers (Carscadden et al. 2001). The omega 3 and omega 6 fatty acids prevalent in capelin were also described by Henderson et al. (1984), though while they found similar high levels of 16.0 and 22.6n3, they did not find high levels of 18:2n6, 20:4n6 or 22:5n6. They found low levels of the fatty acids 20:1n9 and 22:1n11, similar to observations here. Given that their study was on the Balsfjorden capelin population that live exclusively in Balsfjorden Fjord, their fatty acid profiles may reflect different ecosystem-based sources. High levels of the saturated 16 carbon fatty acid (palmitic acid) have been documented in capelin (Ratnayake and Ackman 1979; Henderson et al. 1984; Sargent and Falk-Petersen 1988) and is thought to be incorporated into fish tissues at proportions similar to those in copepods (Ratnayake and Ackman 1979). Palmitic acid in copepods is related to heavy feeding on phytoplankton diatoms (Falk-Petersen et al. 2009b), indicative of pelagic feeding ecology.

The fatty acid 18:2n6 is thought to reflect terrestrial sources (Budge and Parrish 1998) as well as freshwater sources in seals (Smith et al. 1996). Though capelin are marine fish, they return to coastal areas to feed and spawn (Vesin et al. 1981). Higher levels of 18:2n6 in capelin may indicate terrestrial carbon contributions from the coastal marine ecosystem (Dalsgaard et al. 2003); as capelin migrate inshore to spawn, their diet reflects increasingly coastal sources (Davoren and Montevecchi 2003). Higher levels of this fatty acid in beluga from Hudson Bay may suggest capelin consumption or overall greater incorporation of terrestrial carbon in the Hudson Bay ecosystem as compared to the far north ecosystem. Fatty acids 20:4n6 and 22:5n6 are also both synthesized from lineolate (18:2n6) and may be synthesized or modified by zooplankton at the lower levels of the pelagic ecosystem (Budge et al. 2008).

Differences in prevalence of the 20- and 22- monounsaturates in capelin and Arctic cod may reflect differences in their trophic levels, as these fatty acids increase in proportion with trophic level within a food web (Budge et al. 2006; Dalsgaard et al. 2003; Loseto et al. 2009). The low levels of 20- and 22- monounsaturates in capelin compared with Arctic cod may suggest capelin feed at a lower trophic level (i.e. incorporating phytoplankton prey). However, fatty acids may not be the best indicators of trophic level. If Arctic cod and capelin are in fact feeding in different food webs, factors such as prey species and biosynthesis rate variability between the food webs may alter the relative proportions of these fatty acids in cod and capelin

(Dalsgaard et al. 2003). This variability may influence the relative proportion of the 20- and 22-carbon length monounsaturates in the fatty acid profiles of these fish.

The different habitat use and ecology of Arctic cod and capelin may provide a better explanation of differences between fatty acid profiles. Arctic cod are sea-ice specialists, adapted to feeding on prey items under the sea ice (Gradinger and Bluhm 2004), while capelin utilize ice-free water (Carscadden et al. 2001; Vilhjálmsson 2002) supporting different feeding ecologies. The primary producers in sea-ice ecosystems are ice algae (Gosselin et al. 1998; Lizotte 2001). Algal taxa have unique fatty acids compositions compared to pelagic phytoplankton (Dunstan et al. 1993; Viso and Marty 1993). These lower trophic level differences may be identifiable in higher level consumers and indicate important sources of primary productivity in the ecosystem (Budge et al. 2008). The higher levels of 22.6n3 found in capelin may indicate they feed predominantly in a pelagic food web (Copeman and Parrish 2003; Budge et al. 2008). Prevalent fatty acids support different feeding ecologies among capelin and Arctic cod and provide identifiable differences between sea-ice and pelagic-based food webs in consumers further up the food chain.

Beluga Dietary Preference and Site Discrimination

The fatty acids selected for analysis likely undergo little biotransformation between predator and prey (Iverson et al. 2004). Multivariate analyses comparing the relative percent of these fatty acids in predator and prey species have been used to determine important prey species as well as variations in diet among populations of predators (Iverson et al. 1997; Budge et al. 2002; Loseto et al. 2009). The DFA supported significant differences in diet among the three beluga populations investigated here. The overlap of populations along the second axis of the DFA demonstrated some similarity in diet among the beluga populations.

Despite apparent dietary differences among the beluga populations, our analyses cannot determine the percent contribution of individual species to each beluga population. Results from both the PCA and ANOVA support that regional differences found in beluga fatty acid profiles relate well to the differences in Arctic cod and capelin. Beluga from Pangnirtung had a fatty acid profile most similar to Arctic cod and the beluga from Arviat the least similar. Beluga from Arviat appeared to be most similar to capelin, a known prey in the area (Watts and Draper 1986). The regional differences may be due to higher sea ice coverage at higher latitudes where Arctic cod may be the dominant forage species that associates with a sea-ice algae food web base, while the farther south population near Arviat has a higher proportion of fatty acids indicative of pelagic phytoplankton producers.

Hudson Bay is typically ice-covered during the winter and ice-free during the summer months, but recent climate change-induced increases in water temperature in Hudson Bay have resulted in decreasing sea ice coverage (Kerr 2007; Serreze et al. 2007). During the 1980s, beluga in Hudson Bay appear to have been consuming more capelin, a species associated with open Arctic waters, but Arctic cod still comprised part of their diet. Distributional shifts in both capelin and cod to farther northern latitudes is predicted (Tynan and Demaster 1997; Carscadden et al. 2001). How ecosystem trends, beluga diet, and prey species distribution and abundance respond to environmental changes is difficult to assess. The diet of thick-billed murres (*Uria lomvia*) in Hudson Strait changed to consume more capelin between the 1980s and 2000s (Gaston et al. 2003). One might expect beluga to show a similar dietary shift, and change in distribution, or both. Similarly, beluga farther north may increase the proportion of capelin in their diets.

Conclusion

We resurrected archived data to address five objectives. First, Arctic cod and capelin were distinguished from each other based on their fatty acid profiles. Cod were characterized by the 20 and 22 carbon length monounsaturates and capelin by omega 3 and 6 fatty acids. Second, Hudson Bay beluga had a different diet than the Cumberland Sound and High Arctic whales. Third, this difference was due to the prevalence of capelin in the Hudson Bay diet. Capelin was relatively more important than Arctic cod to Hudson Bay beluga, but less important to the Cumberland Sound and High Arctic beluga populations. Fourth, fatty acid profiles in the blubber of whales discriminated statistically between Hudson Bay whales and the other two areas, but not between the latter two. Finally, by characterizing the blubber of beluga sampled 20 years ago, we provide a reference for modern and future studies. The dietary difference is likely due to the extensive open-water season in Hudson Bay. As water temperatures increase and sea ice diminishes, both Arctic cod and capelin will likely shift their range north. The effect this will have on beluga in Hudson Bay is not completely understood requiring more research on predator-prey ecology in Hudson Bay.

References

Aitchison, J. 1986. The statistical analysis of compositional data. Chapman & Hall, London.

Andersen, S.M., Lydersen, C., Grahl-Nielsen, O., Kovacs, K.M. 2004. Autumn diet of harbour seals (*Phoca vitulina*) at Prins Karls Forland, Svalbard, assessed via scat and fatty-acid analysis. Can. J. Zool. 82, 1230–1245.

Budge, S.M., Wooler, M.J., Springer, A.M., Iverson, S.J., McRoy, C.P., Divoky, G.J. 2008. Tracing carbon flow in an arctic marine food web using fatty acid-stable isotope analysis. Oecologia 157, 117–129.

Budge, S.M., Iverson, S.J., Koopman H.N. 2006. Studying trophic ecology in marine ecosystems using fatty acids: A primer on analysis and interpretation. Mar. Mamm. Sci. 22, 759–801.

Budge, S.M., Iverson, S.J., Bowen, W.D., Ackman, R.G. 2002. Among- and within-species variability in fatty acid signatures of marine fish and invertebrates on the Scotian Shelf, Georges Bank, and southern Gulf of St. Lawrence. Can. J. Fish. Aquat. Sci. 59, 886–898.

Budge, S.M., Parrish, C.C. 1998. Lipid biogeochemistry of plankton, settling matter and sediments in Trinity Bay, Newfoundland. II. Fatty acids. Org. Geochem. 29, 1547–1559.

Carscadden, J.E., Frank, K.T., Leggett, W.C. 2001. Ecosystem changes and the effects on capelin (*Mallotus villosus*), a major forage species. Can. J. Fish. Aquat. Sci. 58, 73–85.

Carscadden, J.E., Frank, K.T., Miller, D.S. 1989. Capelin (*Mallotus villosus*) spawning of the southeast shoal: influence of physical factors past and present. Can. J. Fish. Aquat. Sci. 46, 1743–1754.

Carscadden, J.E., Vilhjálmsson, H. 2002. Capelin – what are they good for? ICES J. Marine Sci.: Journal du Conseil 59, 863–869.

Copeman, L.A., Parrish, C.C. 2003. Marine lipids in a cold coastal ecosystem: Gilbert Bay, Labrador. Mar. Biol. 143, 1213–1227.

Dahl, T.M., Lydersen, C., Kovacs, K.M., Falk-Petersen, S., Sargent, J., Gjertz, I., Gulliksen, B. 2000. Fatty acid composition of the blubber in white whales (*Delphinapterus leucas*). Pol. Biol. 23, 401–409.

Dalsgaard, J., John, M., Kattner, G., Muller-Navarra, D.C., Hagen, W. 2003. Fatty acid trophic markers in the pelagic food marine environment. Adv. Mar. Biol. 46, 227–340.

Davoren, G.K., Montevecchi, W.A. 2003. Signals from seabirds indicate changing biology of capelin stocks. Mar. Ecol. Prog. Ser. 258, 253–261.

Dunbar, M.J. 1981. Physical causes and biological significance of polynyas and other open water in sea ice. In: Stirling, I., Cleator, H. (Eds.), Polynyas in the Canadian Arctic. Canad. Wild. Serv. Occasional Paper, pp. 29–43.

Dunstan, G.A., Volkman, J.K., Barrett, S.M., Garland, C.D. 1993. Changes in the lipid composition and maximisation of the polyunsaturated fatty acid content of three microalgae grown in mass culture. J. Appl. Phycol. 5, 71–83.

Falk-Petersen, S., Haug, T., Nilssen, K.T., Wold, A., Dahl, T.M. 2004. Lipids and trophic linkages in harp seal (*Phoca groenlandica*) from the eastern Barents Sea. Pol. Res. 23, 43–50.

Falk-Petersen, S., Haug, T., Hop, H., Nilssen, K.T., Wold, A. 2009a. Transfer of lipids from plankton to blubber of harp and hooded seals off East Greenland. Deep Sea Res. 56, 2080–2086.

Falk-Petersen, S., Mayzaud, P., Kattner, G., Sargent, J.R. 2009b. Lipids and life strategies of Arctic *Calanus*. Mar. Biol. Res. 5, 18–39.

Ferguson, S.H., Stirling, I., McLoughlin, P. 2005. Climate change and ringed seal (*Phoca hispida*) recruitment in western Hudson Bay. Science 21, 121–135.

Fraser, A.J., Sargent, J.R., Gamble, J.C. 1989. Lipid class and fatty acid composition of *Calanus finmarchicus* (Gunnerus) *Pseudocalanus* sp. and *Temora longicornis* (Muller) from a nutrient-enriched seawater enclosure. J. Exper. Mar. Biol. 130, 81–92.

Frost, K.J., Lowry, L.F. 1981. Foods and trophic relationships of cetaceans in the Bering Sea. In: Wood, D.W., Calder, J.A. (Eds.), The eastern Bering Sea shelf: oceanography and resources, Vol. 2, pp. 825–836. University of Washington Press, Seattle.

Gagnon, A.S., Gough, W.A. 2005. Trends in the dates of ice freeze-up and breakup over Hudson Bay, Canada. Arctic 58, 370–382.

Gaston, A.J., Woo, K., Hipfner, M. 2003. Trends in forage fish populations in Northern Hudson bay since 1981, as determined from the diet of nestling thick-billed Murres, *Uria lomvia*. Arctic 56, 227–233.

Gosselin, M., Levasseur, M., Wheeler, P.A., Horner, R.A., Booth, B.C. 1998. New measurements of phytoplankton and ice algal production in the Arctic Ocean. Deep Sea Res. Part II: Top. Stud. Oceanog. 44, 1623–1644.

Gradinger, R.R., Bluhm, B.A. 2004. In-situ observations on the distribution and behavior of amphipods and Arctic cod (*Boreogadus saida*) under the sea ice of the High Arctic Canada Basin. Pol. Biol. 27, 595–603.

Harwood, L.A., Smith, T.G. 2002. Whales of the Inuvialuit settlement region in Canada's Western Arctic: An overview and outlook. Arctic 55(1), 77–93.

Heide-Jorgensen, M.P., Teilmann, J. 1994. Growth, reproduction, age structure and feeding habits of white whales (*Delphinapterus leucas*) in West Greenland waters. Meddr Gronland. Biosci. 39, 195–212.

Henderson, R.J., Sargent, J.R., Hopkins, C.C.E. 1984. Changes in the content and fatty acid composition of lipid in an isolated population of the capelin *Mallotus villosus* during sexual maturation and spawning. Mar. Biol. 78, 255–263.

Hobson, K.A., Fisk, A., Karnovsky, N., Holstd, M., Gagnone, J.-M., Fortier, M. 2002. A stable isotope (d13C, d15N) model for the North Water food web: implications for evaluating trophodynamics and the flow of energy and contaminants. Deep Sea Res. II 49, 5131–5150.

Iverson, S.J., Field, C., Bowen, W.D., Blanchard, W. 2004. Quantitative fatty acid signature analysis: A new method of estimating predator diets. Ecol. Monogr. 74, 211–235.

Iverson, S.J., Frost, K.J., Lowry, L.F. 1997. Fatty acid signatures reveal fine scale structure of foraging distribution of harbour seals and their prey in Prince William Sound, Alaska. Mar. Ecol. Prog. Ser. 151, 255–271.

Jamieson, G.R. 1975. GLC identification techniques for long-chain unsaturated fatty acids. J. Chromatogr. Sci. 10, 491–497.

Kattner, G., Hagen, W. 1995. Polar herbivorous copepods – different pathyways in lipid biosynthesis. ICES J. Mar. Sci. 52, 329–335.

Kattner, G., Hirche, H.-J., Krause, M. 1989. Spatial variability in lipid composition of calanoid copepods from Fram Strait. Arctic Mar. Biol. 102, 473–480.

Kenkel, N.C. 2006. On selecting an appropriate multivariate analysis. Can. J. Plant Sci. 86, 663–676.

Kerr, R.A. 2007. Climate change: Is battered arctic sea ice down for the count? Science 318, 33–34.

Kirsch, P.A., Iverson, S.J., Bowen, W.D. 2000. Effect of a low-fat diet on body composition and blubber fatty acids of captive juvenile harp seals (*Phoca groenlandica*). Phys. Biochem. Zool. 73, 45–59.

Laidre, K.L., Stirling, I., Lowry, L.F., Wiig, Ø., Heide-Jorgenson, M.P., Ferguson, S.H. 2008. Quantifying the sensitivity of arctic marine mammals to climate-induced habitat change. Ecol. Appl. 18, S97–S125.

Lizotte, M.P. 2001. The contributions of sea ice algae to Antarctic marine primary production. Amer. Zool. 41, 57–73.

Lønne, O.J., Gulliksen, B. 1989. Size, age and diet of polar cod, *Boreogadus saida* (Lepechin 1773) in ice covered waters. Pol. Biol. 9, 187–191.

Loseto, L.L., Stern, G.A., Connelly, T.L., Deibel, D., Gemmill, B., Prokopowicz, A., Fortier, L., Ferguson, S.H. 2009. Summer diet of beluga whales inferred by fatty acid analysis of the eastern Beaufort Sea food web. J. Exper. Mar. Biol. Ecol. 374, 12–18.

Moskalenko, B.F. 1964. On the biology of polar cod, *Boreogadus saida* (Lepechin). Voprosy Ikhtiologii (Alaska Department of Fish and Game, translation, p. 18) 4, 433–443.

Parkinson, C.L., Cavalieri, D.J. 2008. Arctic sea ice variability and trends, 1979–2006. J. Geophys. Res. 113, C07003. doi:10.1029/2007JC004558.

Parkinson, C.L., Cavalieri, D.J., Gloersen, P., Zwally, H.J., Comiso, J.C. 1999. Arctic sea ice extents, areas, and trends, 1978–1996. J. Geophys. Res. 104, 837–856.

Ratnayake, W.N., Ackman, R.G. 1979. Fatty alcohols in capelin, herrin, and mackerel oils and muscle lipids: I. Fatty alcohol details linking dietary copepod fat with certain fish depot fats. Lipids 14, 795–803.

Rose, G.A. 2005. Capelin (*Mallotus villosus*) distribution and climate: a "sea canary" for marine ecosystem change. ICES J. Mar. Sci. 62, 1524–1530.

Sargent, J.R., Falk-Peteren, S. 1988. The lipid biochemistry of calanoid copepods. Hydrobiology 167/168, 101–114.

Saucier, F.J., Senneville, S., Prinsenberg, E.S., Roy, E.F., Smith, G., Gachon, E.P., Cava, E.D., Laprise, E.R. 2004. Modelling the sea ice-ocean seasonal cycle in Hudson Bay, Foxe Basin, and Hudson Strait, Canada. Clim. Dyn. 23, 303–326.

Seaman, G.A., Lowry, L.F., Frost, K.J. 1982. Foods of belukha whales (*Delphinapterus leucas*) in western Alaska. Cetology 44, 1–19.

Serreze, M.C., Holland, M.M., Stoeve, J. 2007. Perspectives on the arctic's shrinking sea ice coverage. Science, 315, 1533–1536.

Smith, R.J., Hobson, K.A., Koopman, H.N., Lavigne, D.M. 1996. Distinguishing between populations of fresh-and salt-water harbour seals (*Phoca vitulina*) using stable-isotope ratios and fatty acid profiles. Can. J. Fish. Aquat. Sci. 53, 272–279.

Stewart, R.E.A. 1994. Size-at-age relationships as discriminators of white whale (*Delphinapterus leucas* Pallas 1776) stocks in the eastern Canadian Arctic. Meddr. Grnland. Biosci. 39, 217–225.

Stirling, I., Parkinson, C.L. 2006. Possible effects of climate warming on selected populations of Polar Bears (*Ursus maritimus*). Can. Arctic. 59, 261–276.

Stirling, I., Lunn, N.J., Iacozza, J., Elliot, C., Obbard, M. 2004. Polar bear distribution and abundance on the southwestern Hudson Bay coast during open water season, in relation to population trends and annual ice patterns. Arctic 57, 15–26.

Thiemann, G.W., Iverson, S.J., Stirling, I. 2007. Variability in the blubber fatty acid composition of ringed seals (*Phoca hispida*) across the Canadian arctic. Mar. Mamm. Sci. 23, 241–261.

Tynan, C.T., DeMaster, D.P. 1997. Observations and predictions of Arctic climate change: Potential effects on Marine Mammals. Arctic 50, 308–322.

Vesin, J.P., Leggett, J.C., Able, K.W. 1981. Feeding ecology of capelin (*Mallotus villosus*) in the estuary and western gulf of St. Lawrence and its multispecies implications. Can. J. Fish. Aquat. Sci. 38, 257–267.

Vilhjálmsson, H. 2002. Capelin (*Mallotus villosus*) in the Iceland-East Greenland-Jan Mayen ecosystem. ICES J. Mar. Sci. 59, 870–883.

Viso, A.C., Marty, J.C. 1993. Fatty acids from 28 marine microalgae. Phytochemistry 34, 1521–1533.

Watts, P.D., Draper, B.A. 1986. Note on the behavior of beluga whales feeding on capelin. Arct. Alp. Res. 18, 439.

Welch, H.E., Bergmann, M.A., Siferd, T.M., Martin, K.A., Curtis, M.F., Crawford, R.E., Conover, R.J., Hop, H. 1992. Energy flow through the marine ecosystem of the Lancaster Sound region, Arctic Canada. Arctic 45, 343–357.

Migration Route and Seasonal Home Range of the Northern Hudson Bay Narwhal (*Monodon monoceros*)

K.H. Westdal, P.R. Richard, and J.R. Orr

Abstract The northern Hudson Bay narwhal (*Monodon monoceros*) population gathers in the area of Repulse Bay, Nunavut in the summer season. This population is hunted by local Inuit and co-managed by the Nunavut Wildlife Management Board and the Department of Fisheries and Oceans. There is some uncertainty as to the size of the population, the migration route this population takes to its wintering areas, if its winter range overlaps with that of other narwhal populations, and whether it is hunted by other communities during migrations. In the face of a changing climate, this ecological information is essential to understanding the success of the population in the future.

The main focus of this paper is to provide summer and winter home range data of narwhals, as well as migration routes. This in turn will help to determine if past aerial surveys covered appropriate areas and what boundaries should be considered for future aerial population surveys. Ultimately this information will contribute to written documentation of traditional ecological knowledge and may assist in determining if this population is a separate stock. Finally, this study establishes a baseline to evaluate future impacts of climate change on this Hudson Bay narwhal population.

Nine narwhals were tagged with satellite-linked tracking devices in August 2006 and 2007 in the vicinity of Repulse Bay, Nunavut. Whales were tracked using the ARGOS system for 100 to 305 days with two of the tags transmitting long enough to show the beginning of the migration from wintering areas back to summer areas in early May. The trajectories of the tagged narwhals were estimated from the ARGOS locations using a movement state-space model. Home range size for each data set was calculated using 95% and 50% kernel estimates. In addition, 17 hunters and elders were interviewed in the community of Repulse Bay in order to gather traditional ecological knowledge of the species to add to the scientific analysis.

K.H. Westdal (✉)
Department of Environment and Geography, University of Manitoba,
440 Wallace Building, Winnipeg, MB R3T 2N2, Canada
e-mail: kristinwestdal@hotmail.com

S.H. Ferguson et al. (eds.), *A Little Less Arctic: Top Predators in the World's Largest Northern Inland Sea, Hudson Bay*, DOI 10.1007/978-90-481-9121-5_4,
© Springer Science+Business Media B.V. 2010

Results of local and scientific knowledge suggest that a portion of the summer home range falls to the east of past aerial survey coverage and that winter range does not overlap with that of other narwhal populations. Migration route of tracked animals coincide with traditional ecological knowledge of narwhal migration and suggests that this population is probably rarely hunted by other communities en-route between summer and winter areas.

Keywords Aerial survey • ARGOS • Kernel home range • Movement state-space model • Repulse Bay • Satellite telemetry • Subsistence hunt • Traditional ecological knowledge

Introduction

The narwhal (*Monodon monoceros*) is an Arctic cetacean known to travel between bays and fjords in the summer and deep offshore areas of heavy pack ice in the winter (Laidre et al. 2002). Ice, water depth and the presence of upwellings are believed to play key roles in habitat selection (Heide-Jørgensen et al. 2002a; Laidre et al. 2004), with narwhals presumably returning to the same locations in the Arctic year after year (Laidre et al. 2004). However, information is limited on narwhal migration routes as well as summer and winter home ranges due to the difficulty and expense in gathering such data (Heide-Jørgensen et al. 2003).

Changes in climate and decreasing sea ice extent in the Arctic, may cause narwhals to be more vulnerable than other cetaceans to these disturbances such as range restriction and predation, due to a strong association with the sea ice (IWC 1997; Tynan and DeMaster 1997; Higdon and Ferguson 2009). Laidre et al. (2008) further suggest that narwhals are among the most vulnerable species to climate and habitat change due to their specialization in feeding, narrow distribution, seasonal dependence on ice, and reliance on pack ice for predator avoidance. Gathering basic ecological data such as distribution and migration of the northern Hudson Bay (NHB) narwhals is useful in understanding habitat use, and can aid in understanding future ecological challenges to this population in response to climate change.

The NHB narwhals comprise one of three narwhal populations that inhabit Arctic waters (Strong 1988). Although there is some uncertainty about their exact summer home range and migration routes, it was thought that the NHB narwhal migrates between northern Hudson Bay, largely Repulse Bay, Lyon Inlet, Frozen Strait, and western Foxe Channel, in the summer (Richard 1991; Bourassa 2002) to eastern Hudson Strait in the winter (Richard 1991). There is uncertainty on how vulnerable this population is to hunting by other Hudson Bay and Strait communities during its migration. The NHB narwhal population is known to be hunted by Inuit primarily from Repulse Bay (Naujaat), and occasionally from five other communities in Nunavut: Chesterfield Inlet (Igluligaarjuk), Coral Harbour (Salliq),

Rankin Inlet (Kangiqliniq), Whale Cove (Tikirarjuaq), and Cape Dorset (Kingait) (Fisheries and Oceans Canada 1998). The harvest of this population is currently co-managed by the local Hunters and Trappers Organizations (HTOs), the Nunavut Wildlife Management Board (NWMB) and the Department of Fisheries and Oceans (DFO).

To gain a more complete understanding of the NHB narwhal population, Inuit traditional ecological knowledge (TEK) was also incorporated in the study. TEK, as defined by Berkes (1999:8) is "a cumulative body of knowledge, practice, and belief, evolving by adaptive processes and handed down through generations by cultural transmission, about the relationship of living beings, including humans, with one another and with their environment." Combining TEK and science is useful in that they each provide different types and spatial extents of information (Usher 2000). TEK can be especially valuable in remote locations where scientific data can only be gathered for short periods of time (Ferguson et al. 1998). Furthermore, engaging all stakeholders in the research process is invaluable in the process of creating sustainable management plans (Gilchrist et al. 2005).

In this study, results from satellite tracking data and traditional ecological knowledge, with a focus on empirical observations, from Repulse Bay community members on narwhal seasonal occurrence and movement were used to describe the population's summer range, migration routes and winter range. The summer range information was used to establish if it is different from past survey coverage. The autumn migration information was used to assess whether this population is vulnerable to hunting in other areas en route to and within its winter range. Finally, winter range information was compared with other documented Canadian narwhal populations' winter ranges.

Methods

Scientific

Whale Capture and Instrumentation

In order to deploy the satellite-linked tags with safe and effective capture and handling of all animals, tagging techniques adhered to DFO scientific permits and Animal Care Committee (ACC) protocols (following DFO ACC standard operating procedures, AUP #: FWI-ACC-2007-2008-037), and were supported by the local Hunter and Trapper Organizations.

In 2006 and 2007, nine narwhals were caught and fitted with satellite-linked transmitters in the Repulse Bay area (Fig. 1). Five were tagged in August of 2006 in Lyon Inlet (66°29'54"N, 83°58'20"W) and four in August of 2007 in Repulse Bay (66°31'19"N, 86°14'06"W). All nine whales were caught using the same

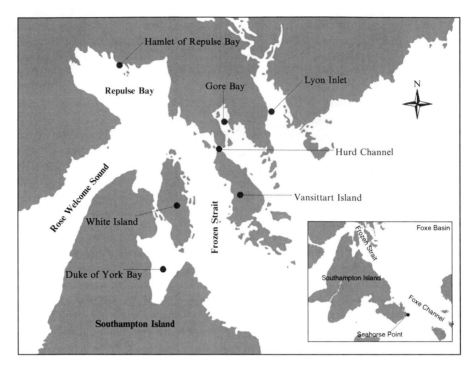

Fig. 1 Local area map showing the region where the NHB narwhals are known to summer with map insert showing Seahorse Point, Southampton Island

stationary net technique, where a mesh net is anchored to land and set out to sea perpendicular to the shoreline. The net was dotted with buoys along its length to keep it afloat and the bottom contained a lead line to keep it perpendicular in the water column. The net was under watch 24 h a day for any sign of whales near or in the net (Orr et al. 2001).

Once narwhals swam into the net and were caught, they were pulled to the surface, freed from any net material and secured with large straps and a hoopnet before attaching the satellite-linked transmitter ("tag"). Tags were attached to the whales with nylon pins inserted through the skin and fat below their dorsal ridge (Fig. 2). Standard body length, tusk length (where applicable), fluke width, sex, and scarring patterns, were recorded prior to release (Table 1). The interval between capture and release averaged 30 min per animal, with none lasting more than an hour.

The instrument used both years were SPLASH tags (Wildlife Computers, Redmond, WA, USA), which were programmed to transmit when surfaced. They were programmed to transmit a maximum 400 times per day in August and September. The rest of the year, to prolong battery life, tags were programmed to transmit every fourth day and to a maximum of 100 times on those days.

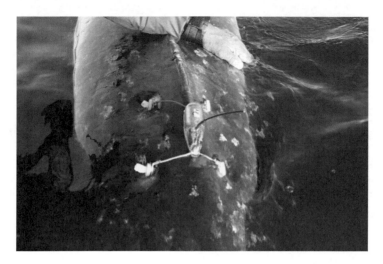

Fig. 2 SPLASH satellite transmitter attached to narwhal and ready for release in 2007 (Photo Credit: K. Westdal)

Table 1 Individual narwhal satellite-tagged over 2 years used in this study

TAG ID	Transmitter type	Year tagged	Date tagged in August	Body length (cm)	Fluke width (cm)	Tusk length (cm)	Sex	Date of last position	Transmission duration (days)
57595	SPLASH	2006	11	437	112	152(broken)	M	June 11/07	305
57596	SPLASH	2006	11	396	86	NA	F	June 3/07	297
57597	SPLASH	2006	11	262	66	27	M	Dec 24/06	146
57598	SPLASH	2006	11	353	ND	91	M	Apr 20/07	253
57599	SPLASH	2006	11	396	89	NA	F	Apr 4/07	237
36641	SPLASH	2007	8	400	90	NA	F	Nov 15/07	100
40152	SPLASH	2007	9	375	89	98	M	Dec 9/07	123
37024	SPLASH	2007	9	385	92	110	M	Dec 17/07	131
40622	SPLASH	2007	10	364	91	101	M	Nov 22/07	105

ND = no data available; NA = not applicable.

Data Collection and Analysis

Instrumented narwhals were tracked using the ARGOS System (http://www. clsamerica.com) from 100 (tag 36641) to 305 days (tag 57595). This satellite tracking system uses an array of polar-orbiting satellites, and ground-based receiving and processing stations that receive the satellite-linked data and calculate locations (CLS America Inc. 2007). ARGOS assigns location accuracy indices (or classes) to each location received, based on a proprietary algorithm. These range from high to low precision. ARGOS guarantees the minimal accuracy of location quality classes 3, 2, and 1 to be respectively 250 m, 500 m and 1.5 km. Location classes A and B have no guaranteed measure of positional accuracy but A is considered more accurate than B (CLS America Inc. 2007). In addition, Vincent et al. (2002) determined that, while location accuracy does indeed vary between location classes similarly to ARGOS accuracy ratings, it also varies within each location class. Locations are also irregularly spaced in time, due to irregular narwhal surface activity and punctuated satellite coverage. Irregular sampling can bias home range estimation because of heavy weighting in areas where more data were available (De Solla et al. 1999). This problem has been addressed, either by taking a daily best or a mean or median location to calculate home range (e.g., Heide-Jørgensen et al. 2003; Laidre et al. 2004). The problem with such an approach is that a large part of the information on the dataset is not used when estimating home range. Jonsen et al. (2005) developed a state-space model that uses all the data, while accounting for the variance of the accuracy of various location classes (Vincent et al. 2002), and their irregularly-spaced time intervals. The state-space model uses all location data (B to 3) and estimates each animal's trajectory by estimating locations that are evenly spaced in time. The state-space model was used here in order to prevent the loss of any data which could be valuable ecological information (Jonsen et al. 2005). Estimated locations generated by the model were plotted in ArcView® 3.3 (Environmental Systems Research Institute Inc., Redlands, California) and used to calculate winter and summer home ranges and to display estimated migration routes.

Home ranges were calculated for all animals in each year using the Animal Movement (AM) extension for ArcView (Hooge and Eichenlaub 2000). The extension implements the Worton (1989) kernel method to generate a probability of finding an animal in portions (or kernels) of the study area, given the location data provided (De Solla et al. 1999). Fifty percentile and 95 percentile probability contours were generated by the AM extension. The 50 percentile contours represent areas with at least a 50% probability of occurrence, or the median home range of the sample of tracked animal (Hooge et al. 1999). The 95% probability contours represent a probability of 95% of occurrence of the sampled animals. Following Worton (1989) and Hooge et al. (1999), the home range was calculated using the default kernel parameter generated from the dataset by the AM tool's algorithm. The model estimated locations at different intervals in winter than in summer because the tags were duty-cycled to transmit every 4 days after September 30th. The location interval was therefore every 4 h in summer (August–September) and every 4 days in winter.

Traditional Ecological Knowledge

Narwhals historically and still today have an important role in the community of Repulse Bay culturally and economically. Hunting practices and use of the animals have changed over time in Repulse Bay, but the importance of the animal in the community remains the same (Gonzalez 2001). Narwhals are hunted for local food consumption as well as a way to generate income within the community. They are hunted for their skin, or maktaq, as well as meat (Nuttall et al. 2005) and ivory tusks.

Traditional ecological knowledge documentation of the NHB narwhal population in the community of Repulse Bay took place with the direction and consent of community leaders and willing participants. We received consent from the Nunavut Research Institute and the local Hunters and Trapper Organization (Arviq Hunters and Trappers Organization) to come to the community and talk with elders and hunters about narwhals prior to the start of the research.

Semi-structured (or semi-directive) interviews were conducted to gather information about this narwhal population from participants within the community of Repulse Bay. The interviews required that we have specific topics in mind prior to the interview from which open ended questions were able to develop during the interview. The semi-structured interview has been effective in other projects involving TEK as it allows flexible guidelines, encourages the participant to elaborate on certain topics, and can take place in a variety of community settings. This method has been previously used to gain knowledge on the TEK of northwestern Alaskan beluga on topics such as migration, population and feeding (Huntington 1998). Research by Gonzalez (2001) on the TEK of the NHB narwhals from people in the community of Repulse Bay was used as a reference point for topics previously covered.

Seventeen community members were interviewed (Table 2). Participants interviewed were drawn from a list of potential interviewees created by the Arviq Hunters and Trappers Organization and the project interpreter, Marius Tungilik. Participants agreed to have their names acknowledged in the text as the source of information but not have their names attached to specific comments.

Most participants have spent the majority of their lives in the Repulse Bay area and were still actively involved in hunting at the time of the interview. Information relayed by the participants during the interviews was openly expressed as knowledge from personal experience and observations as well as knowledge passed down through generations and from others in the community. Qualitative data was

Table 2 Repulse Bay community participants

Honore Aglukka	Quviq Malliki
Michel Akkuardjuk	Seemee Malliki
Louis Angogingoar	Donat Milortok
John Ignerdjuk	Luky Putulik
John Ivalutanar	Pie Sanertanut
Sata Kidlapik	Antoine Siatsiak
Laurent Kringayark	Mark Tagornak
Laimmiki Malliki	Charlie Tinashlu
	David Tuktudjuk

analyzed using an interpretive approach that aims to connect ideas and categorize results (Kitchin and Tate 2000). Results were grouped into related categories and from there comments were scanned to look for patterns in the data in order to summarize results. Results paraphrase responses given by participants with no interpretation.

Results

Satellite Telemetry

All nine tags on the whales captured in 2006 and 2007 provided locations throughout the month of August. Five tags from 2006 provided location data well into the winter season (first ceased transmitting December 24th), two were still transmitting early May when both began migrating west back into Hudson Strait (Fig. 3). None of the four tags from 2007 transmitted past December, and thus did not provide any information on wintering areas.

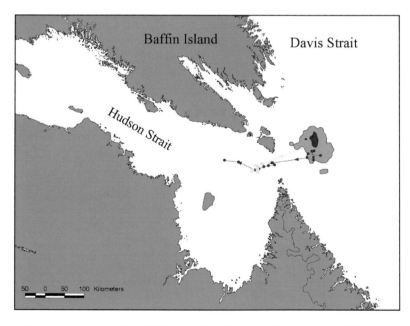

Fig. 3 Winter home range of the NHB narwhal in blue; dark blue inner ring (50% kernel probability), light blue outer (95% kernel probability) and return migration tracks from two whales tagged in 2006

Home Range August 2006 and 2007

Summer (August) home range in 2006 was estimated using the 95% kernel probability contour to occupy approximately 7,900 km², extending from Lyon Inlet around the southern tip of Vansittart Island and into Frozen Strait (Fig. 4). Figure 4 shows the extent of the 2007 aerial photo survey, from which approximately 2,700 km² (34%) of the 2006 home range was left uncovered. However, the 2000 visual survey did cover the 95% kernel area estimate. Using the 95% kernel probability contour, summer (August) home range in 2007 was approximately 4,600 km² (Fig. 5). That range fell entirely within the extent of the 2000 aerial photo and visual surveys (Fig. 5).

Migration Routes

In both 2006 and 2007 whales began to move out of Lyon Inlet and Repulse Bay in September and by early November, those that were still transmitting, were off the north east coast of Southampton Island. The same migration route was used both years: starting north of Southampton Island, they passed through

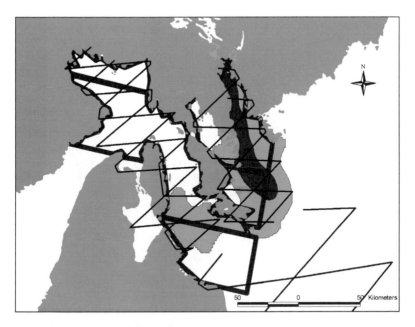

Fig. 4 August 2006 NHB narwhal summer home range (outer ring (*light blue*) 95% kernel probability and inner ring (*dark blue*) displaying the 50% kernel probability) overlayed with August 2000 photographic aerial survey coverage boundaries in red and visual survey flight lines in black (Bourassa 2002)

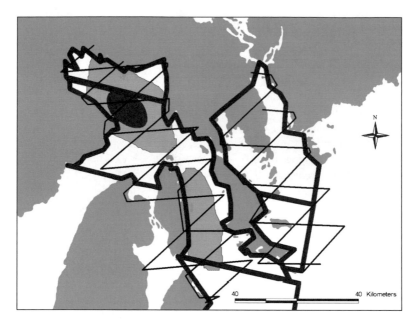

Fig. 5 August 2007 NHB narwhal summer home range (outer ring (*light blue*) 95% kernel probability and inner ring (*dark blue*) displaying the 50% kernel probability) overlayed with August 2000 photographic aerial survey coverage boundaries in red and visual survey flight lines in black (Bourassa 2002)

Foxe Channel into Hudson Strait (Fig. 6). All animals moved into Hudson Strait, passing north of Nottingham Island but on both sides of Salisbury and Mill Islands. Narwhals arrived in their wintering area by late December, as noted by the reduction in travel in an easterly direction and clustering of locations.

Of the nine tagged animals, one travelled farther north than the others passing by Big Island but remaining approximately 50 km offshore of the community of Kimmirut. The other four whales followed a very similar track off Cap de Nouvelle France on Quebec's Ungava Peninsula, later passing between 40 and 100 km offshore of the community of Kangiqsujuaq.

Winter Home Range

Winter home range of the NHB narwhals was calculated using winter modeled locations from five whales tagged in August 2006, all August 2007 tags ceased transmittion by December 17th. The winter range was defined by clustering of locations and began January 2nd and ended on May 2nd when narwhals began traveling in a westerly direction. The 95% kernel probability winter home range was approximately 7,000 km² (Fig. 3).

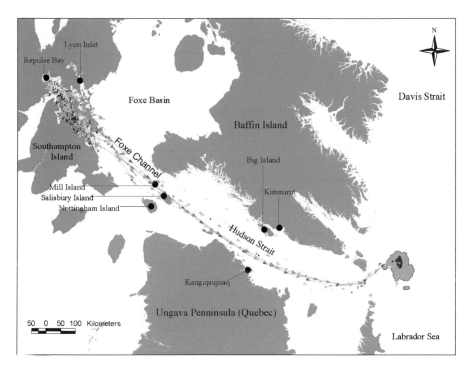

Fig. 6 Migration route of narwhal tagged in 2006 and 2007 ($n = 9$) summering in Repulse Bay area and wintering in the Labrador Sea overlaid with fall migration route drawn by two participants in separate interviews

Traditional Ecological Knowledge

Results of the community interviews as they relate to migration and home range are presented below as summarized from interviews with community members in the following categories: migration route, seasonal locations, density, and change in distribution.

Migration Route

Their main road is Frozen Strait

Narwhals migrate into Repulse Bay in June and July and out in August and September through Frozen Strait (Figs. 1 and 6). Frozen Strait is known to be very deep with strong currents and a preferred place for narwhals to feed. They use the same route in and out and are not known to use Roes Welcome Sound to migrate. It is known that the whales do not go north to Igloolik or south to Coral Harbour (although narwhals have been seen there on rare occasions) and that the large numbers of whales seen further north in Hecla Strait, Lancaster Sound, Prince Regent Inlet, and Admiralty Inlet are part of separate populations.

In the fall, when migration begins, narwhals will spend more time in the middle of Repulse Bay and less time along the shore, according to one participant. Once they decide to move, migration out of the bay happens very quickly. Once the whales leave Repulse Bay they can be found in large numbers in Lyon Inlet where the water is deep.

Long after the narwhals have left Repulse Bay you can still find them in the Inlet

They will make their way to Lyon Inlet through Hurd Channel where there is a strong current. Narwhals will spend some time here before moving on. Once they have left Repulse Bay they can also be found along both sides of White Island, in Duke of York Bay (south of White Island), and along Vansittart Island. Sometimes they will also go between Repulse Bay and these areas. From here, in the later part of the season, they will move through the deep part of Foxe Basin and Foxe Channel out through Hudson Strait. It is not known exactly where they overwinter but interviewees noted that community members have heard from DFO that they may be near Iqaluit in winter. Narwhals have been seen along the north coast of Southampton Island (East Bay area) in the spring and fall.

Seasonal Locations

Spring

Narwhals can be seen in great numbers in the spring at the floe edge during break up. In the past, according to one participant, narwhals were not seen until the first or second week in July. Today they can be seen at the floe edge as early as the first week of June.

Narwhals are first seen near small islands on north end of White Island, along the north coast of Southampton Island, and along the floe edge in June and early July, and between Beach Point and Cape Clarke where people are hunting. One participant mentioned being camped on the north end of White Island in the spring and observed the narwhals moving back and forth with the tide. Two participants mentioned that they had seen and or heard from elders that in the middle of June narwhals can also be seen in Lyon Inlet in holes in the ice feeding.

The whales follow the open water and move into Repulse Bay as cracks in the ice form and food becomes available in the bay.

Summer

In August, narwhals move back and forth in Repulse Bay. They also move in and out of the area with the tides. One participant said that towards the end of August one would see more narwhals down into Frozen Strait.

Fall

Narwhals are last seen in September or October in Repulse Bay (although there was one report in November before the ice was formed in the bay). Two participants noted that in the past they did not see them that late in the season. It was suggested by one that the reason for the late sightings could be due to the presence of killer whales (*Orcinus orca*) in the area.

They are last seen in various spots around the bay: close to town, deep water on the east side of the bay, near Harbor Islands, and in the middle of the bay. One participant suggested that they are last seen in these areas because people are not boating as far due to weather and no narwhal hunting tags are left at this time. When narwhals leave the Repulse Bay area they can be found in Hurd Channel, near the Sturges Islands, around White Island, in Duke of York Bay, along Vansittart Island, and in Lyon Inlet. One participant mentioned that when he was growing up in Lyon Inlet he would "see narwhals there until the snow started falling in October". He mentioned that they would stay in this area until they moved on to overwinter elsewhere.

Winter

Narwhals are not known to inhabit the Repulse Bay area in the winter. However, narwhals have been sighted occasionally in the area at the floe edge and in open water. Two participants noted that there is open water in Roes Welcome Sound, Frozen Strait, and Hurd Channel all year round due to strong currents. Narwhals have been spotted at the floe edge in November, February and May but it was also noted that the location of the floe edge changes every year. Narwhals have also been spotted in Lyon Inlet and specifically Ross Bay and Qariaq Inlet in the winter. These narwhals appeared to have been trapped as the ice had already formed in the area. Most participants agreed that if not stranded, narwhals do not winter in this area.

One participant mentioned that he thought that the narwhals from Repulse Bay winter in the Baffin area. Another mentioned that narwhals winter in the deep part of the Arctic Ocean where their preference for cold waters year round could be fulfilled. It was also mentioned that they may winter in the deep part of southern Hudson Bay. Another participant had heard stories of belugas (*Delphinapterus leucas*) wintering near Coral Harbour but not narwhals. However, it was also noted that people from Coral Harbour have seen narwhals in winter as they are known to pass Seahorse Point on Southampton Island on their migration out through Hudson Strait (Fig. 1, inset).

Whale Density

All participants had some information to offer on the topic of narwhal density. Most participants thought that the highest densities of narwhals in the summer were in Repulse

Bay and Lyon Inlet (specifically near Naujaarjuat Head adjacent to the community in the Inlet according to one participant). One participant said that these two congregations of narwhals (Repulse Bay and Lyon Inlet) were similar in size and two said that there are more in Lyon Inlet than in Repulse Bay in the summer. It was also mentioned that when there are no narwhal in Repulse Bay you can find them in Lyon Inlet and vice versa. Lyon Inlet is known to have more females and young than Repulse Bay.

There were mixed opinions on densities within Repulse Bay during the time narwhals spend in the area. Two participants noted that the largest concentrations in the bay were on the west side of the bay whereas others felt that they were in the greatest numbers on the top of the bay (community side) and near the Harbour Islands. One participant stated that there was a large concentration in the middle of the bay when killer whales were not in the area. If there are killer whales, narwhals come close to the community and into the shallow water. In the spring, when there is still lots of ice and a floe edge, large concentrations of narwhals can be found along the east side of White Island.

On the west side of Foxe Basin, from Lyon Inlet to Igloolik, there are normally no narwhal according to two participants. South of Beach Point (Roes Welcome Sound, Wager Bay) and north of Lyon Inlet narwhals are seen infrequently and in low numbers. Smaller concentrations of narwhals can be found along the coast towards Lyon Inlet (including Hurd channel and into Gore Bay), around both sides of White Island and down into Duke of York Bay in the summer months.

Participants said that it is hard to tell how many whales there are in the area as they are spread out and can travel far. It is also hard to tell if it is the same ones returning year after year as they come in such great numbers. One can see huge numbers of whales together in the summer, numbering in the hundreds. One partici- pant said that he had seen thousands of whales moving from White Island into Repulse Bay in late July.

Changes in Distribution

Five participants remarked that distribution of narwhals in the area has changed through personal experience and stories from others. In the past there were more narwhals in the bay, they were closer to shore, and would feed in the shallows. Possible reasons for the change are thought to be an increase in water temperature in the bay, an increase in motor boat activity, an increase in noise and an increase in hunting activity.

> With the increase in motorboats the narwhals started to move away from the community.
> Narwhals will move away from noise pollution like other animals.

Killer whales are also having a noticeable effect on narwhal distribution in relation to their traditional range. All 17 participants had something to say on this topic. It was mentioned that narwhal movement south into Wager Bay has been observed. Normally narwhals do not travel down Roes Welcome Sound and are not seen in Wager Bay. Participants had both heard and personally saw narwhals in Wager Bay in the last 2 years when killer whales were in the area.

Discussion

Data Processing

There is no consensus in the literature on how to use satellite-linked location data in ecological research. Some researchers preferred to use only location quality 1 through 3 (Heide-Jørgensen et al. 2003), others argued that 0, A and B might be used as well (Britten et al. 1999). In some studies data was also filtered by removing locations that were biologically unrealistic in terms of distance, time and angle from points (Austin et al. 2003; Heide-Jørgensen et al. 2002a; Vincent et al. 2002). In their attempt to reduce location bias, these methods lose some information available in the location data.

The Jonsen et al. (2005) estimation method was used here because it is able to utilize all location data, take into account their accuracy, as determined by Vincent et al. (2002), and because it generates locations that are evenly spaced in time. This method is particularly valuable because it has the ability to use the information of both high and low accuracy locations. High quality locations are relatively few for satellite-linked tracks of marine animals because they spend a considerable amount of time below the surface, a factor that limits the number of repeated transmissions in a satellite pass, hence the accuracy of many ARGOS locations. This method of location modeling is also preferable because it produces a regular time interval between locations and deals with the bias that irregular interval locations could cause in home range analysis.

Home Range Analysis

Performing the Worton (1989) kernel analysis of the home range was preferable over other home range analyses because the method is less sensitive to sample size than other home range estimators such as minimum convex polygons (Powell 2000). It also does not require the home range to be symmetrical such as an ellipse home range and it has the ability to account for 95% of the animal's movement which allows for occasional trips outside of what would be considered the animal's normal range (Powell 2000).

Stock Segregation

Data on stock segregation according to summer range is available from traditional ecological knowledge as well as tracking data. Home ranges in 2006 and 2007 show no use of areas north of Lyon Inlet. Participants involved in the traditional ecological knowledge research of this study stated that narwhals are not found much east of Lyon Inlet nor on the west coast of Foxe Basin in summer and that narwhals summering in Fury and Hecla Strait, Lancaster Sound, Prince Regent Inlet, and Admiralty Inlet are part of separate populations. This information is similar to records from previous traditional ecological knowledge studies (Gonzalez 2001; Stewart et al. 1995).

Summer Home Range

The summer home range of the NHB narwhal did not fall entirely within the coverage of some past population surveys. A portion of the August home range in 2006 was outside the 2002 survey area (Fig. 4). Had additional survey coverage been conducted in that area in August 2000, it might have resulted in more narwhal sightings and consequently a larger population estimate. This home range result should be used to delimit future aerial survey boundaries.

Most participants in the traditional ecological knowledge interviews agreed that the highest densities of narwhals in the summer were in Repulse Bay and Lyon Inlet although there was no agreement on which area had the highest density. This is similar to home range data gathered in this study and to past visual surveys of densities of animals (Richard 1991; Bourassa 2002).

Participants also agreed that narwhals are not known to be common south of Beach Point (Roes Welcome Sound, Wager Bay) and north of Lyon Inlet towards Igloolik (Fig. 1). This is also in agreement with home range data and past surveys indicated previously.

Participants also noted that narwhals can be found along the coast towards Lyon Inlet, around both sides of White Island and down into Duke of York Bay in the summer months although in concentrations smaller than that of Repulse Bay and Lyon Inlet. This is in agreement with 95% kernel densities as well as past survey data. However, large concentrations were seen further south in the 2000 visual survey which could be due to killer whale activity in the area as participants were confident that killer whales altered the distribution of narwhals in the area. Alternatively, there may have been fewer hunters in this area due to strong currents and dangers associated with ice.

Winter Home Range

Studies examining migration routes, winter location, behaviour and stomach contents of narwhal populations that summer in the Canadian High Arctic suggest that the same winter ranges and migration routes are used every year (Heide-Jørgensen et al. 2003). Winter feeding may be the primary reason that narwhals are congregating in those same locations each winter (Laidre and Heido-Jørgensen 2005).

The winter range of tagged Baffin Bay narwhals that summer near Somerset Island was approximately 26,000 km^2 and did not extend further south than 69°N

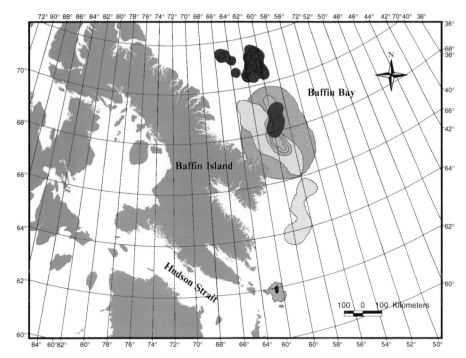

Fig. 7 Winter home range of the tagged NHB narwhals in blue (dark blue inner ring = 50% kernel probability, light blue outer = 95% kernel probability; and from the tagged Baffin Bay narwhals, animals tagged in: Admiralty Inlet in yellow (Dietz et al. 2008), Somerset Island in red (Heide-Jørgensen et al. 2002a, 2003); Melville Bay in purple (Dietz and Heide-Jørgensen 1995), Eclipse Sound in pink (Dietz et al. 2001)

(Heide-Jørgensen et al. 2003). The winter range of tagged Baffin Bay narwhals that summer in Eclipse Sound and Melville Bay (West Greenland) was further south in an area of approximately 22,500 km² that did not extend further south than 68°N (Dietz et al. 2001; Heide-Jørgensen et al. 2002a). The winter range of the 2006 NHB narwhals did not extend further north than 62°N. This winter home range adds to data that suggests that the NHB narwhal is a separate population (Fig. 7). Contaminant and genetic studies of de March and Stern (2003) and de March et al. (2003) point to the same conclusion: that NHB narwhals are probably a separate population. Koski and Davis (1994) observed narwhals in Hudson Strait and in southern Davis Strait in late winter in 1981 near the 2006 tracked narwhal winter home range. Together, these results strongly suggest that NHB narwhals should be managed separately from other Canadian populations.

Tracking data from this study suggest that the size of the tracked NHB narwhal winter range was at least 7,000 km². In comparison, the size of the winter home range for animals considered to be part of the Baffin Bay population is more than five times larger. The range of the animals tagged in Tremblay Sound wintering in northern Davis Strait is approximately 12,000 km² (Heide-Jørgensen et al. 2002a) and Somerset Island wintering range in central Baffin Bay is approximately 26,000 km² (Heide-Jørgensen et al. 2003). This could be due to a large difference in population size; the NHB narwhal population is estimated at 5,600 whales (Westdal 2008), whereas the total Baffin Bay stock is estimated to be in excess of 50,000 (Innes et al. 2002). Less likely in terms of the differences in area usage, but worth noting, is the sample size used to calculate home range in each of the three studies. Sample size for the NHB narwhals was five whales (four lasting the whole season) whereas the sample size was fifteen for narwhals tagged in Admiralty Inlet, nine in Melville Bay and seventeen in Eclipse Sound. Home range kernels, although proven to be the most robust of area usage models, are still sensitive to sample size (Hooge et al. 1999). A larger number of tagged narwhals is needed to verify if the estimate presented here represents the whole population's winter home range, considering that narwhal sightings have been made in eastern Hudson Strait in March in the past (Koski and Davis 1994).

Migration

Satellite tracking results reinforce traditional ecological knowledge on narwhal migration route. Both sets of data suggest that this population may migrate past Baffin Island and northern Quebec communities. Nevertheless, the tracking shows that most tagged narwhals migrated offshore of those coastlines. Changes in seasonal ice conditions in this region may impact the timing and pattern of narwhal migration in the future. The unpredictability of autumn ice formation would make narwhals, which migrate in large herds at that time of year, more susceptible to large ice entrapments (Heide-Jørgensen et al. 2002b).

Conclusions

Satellite tracking and traditional ecological knowledge lend support to the previous assumptions about the population's seasonal home ranges and migration routes (Richard 1991; Bourassa 2002).

It was found that past summer aerial survey coverage for the purpose of estimating population size did not cover a portion of the estimated summer home range of the tagged animals. This is an important finding as inclusion of this area could potentially result in additional sightings and consequently in an increase of the population estimate. Tracking results also suggest that this population of narwhals does not overlap with that of other narwhal populations in summer or winter. This is significant because it reinforces the current management assumption that the population is separate from other narwhal populations. The NHB narwhal migration routes defined by satellite-linked data and community participants also suggest that this population is not hunted extensively by communities en route to and from its winter area. This data reinforces current management plans that do not take communities in South Baffin and Quebec into account as catches in communities outside of the Kivalliq region are rare and not considered significant.

This work provides a legacy to test assumptions that this population of narwhals is susceptible to climatic and sea ice changes. Sea ice in its seasonal range, from northern Hudson Bay to southern Davis Strait is predicted to decline. This sets the stage for future studies to measure if the population has experienced a shift in distribution or a change in population size subsequent to sea ice and climatological change.

Lastly, and arguably most significant, an important finding in this research was the value found in incorporating western science and traditional ecological knowledge in addressing the research questions. There was general agreement between science and traditional ecological knowledge on all important issues examined in this research. The complement of long term traditional ecological knowledge and relatively short duration but intense scientific data provides a robust data set in time and space. When science agrees with long term community observation, it provides greater confidence in scientific results (Johannes et al. 2000). In this case, and in the face of climate change where long term scientific data is not available, the incorporation of TEK to address research questions is relevant and valuable.

References

Austin, D., McMillan, J.I., & Bowen, W.D. (2003). A three-stage algorithm for filteringerroneous Argos satellite locations. Marine Mammal Science, 19(2), 371–383.

CLS America Inc. (2007). ARGOS User's Manual. vi + 58 p. http://www.argos-system.org/documents/userarea/argos_manual_en.pdf

Berkes, F. (1999). Sacred ecology: traditional ecological knowledge and resource management. Philadelphia: Taylor & Francis.

Bourassa, M.N. (2002). Inventaires de la population de narvals (Monodon monoceros) du nord de la baie d'Hudson et analyse des changements démographiques depuis 1983. Unpublished M.Sc. Thesis, Université du Québec à Rimouski, Rimouski, QC, Canada. xxii + 69 p.

Britten, M.W., Kennedy, P.L., & Ambrose, S. (1999). Performance and accuracy evaluation of small satellite transmitters. Journal of Wildlife Management, 63(4), 1349–1358.

de March, B.G.E., & Stern, G. (2003). Stock separation of narwhal (Monodon monoceros) in Canada based on organochlorine contaminants. Canadian Science Advisory Secretariat Research Document 2003/079. Fisheries and Oceans Canada: Science.

de March, B.G.E., Tenkula, D.A., & Postma, L.D. (2003). Molecular genetics of narwhal (Monodon monoceros) from Canada and West Greenland (1982-2001). Canadian Science Advisory Secretariat Research Document 2003/080. Fisheries and Oceans Canada: Science.

De Solla, S.R., Bondurianski, R., & Brooks, R.J. (1999). Eliminating autocorrelation reduces biological relevance of home range estimates. Journal of Animal Ecology, 68, 221–234.

Department of Fisheries and Oceans Canada. (1998). Hudson Bay narwhal. Canada Fisheries and Oceans Canada, Central and Arctic Region, DFO Science Stock Status Report E5-44, p. 5.

Dietz, R., & Heide-Jørgensen, M.P. (1995). Movements and swimming speeds of narwhals (Monodon monoceros) instrumented with satellite transmitters in Melville Bay, Northwest Greenland. Canadian Journal of Zoology, 73, 2106–2119.

Dietz, R., Heide-Jørgensen, M.P., Richard P.R., & Acqaurone, M. (2001). Summer and fall movements of narwhal (*Monodon monoceros*) from northeastern Baffin Island towards northern Davis Strait. Arctic, 54, 244–261.

Dietz, R., Heide-Jørgensen, M.P., Richard, P., Orr, J., Laidre, K., & Schmidt, H.C. (2008). Movements of narwhals (*Monodon monoceros*) from Admiralty Inlet monitored by satellite telemetry. Polar Biology, 31, 1295–1306.

Ferguson, M.A.D., Williamson, R.G., & Messier, F. (1998). Inuit knowledge of long-term changes in a population of Arctic tundra caribou. Arctic, 51, 201–219.

Gilchrist, G., Mallory, M., & Merkel, F. (2005). Can local ecological knowledge contribute to wildlife management? Case studies of migratory birds. Ecology and Society, 10(1), 20.

Gonzalez, N. (2001). Inuit traditional knowledge of the Hudson Bay narwhal (tuugaalik) population, p. 26. Fisheries and Oceans Canada, Iqaluit Nunavut.

Heide-Jørgensen, M.P., Dietz, R., Laidre, K.L., & Richard, P. (2002a). Autumn movements, home ranges, and winter density of narwhals (*Monodon monoceros*) tagged in Tremblay Sound, Baffin Island. Polar Biology, 25(5), 331–341.

Heide-Jørgensen, M.P., Richard, P., Ramsay, M., & Akeeagok, S. (2002b). Three recent ice entrapments of Arctic cetaceans in West Greenland and the eastern Canadian High Arctic. NAMMCO Scientific Publications 4, 143–148.

Heide-Jørgensen, M.P., Dietz, R., Laidre, K.L., Richard, P., Orr, J., & Schmidt, H.C. (2003). The migratory behaviour of narwhal (*Monodon monoceros*). Canadian Journal of Zoology, 81, 1298–1305.

Higdon, J.W., & Ferguson, S.H. (2009). Loss of Arctic sea ice causing punctuated change in sightings of killer whales (*Orcinus orca*) over the past century. Ecological Applications, 19(5), 1365–1375.

Hooge, P.N., Eichenlaub, W.M., & Solomon, E.K. (1999). Using GIS to analyze animal movements in the marine environment. Retrieved January, 2008, from http://www.absc.usgs.gov/glba/gistools

Hooge, P.N., & Eichenlaub, B. (2000). Animal movement extension to ArcView ver. 2.0. Alaska Science Center-Biological Science Office, U.S. Geological Survey, Anchorage, AK.

Huntington, H.P. (1998). Observations on the utility of the semi-directive interview for documenting traditional ecological knowledge. Arctic, 51(3), 237–242.

Innes, S., Heide-Jørgensen, M.P., Laake, J.L., Laidre, K.L., Cleator, H.J., Richard, P., & Stewart, R.E.A. (2002). Surveys of beluga and narwhal in the Canadian High Arctic in 1996. NAMMCO Science Publications, 4(1), 69–190.

IWC (International Whaling Commission). (1997). Report of the IWC workshop on climate change and cetaceans. Report of the International Whaling Commission, 47, 291–320.

Johannes, R.E., Freeman, M.M.R., & Hamilton, R.J. (2000). Ignore fishers' knowledge and miss the boat. Fish and Fisheries, 1(3), 257–271.

Jonsen, I.D., Flemming, J.M., & Myers, R.A. (2005). Robust state-space modeling of animal movement data. Ecology, 86(11), 2874–2880.

Kitchin, R., & Tate, N. (2000). Conducting research in human geography: theory, methodology and practice. Harlow: Pearson Education Limited.

Koski, W.R., & Davis, R.A. (1994). Distribution and numbers of narwhals (*Monodon monoceros*) in Baffin Bay and Davis Strait. Meddr Grønland. Bioscience, 39, 15–40.

Laidre, K.L., Heide-Jørgensen, M.P., & Dietz, R. (2002). Diving behaviour of narwhal (*Monodon monoceros*) at two costal localities in the Canadian High Arctic. Canadian Journal of Zoology, 80, 624–635.

Laidre, K.L., Heide-Jørgensen, M.P., Logdson, M.L., Hobbs, R.C., Heagerty, P., Dietz, R., Jorgensen, O.A., & Treble, M.A. (2004). Seasonal narwhal habitat associations in the high Arctic. Marine Biology, 145, 821–831.

Laidre, K.L., & Heide-Jorgensen, M.P. (2005). Winter feeding intensity of narwhals (*Monodon monoceros*). Marine Mammal Science, 21(1), 45–57.

Laidre, K.L., Stirling, I., Lowry, L.F., øystein, W., Heide-Jørgensen, M.P., & Ferguson, S.H. (2008). Quantifying the sensitivity of Arctic marine mammals to climate-induced habitat change. Ecological Applications, 18(2), S97–S125.

Nuttall, M., Berkes, F., Forbes, B., Kofinas, G., Vlassova, T., & Wenzel, G. (2005). Hunting, herding, fishing and gathering: indigenous peoples and renewable resource use in the Arctic. In: Symon, C., Arris, L., & Heal, B. (Eds.), Arctic climate impact assessment (pp. 661–702). New York: Cambridge University Press.

Orr, J.R., Joe, R., & Evic, D. (2001). Capturing and handling of White Whales (Delphinapterus leucas) in the Canadian Arctic for instrumentation and release. Arctic, 54(3), 299–304.

Powell, R.A. (2000). Animal home ranges and territories and home range estimators. In: Boitani, L., & Fuller, T.K. (Eds.), Research techniques in animal ecology: controversies and consequences (pp. 65–110). New York: Columbia University Press.

Richard, P.R. (1991). Abundance and distribution of narwhals (Monodon monoceros) in northern Hudson Bay. Canadian Journal of Fish Aquatic Science, 48, 276–283.

Stewart, D.B., Akeeagok, A., Amarualik, R., Panipakutsuk, S., & Taqtu, A. (1995). Local knowledge of beluga and narwhal from four communities in Arctic Canada. Canadian Technical Report on Fish Aquatic Science 2065: viii + 48p. + Appendices on disk.

Strong, J.T. (1988). Status of narwhal (Monodon monoceros) in Canada. Canadian Field-Naturalist 102, 391–398.

Tynan, C.T., & DeMaster, D.P. (1997). Observations and predictions of arctic climate change: potential effects on marine mammals. Arctic, 50(4), 308–322.

Usher, P.J. (2000). Traditional ecological knowledge in environmental assessment and management. Arctic, 53(2), 183–193.

Vincent, C., McConnell, B.J., Ridoux, V., & Fedak, M.A. (2002). Assessment of Argos location accuracy from satellite tags deployed on captive gray seals. Marine Mammal Science, 18(1), 156–166.

Westdal, K. (2008). Movement and diving of northern Hudson Bay narwhals (Monodon monoceros): relevance to stock assessment and hunt co-management. M.Env. Thesis. Department of Environment and Geography, University of Manitoba, p. 103.

Worton, B.J. (1989). Kernel methods for estimating the utilization distribution in home-range studies. Ecology, 70(1), 164–168.

Polar Bear Ecology and Management in Hudson Bay in the Face of Climate Change

E. Peacock, A.E. Derocher, N.J. Lunn, and M.E. Obbard

Abstract Hudson Bay, Canada has been a region of intensive research on polar bear population ecology dating back to the late 1960s. Although the impacts of climate change on sea ice habitat throughout the circumpolar range of the species is of concern, Hudson Bay is the only region where the duration of sea ice cover has been linked empirically with declines in a suite of parameters: polar bear body condition; individual survival; natality; and population size. Research in Hudson Bay has also focused on contaminants in polar bear tissues, population genetics, behaviour and denning, as well as predator-prey interactions. These decades of research in Hudson Bay provide important baseline information with which to monitor the rate and extent of the impact of climate change on polar bear ecology. Climate change has already become a critical issue for polar bear management in the region; human–bear conflicts in Nunavut have increased, which had been an explicit prediction of an effect of climate change. In addition, relative to polar bear numbers in the early 1960s – before government-based harvest management – polar bear abundance has also increased. The recent empirical data demonstrating a decline in the Western Hudson Bay polar bear subpopulation has been interpreted as incongruous with observations of respected Inuit elders of the marked increase in polar bears from historical numbers, catalyzing divergent views on polar bear management. Lastly, should the duration of the ice-free season continue to increase, industrial shipping and future mining and oil and gas developments will affect polar bears in the region in ways that are not well understood. We review current knowledge of polar bear ecology in Hudson Bay, as it relates to climate change, and present an overview of future research needs and management challenges.

E. Peacock (✉)
Wildlife Research Section, Department of Environment, Government of Nunavut,
Igloolik, NU, Canada
Present address: US Geological Survey, Alaska Science Center, Anchorage, AK 99516, USA
e-mail: lpeacock@usgs.gov

S.H. Ferguson et al. (eds.), *A Little Less Arctic: Top Predators in the World's Largest Northern Inland Sea, Hudson Bay*, DOI 10.1007/978-90-481-9121-5_5,
© Springer Science+Business Media B.V. 2010

Keywords Polar bear • Hudson Bay • Climate change • Harvest management • Human-bear interactions • Denning • Protected areas • Shipping • Tourism

Introduction

The past, present, and future status of the polar bear (*Ursus maritimus*) in the Hudson Bay complex (hereafter, Hudson Bay) is a microcosm of the worldwide circumstance of the species. Thus, we have the possibility of learning conservation impacts from a logistically tractable group of polar bears that may be applicable to the circumpolar region. Polar bears here exist throughout a range of ecological conditions, have a long history of interaction with both aboriginal and non-aboriginal people, and are subject to a broad suite of threats to their viability. The polar bears in Hudson Bay occur from the treed boreal ecosystem of northern Ontario to the Arctic tundra of Fury and Hecla Strait. Polar bears occur in James Bay, Hudson Bay proper, Hudson Strait, and Foxe Basin (Fig. 1). In the southern reaches, polar bears spend up to 5 months in a hypophagic condition (fasting) during the ice-free season. In the northern reaches of Foxe Basin, polar bears can forage year round on ice floes, though this ecological condition is changing (Stirling and Parkinson 2006). The primary prey of the polar bear throughout both Hudson Bay and the range of the species is the ringed seal (*Pusa hispida*), though prey diversity in the region includes the entire range of important prey species for polar bears: bearded seals (*Erignathus barbatus*); harbour seals (*Phoca vitulina*); harp seals (*Pagophilus groenlandicus*); hooded seals (*Cystophora cristata*); beluga (*Delphinapterus leucas*), and notably walrus (*Odobenus rosmarus*) in Foxe Basin (Thiemann et al. 2008).

Both Cree First Nations, in Manitoba, Québec, and Ontario, and Inuit communities throughout Hudson Bay have long used polar bears for sustenance, clothing, and economic opportunity (see Henri et al. this volume). European-Canadians have interacted with polar bears in Hudson Bay since the 1600s through exploration, fur trading, military activity, government management, sport hunting, research, and tourism. Both aboriginal and non-aboriginal communities have also had long-standing negative interactions with polar bears as a result of kills in defence of life and property and improper waste management. Anthropogenic influences on polar bears throughout Hudson Bay include subsistence and sport harvest, contaminant accumulation, increasing shipping and development, and most notably the impact of climate change on the loss of sea ice and ecosystem shifts. These threats to polar bear viability in Hudson Bay are those that the species faces throughout its range. In contrast to most other polar bear populations, there has been almost 40 years of research on polar bears in Hudson Bay, resulting in large and continuous datasets and broad ecological inquiry. Indeed, the first empirical evidence of impacts of climate change on polar bears has come from research initiatives in Hudson Bay. In this chapter, we review the breadth of research on polar bears throughout Hudson Bay and comment on the future ecological and conservation circumstance of the species in Hudson Bay.

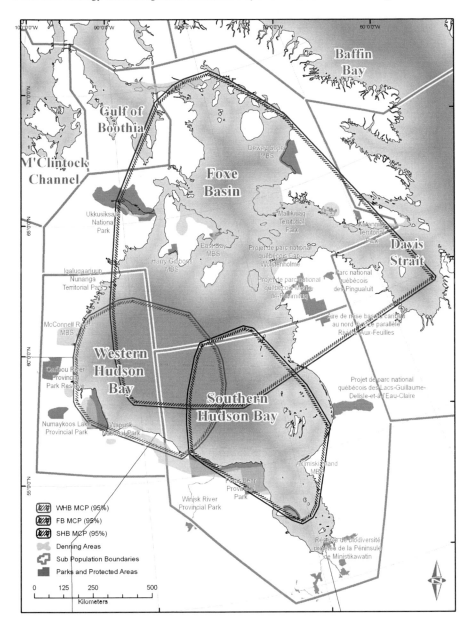

Fig. 1 Boundaries of polar bear subpopulations (management zones) in Hudson Bay. Composite home-ranges (95% minimum convex polygons) are derived from movements of satellite-collared adult female polar bears, 2007–2009. Existing protected areas and known denning areas are also shown

Delineation of Subpopulations and Important
Seasonal Areas of Concentration

Since the 1960s, capture and tagging programs (Jonkel et al. 1976; Stirling et al. 1977b; Lunn et al. 1987; Taylor et al. 2006), records of recapture and harvest of tagged polar bears (Taylor and Lee 1995), and population genetics (Paetkau et al. 1995, 1999; Crompton et al. 2008) have elucidated boundaries among three semi-discrete groups of polar bears in Hudson Bay (Fig. 1): Southern Hudson Bay; Western Hudson Bay; and Foxe Basin. The boundaries between subpopulations reflect learned patterns of ice movement relative to food availability and accessibility. In addition, high levels of philopatry are created by the patterns of ice melt in Hudson Bay that result in discrete summertime on-shore refugia (Stirling et al. 2004). These summering areas are used to define units for polar bear harvest management. There is low genetic differentiation among the subpopulations (Paetkau et al. 1995). However, abundant groups of large-ranging animals may be genetically similar but demographically independent (Mills and Allendorf 1996). Hudson Bay polar bears show evidence of demographic independence among these three groups (Taylor and Lee 1995), and movement data reflect these core areas (Fig. 1). More recent evidence (Crompton et al. 2008) suggests the polar bears in James Bay may be more genetically distinct than previously considered.

During the ice-free season, polar bears belonging to the Western Hudson Bay subpopulation occur along the western shore of Hudson Bay from the Manitoba-Ontario border north to Chesterfield Inlet, Nunavut (Derocher and Stirling 1990). Bears belonging to the Southern Hudson Bay subpopulation are found along the Hudson Bay coast from the Manitoba-Ontario border east to Cape Henrietta Maria, and south down the James Bay coast to the vicinity of the Ekwan River. Bears are also found on the islands in James Bay, primarily Akimiski Island and North and South Twin Islands (Jonkel et al. 1976; Crête et al. 1991; Obbard and Walton 2004). The large and less-studied Foxe Basin subpopulation extends west and north from Hudson Strait and the northwestern Québec coast of Hudson Bay to Fury and Hecla Strait between the Melville Peninsula and Baffin Island. Though the northern extent of Foxe Basin is rarely, though increasingly, completely ice-free in the summer, important summer resting areas include all islands of Foxe Basin and Hudson Strait, most notably Rowley, Koch, Southampton, Coats and Mansel islands (Stenhouse and Lunn 1987; Taylor et al. 1990a; Crête et al. 1991; Peacock et al. 2009b). Important mainland areas of summertime concentration include the northern Foxe Peninsula near Cape Willoughby of western Baffin Island, the southern reaches of Lyon Inlet, Wager Bay and the area of Cape Penryn (Peacock et al. 2009b). Few bears summer along the coast of Québec, whether in the Foxe Basin region (Peacock et al. 2009a) or Southern Hudson Bay.

In anticipation of sea ice freeze-up, which ranges from early November to early December across Hudson Bay, bears move towards regions where ice appears first. Similarly at smaller scales, polar bears move to heads of inlets or shallow-water bays as ice forms. For example, polar bears from the summer concentration area near Churchill, Manitoba predictably migrate northwards along the coast, into Nunavut in October and November. Some bears (largely subadults and some adult males) follow the first sea ice that forms nearshore north of Churchill whereas others remain on land longer and then head northeast from the Cape Churchill area. Likewise, polar bears on Southampton Island predictably move to the southern bays of the island during freeze-up. Freeze-up is about 2 weeks later in southern than in western Hudson Bay, and these bears are the last to return to the ice.

Maternity denning occurs when pregnant females reside in a den from approximately August (in Foxe Basin, likely, October) to March during pregnancy, birth, and early lactation. In Hudson Bay denning occurs exclusively on land and throughout the region (Urquhart and Schweinsburg 1984), although there are some important areas of den concentration (Fig. 1). Maternity denning in Ontario is known to occur up to 120 km inland (Kolenosky and Prevett 1983; Obbard and Walton 2004). Dens are sparsely distributed between the Manitoba-Ontario border and the James Bay coast with most dens occurring within 40 km of the coast. Aerial surveys conducted periodically since the 1970s suggest that approximately 33% of dens in the Southern Hudson Bay region are found within Polar Bear Provincial Park (Obbard and Walton 2004). Scientific researchers have known that polar bears den on the Twin Islands in James Bay since the 1930s (Doutt 1967), on Akimiski Island in James Bay (Prevett and Kolenosky 1982; Kolenosky and Prevett 1983), and the Belcher Islands (Jonkel et al. 1976; Urquhart and Schweinsburg 1984). Well-delineated denning areas for the Western Hudson Bay subpopulation include areas partially protected by the Wapusk National Park in the vicinity of the Owl River south of Churchill and south of Cape Tatnam in Manitoba's Cape Tatnam Wildlife Management Area (Stirling et al. 1977b; Richardson et al. 2005) with a few scattered along the Nunavut coast, west of the community of Arviat (Urquhart and Schweinsburg 1984). Pregnant polar bears in western Hudson Bay dig dens in peat deposits along the sides of rivers, creeks and lakes as early as August and remain there until covered by snow to reduce energy expenditure (Clark et al. 1997; Clark and Stirling 1998; Lunn et al. 2004; Richardson et al. 2005). Parturition can occur in western Hudson Bay as early as November (Derocher et al. 1992). In southeastern Hudson Bay, bears construct maternity dens in palsas (oval or round elevated extents of soil on top of permafrost, associated with bogs), river banks, gravel ridges, and in the lee of clumps of spruces (Kolenosky and Prevett 1983; Obbard and Walton 2004). Because polar bears have a reduced on-shore period in Foxe Basin, pregnant females likely do not seek inland denning sites until late September or October. Denning areas in Foxe Basin are less studied and appear more dispersed. However likely areas include inland and higher topography parts of eastern Southampton Island, Mansel Island (Stenhouse and Lunn 1987), Koch and Rowley islands (T. Ikummaq, pers. comm.), southern Wager Bay (Ayotte 1977), Foxe Peninsula, and the Baffin Island shore of Fury and Hecla Strait (Urquhart and Schweinsburg 1984). Females and young cubs typically return to the sea ice between late February and late March in southern reaches of Hudson Bay and Foxe Basin, but not until April in the northern extent of Foxe Basin.

During the winter and spring, polar bears concentrate at the floe-edge (boundary between landfast and moving ice), ice leads and pressure ridges, though they range across all sea ice habitats. Early spring, during ringed seal pupping, is the crucial peak feeding period of the year for polar bears. In Hudson Bay this season ranges from March to May, although little research has been conducted at this time of year. As ringed seals give birth in dispersed areas, wherever the ice is sufficiently stable and with sufficient snow to dig a lair, there does not appear to be particular regions of importance for springtime foraging for polar bears. However, some regions of Hudson Bay are consistently important for polar bears in the springtime: Markham Bay in Hudson Strait; Wager Bay; between Lyon Inlet and Winter Island; Steensby Inlet; East Bay of Southampton Island (V. Sahanatien personal communication 2009); northern James Bay; and a large region of south-central Hudson Bay lying between 80–84° W and 57–60° N (M. Obbard 2009). From April through June, polar bears mate on the sea ice. During this time, some individuals from all subpopulations co-mingle on the sea ice in parts of Hudson Bay proper (Fig. 1; Taylor et al. 1990b), which explains the lack of genetic differentiation among the three subpopulations.

Population Ecology

Indications of polar bear numbers in Hudson Bay are available from historic times through harvest records of furriers. Most records come from the Hudson's Bay Company and the Northwest Company. The first permanent Hudson's Bay Company trading post was established in 1682 at the mouth of the Hayes River (Stirling et al. 1977b). Polar bears were said to be common, and aboriginal people incorporated polar bear meat and fat in their diets and traded furs. Until recently in western Hudson Bay (Regehr et al. 2007), human harvest of polar bears has been the primary population-limiting factor (Taylor et al. 1987; Prestrud and Stirling 1994). Thus, unregulated harvest and the business of the fur trading companies reduced polar bear numbers. Prestrud and Stirling (1994) argued that recorded polar bear harvest across Canada increased perceptibly in the late 1960s. This was likely due to increased price of furs, increased sport hunting, use of aircraft and ships for hunting, and ubiquitous use of rifles and snow machines by aboriginal hunters. This trend in Canada, including in Hudson Bay, was occurring throughout the circumpolar Arctic, depressing polar bear numbers. Declining polar bear numbers prompted the signing of the international Agreement on the Conservation of Polar Bears in 1973 (hereafter, the International Agreement). As a result, since the early 1970s and up until recently, harvest of polar bears in Hudson Bay was reduced (with intermittent increases) and population numbers increased, which is consistent with local ecological knowledge (Tyrrell 2007). Derocher and Stirling (1995) report that the abundance in Western Hudson Bay increased until about 1978, and thereafter had been approximately stable (Lunn et al. 1997b). Prevett and Kolenosky (1982) also recorded an increase in population size in Southern Hudson Bay from the 1960s to early 1980s. However, using population estimates from early scientific surveys can be problematic because of changes in effort and study design over time.

Recent population research in Western Hudson Bay has empirically tied a recent population size decrease of 22% (1984–2004), from approximately 1,200

Table 1 Demographic population parameters of polar bears in Hudson Bay

Management zone	Mean litter size (SE)[a]	Adult female survival[b]	Abundance (year of estimate)
Southern Hudson Bay	1.68 (0.09)	0.90	900[c] (2005)
Western Hudson Bay	1.54 (0.11)	0.93	935 (2004)
Foxe Basin	1.64 (0.15)	Unknown	2,300 (2004)

[a] Cubs-of-the-year.

[b] Annual total apparent survival (includes harvest rate).

[c] 673 (396–950, 95% CI) for southern Hudson and James Bay (Obbard et al. 2007) plus an approximated number of bears unavailable for capture and on the Twins and Akimiski islands.

(Lunn et al. 1997b) to 935 animals, to earlier break-up of sea ice (Regehr et al. 2007). Concomitantly, natural survival rates (excluding mortality due to human harvest) of subadult (independent bears ages 2–4) and senescent bears (>20 years old) and natality rates (mean cub-of-the-year litter size multiplied by litter production rates of adult females) have decreased (Table 1), resulting in a population growth rate (λ) < 1 (for a stable population, $\lambda = 1$), which indicates that the subpopulation is currently declining, even without considering the effects of harvest. The Southern Hudson Bay subpopulation is thought to have been stable for the past 20 years, at approximately 900 animals (Obbard et al. 2007), although biologically important declines in survival rates (Table 1) and body condition (Obbard et al. 2006) since the 1980s suggest the population is at a tipping point and that abundance may begin to decline.

The more northerly Foxe Basin region of Hudson Bay differs in that the summers are not completely ice-free, and polar bears enjoy a longer foraging season. However, research has not been continuous in Foxe Basin (Lunn et al. 1987; Taylor et al. 2006; Peacock et al. 2008, 2009b), and as such, there is no scientific evidence of any changes in population demography. Suggestions of increasing observations of polar bears in Foxe Basin by local Inuit led to an increase in hunting quotas in Nunavut in 2005 (Stirling and Parkinson 2006). Current demographic parameters suggest the subpopulations in Foxe Basin (Peacock et al. 2009b) and Southern Hudson Bay are more productive than Western Hudson Bay with larger litter sizes (Table 1). Aerial survey research in 2009 (Peacock et al. 2009b) suggests an abundant population of polar bears in Foxe Basin (S. Stapleton 2009), possibly higher than the abundance estimate from 1994 (Taylor et al. 2006).

Polar Bears in the Ecosystem

Polar bears are currently the top predator in the Hudson Bay ecosystem and primarily prey upon ringed seals and bearded seals although direct observations of predation are uncommon because fieldwork in the area has been conducted when bears are ashore and relying largely on stored fat reserves. The success of polar bears as an Arctic marine predator is driven largely by their seasonal access to the fat-rich blubber layer of seals and possibly other marine mammal prey.

Little is known about the distribution of prey throughout Hudson Bay (see Smith 1975; Lunn et al. 1997a and Chambellant, this volume), especially in Foxe Basin. Still, ringed and bearded seals are widely distributed and available to bears. Using fatty acid signature analysis, the diet in Western Hudson Bay bears was estimated to be largely (ca. 70%) ringed seal, followed by equal proportions of bearded and harbour seals (ca. 10–15%) and about 5% from harp seals (Thiemann et al. 2008). Southern Hudson Bay bears had a similar diet although ringed seals were more common and harp seals less so, reflecting the scarcity of harp seals in this area. The Foxe Basin population had about 60% ringed seal in their diet with about 40% divided amongst bearded seals, harp seals, and walrus. However, not all possible food items of polar bears, such as bowhead whales (*Balaena mysticetus*), were assessed using fatty acid signature analysis. Large prey such as walrus and bearded seals are taken most often by adult males (Thiemann et al. 2008).

It is likely that during the commercial whaling period, polar bears had access to bowhead carcasses (Higdon et al. this volume). In the ice-free season, we have also observed polar bears scavenging on bowhead, seal, walrus and whale carcasses, left by aboriginal hunters or otherwise. In other regions, it has been shown that bowhead whale, available as carcasses left by hunters, composed up to 26% of polar bear diets (Bentzen et al. 2007). Similarly, as killer whales (*Orcinus orca*) have increased in Hudson Bay in recent years, their predation on bowheads results in increasing carcasses available to bears during the ice-free period (Ferguson et al. this volume). It is unknown how much summertime scavenging of these foods, in potentially increased volumes, accounts for diets of polar bears in Hudson Bay.

Polar bears are opportunistic foragers and while absolutely dependent upon the marine ecosystem for sustenance, they will feed opportunistically on a variety of terrestrial species (Russell 1975; Smith and Hill 1996) ranging from blueberries to kelp to seabirds to geese to caribou (*Rangifer tarandrus*). However, the importance of terrestrial food items is considered to be minimal overall (Ramsay and Hobson 1991; Hobson et al. 2009). Some researchers have suggested that polar bears may make greater use of terrestrial resources if the sea ice conditions undergo further deterioration in Hudson Bay (Dyck and Kebreab 2009; Rockwell and Gormezano 2009). Unfortunately, there is little indication that there are sufficient resources, in terms of abundance and nutritive value, to support viable populations of polar bears forced to live on land (Rode et al. in press).

Hunting of Polar Bears

The Cree and Inuit have traditionally harvested polar bears in Hudson Bay for millennia for sustenance and clothing (Fast and Berkes 1994; also see Henri et al. this volume). Historically, 2,000–4,000 aboriginal people (Sturtevant 1984;

Oswalt 1999) lived nomadically across greater Hudson Bay, and polar bear density was likely at an equilibrium with humans as their principal predator. As noted above, the Hudson's Bay Company and other furriers became established in the late 1600s and existed until the 1950s throughout the region, providing firearms and cash incentives for polar bear harvest. Reports of single aboriginal harvesters taking tens of bears in a single season in Foxe Basin during this period are common (Peacock unpublished data 2006–2009). Elders in Arviat, Whale Cove, and Rankin Inlet, Nunavut, report seeing very few polar bears on the Kivalliq coast of Hudson Bay in their earlier hunting years (1930s–1960s, NTI 2009; Freeman and Foote 2009). We conclude that few polar bears were seen as a result of unregulated hunting and the fur trade depressing polar bear abundance during this period.

Provincial and territorial jurisdictions in Hudson Bay implemented quota systems or limited harvest by agreement (Ontario Ministry of Natural Resources 1980; James Bay and Northern Québec Agreement 1975). Québec does not limit the number of bears taken annually in Hudson Bay, although the treaty allows for limiting harvest of a species, which can be "from time to time completely protected to ensure the continued existence of that species or a population" (James Bay and Northern Québec Agreement 1975). Harvest limits were adopted partly in anticipation of the signing of the International Agreement and partly with the settlement of aboriginal land claims. The Northwest Territories (now the Nunavut Territory) established a quota system in 1968 (Prestrud and Stirling 1994) and also allowed Inuit to sell the right-to-harvest to non-aboriginal hunters. However, Inuit and Cree take the majority of polar bears harvested in Hudson Bay for their own economic, nutritional, and cultural subsistence (Table 2). In 2008 and 2009 (calculated only

Table 2 Harvest and harvest limits for polar bears in Hudson Bay for the 2006–2007, 2007–2008 and 2008–2009 harvest years

Subpopulation	Jurisdiction	Three-year mean reported harvest (SE)	Proportion, defence kills (SE)	Proportion, sport harvest (SE)	Annual harvest limits
Southern Hudson Bay	Nunavut[a]	25 (0.0)	0.01 (0.01)	0	25
	Ontario	4 (0.6)	0.50 (approx.)	None	30
	Québec	2.5 (1.5)	Unknown	None	No numeric limits[b]
Western Hudson Bay	Manitoba	2.7 (0.9)	All are defence kills, as no regular harvest	None	Defence kills only
	Nunavut	39.7 (7.7)	0.29 (0.23)	0.19 (0.10)[b]	8
Foxe Basin	Nunavut	101.3 (0.9)	0.22 (0.01)	0.16 (0.06)	105
	Québec	1.7 (0.3)	Unknown	None	No numeric limit

[a] Nunavut Total Allowable Harvest (TAH), annual quotas fluctuate slightly around the TAH depending on prior years' actual harvest.

[b] Québec does not limit the number of bears taken annually in Hudson Bay. See text for details.

for 2008 and 2009 harvest years, as the US *Endangered Species Act* ban on import of bear hides came into effect in May 2008), 83% of the harvest in Nunavut, the only jurisdiction where sport hunting is permitted, were taken by Inuit (Government of Nunavut data). In 1954, legislation was passed that prohibited non-aboriginals from hunting polar bears or for anyone to trade or barter in hides from bears in Manitoba. In other jurisdictions, government fur-purchasing programs still exist today, providing access to fur markets worldwide and incentive to harvest polar bears, beyond nutritional sustenance. Fur sales are regulated by territorial/provincial, federal, and international import-export policies. Both sport hunting and the export of polar bear hides are likely to decrease in coming years. Since the 2008 listing of the polar bear as *threatened* under the *Endangered Species Act* of the United States, sport hunters are unable to import polar bear hides into the once substantial US market.

The harvest of polar bears from the Western Hudson Bay management zone (Government of Nunavut data) is dominated by subadult animals (53 ± 2% SE from 1999–2008) with a strong bias towards males (63 ± 5%) resulting in females dominating the population (Derocher et al. 1997). Male biased harvest maintains the reproductive core of the population. However, excessive harvest of males could result in negative demographic effects such as males becoming too scarce to mate with available females (Coltman et al. 2003; Molnár et al. 2008).

People Living with Polar Bears

Thirty-four thousand people now live in polar bear habitat on the shores of Hudson Bay (Henri et al. this volume). In Manitoba, the well-established Polar Bear Alert Program contends with polar bears in the community of Churchill, and defence kills are kept to a minimum (3.0 ± 0.9 per year over last 5 years; Peacock et al. 2009a). It has also been suggested that some public education campaigns have resulted in a decline in defence kills in Nunavut from 1980–2000 (Dyck 2006). However, defence kills have recently increased in Nunavut in Hudson Bay, most significantly in Foxe Basin (S. Medill, Government of Nunavut unpublished data, 1980–2008; Fig. 2). Still, inconsistent reporting among communities, definition of defence kills, and tactics of deterrence make conclusions difficult.

Examining trends of the number of problem bears in Churchill can also be difficult, due to variable effort (Towns et al. 2009). With the beginning of the deterrence-based Polar Bear Alert Program in 1984, the number of defence kills has decreased while the number of problem bears captured has increased (Towns et al. 2009). Towns et al. (2009) suggest the number of problem polar bears in Churchill has increased from 1970 to 2004 due to: (1) increased nutritional stress of bears; (2) increased emphasis on deterrence as opposed to harvest since 1984; and (3) earlier break-up date. In addition, the abundance of bears in Western Hudson Bay increased until the 1980s (Lunn et al. 1997b),

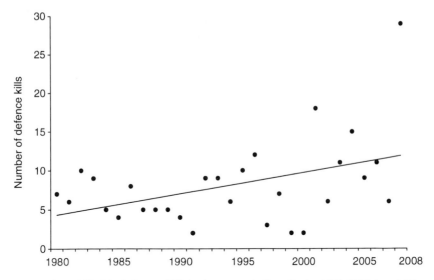

Fig. 2 Polar bears killed in defence of life and property in Foxe Basin, 1980–2008 (y = 0.27x + 40.6; R^2 = 0.17; p = 0.03)

A mother polar bear with two cubs-of-the-year in April 2009, near East Bay, Southampton Island in the Foxe Basin subpopulation. Photo: E. Peacock

which likely related to the increasing trend in problem bears at that time. Further exacerbating the problem of human–bear conflicts in Nunavut is a change in polar bear distribution over time with more bears spending the

Polar bears near Churchill, Manitoba. Photo: A. Derocher

ice-free period farther north (Nunavut Wildlife Management Board, Public Hearings 2007; Towns et al. 2010).

Polar Bear Tourism

The shore of Hudson Bay near Churchill is the site of the most-viewed, most-photographed polar bears in the world. Polar bears spend the autumn in large numbers in and around Churchill and the coastal areas to the east of the town providing reliable viewing for tourists. The reasons for polar bears using this area include patterns of melting ice that result in the last sea ice in the western region of Hudson Bay to be located offshore, a nearby denning area and until 2005 (Towns et al. 2009), an unmaintained garbage dump. Large-scale tourism built around polar bear viewing has occurred in Churchill since the 1980s (Dyck and Baydack 2004), although smaller-scale tourism has occurred in association with the military operations since the 1960s (Stirling et al. 1977a). Early operations were fraught with feeding bears and unregulated viewing activities. With regulation of tourist operators and the closure of garbage dumps, polar bear tourism has grown into a business of CDN$10 million/year (Manitoba Department of Conservation). Ecotourism in Ontario is limited to a few small guiding operations based in Peawanuck and also in Fort Severn. These guides accompany small parties of tourists to view bears from ATVs along beach ridges. There are no significant tourist operations in other parts of greater Hudson Bay.

Some authors (Dyck et al. 2007) have argued that the tourism in Churchill negatively influences polar bears, due to increased vigilance activity when tourist tundra vehicles are present, and therefore presumably increased energy expenditure. Stirling et al. (2008) argue that, as polar bears are in a fasting state during the tourist season and only a small portion of the population is in the viewing area, there is no negative trade-off (i.e., decreased foraging activity) associated with increased vigilance at the population level. Local residents in Nunavut suggest that human-habituated and possibly food-conditioned bears from Churchill, which move north to Nunavut in anticipation of freeze-up, are pre-disposed to bother Inuit communities and camps. However, there are poor or no controls of attractants in communities and camps in Nunavut (e.g., unfenced garbage dumps). Though food conditioning is known to contribute to increased human–bear interactions, it is likely now rare in Churchill since garbage is now managed and unavailable, and tourist operators are prohibited from feeding bears. In some instances, available dog team food (D. Hedmann, personal communication 2008) and kitchen grey water (Herrero and Herrero 1997) may contribute to some food conditioning.

It is unknown whether the tourist activity in Churchill has any population level effect that would result in decreases in survival or recruitment of polar bears (Dyck et al. 2008, Stirling et al. 2008). Both total (natural mortality plus harvest) and natural survival of polar bears specifically caught by researchers in the Churchill area are lower than those rates of polar bears caught in other areas of the Western Hudson Bay subpopulation (Regehr et al. 2007). However, the authors suggest that these Churchill bears have lower survival because of their lower nutritional status and are thus predisposed to come into Churchill in search of food. Secondly, bears that are relocated from Churchill are flown north to Nunavut, where they may be more susceptible to hunting (Regehr et al. 2007).

Climate Change and Polar Bears

A 30-year data set of polar bears in Western Hudson Bay has empirically tied ice break-up with declining demographic parameters of polar bears (Stirling et al. 1999; Regehr et al. 2007). The population declined from 1200 (Lunn et al. 1997b) to 935 as a result of declining survival rates (Regehr et al. 2007). Over this period, declines in body condition and natality (Stirling et al. 1999) were related to increasing length of the ice-free period (Stirling et al. 1999). Similar significant reductions in body condition have been detected in Southern Hudson Bay (Obbard et al. 2006). Demographic changes were not as strongly linked with changing ice conditions, which have been less dramatic in Southern Hudson Bay. In Western Hudson Bay, detectable changes in body condition (Stirling et al. 1999) preceded detectable changes in survival and population size (Regehr et al. 2007). Though the subpopulation in Southern Hudson Bay is currently considered to be stable, it is expected (Stirling and Parkinson 2006; Obbard et al. 2007) that it will mirror the events in Western Hudson Bay. Declining polar bear ice habitat in Foxe Basin (Sahanatien and Derocher 2007) has generated similar predictions (Stirling and Parkinson 2006) for the polar bears in Foxe Basin.

Based on the Western Hudson Bay dataset, Derocher et al. (1992) found a threshold of maternal weight of 189 kg in the autumn for successful reproduction (i.e., a female gives birth to a cub that survives to the following autumn). On-going work suggests that successful reproduction could cease if break-up was to occur in early June which would be a 2-month increase in the fasting period (Molnár 2009). Consequently, if ice changes occur as predicted and the on-shore period continues to increase, the subpopulation in western Hudson Bay is critically threatened. The ice-free season in Foxe Basin is currently much shorter (0–2 months) than in the southern reaches of Hudson Bay (4–5 months). Also, the lengthening ice-free season in this area is not as limiting, as ice is predicted to continue to form in winter. As a result it is unknown when or whether minimum physiological constraints and thresholds for reproduction will be met in Foxe Basin, but this subpopulation is less at risk than those in the more southern areas.

Climate Change in the Ecosystem

The projected changes in the Hudson Bay ecosystem are complex. Fundamentally, the major concern for polar bears is loss of access to their primary habitat: the sea ice. Ongoing declines in the sea ice have already put the Western Hudson Bay sub-population in a negative growth trajectory and this decline is compounded by harvest. There are early indications of change in the Hudson Bay predator-prey relationship. Iverson et al. (2006) and McKinney et al. (2009) have indicated that there has been an increase of harbour seals and bearded seals and a decrease in ringed seals in polar bear diets in Western Hudson Bay (but see Thiemann et al. 2008). It is likely that harp seals will continue to advance into Hudson Bay and this species may become an important prey source for polar bears. The expansion of killer whales into the area is an early indication of a possible regime shift with a new top predator (Higdon and Ferguson 2009; Ferguson et al. this volume). Predicting the future is difficult at the best of times but predicting future ecosystem structure is particularly challenging. There are many unknowns in the system but it is clear that as a highly specialized predator of seals that relies on sea ice as a hunting platform, the future does not look promising for polar bears in the Hudson Bay region. Ultimately, the persistence and dynamics of sea ice will determine the nature of the Hudson Bay top predator community.

The Future for Polar Bears in Hudson Bay

On-going polar bear harvest, changing ice conditions (Gough et al. 2004; Gagnon and Gough 2005b), declining polar bear ice habitat (Sahanatien and Derocher 2007), demonstrated links between ice habitat and polar bear productivity, and increases in contaminant burdens (McKinney et al. 2009) all converge upon a bleak outlook for polar bears in Hudson Bay (Stirling and Parkinson 2006).

The ultimate threat to polar bear persistence in Hudson Bay, and throughout the species' range is reduction in ice-habitat expanse, duration and quality (Stirling and Derocher 1993; Derocher et al. 2004). Polar bear habitat protection introduces novel concepts in conservation biology: How does one protect temporally and spatially dynamic habitat; and how does one safeguard habitat that park boundaries cannot protect? *Marine Protected Areas*, designated under the *Oceans Act* of Canada, could presumably protect important foraging and mating areas for polar bears and their prey in Hudson Bay. Such protected areas could even be designed to accommodate climate-induced spatial change (Prowse et al. 2009). Yet ultimately, protected areas cannot ensure that ice forms predictably and with the quality and quantity appropriate for polar bear population viability. Ragen et al. (2008) correctly state that "short of actions to prevent climate change, there are no known conservation methods that can be used to ensure the long-term persistence" of polar bears. These authors conclude that reducing harvest and protecting essential areas (i.e., ice habitat in the intermediate and terrestrial habitat) can only mitigate declines and do not address the primary conservation concern for the species.

Harvest Management into the Future

The aboriginal right to harvest polar bears in Hudson Bay is inherent in aboriginal land claims agreements (Nunavut Land Claim Agreement 1993) and treaties (Ontario Ministry of Natural Resources 1980; James Bay and Northern Québec Agreement 1975; Natural Resources Transfer Agreement 1930 [Manitoba]), and is consequently protected by the Constitution of Canada. As a result, the harvesting of polar bears for nutritional and cultural reasons will continue in Hudson Bay for the foreseeable future. Nonetheless, the land claims are subject to all treaties and agreements to which Canada as a nation is signatory and thus the provisions of the International Agreement must be embraced in spirit by the jurisdictions managing polar bears in Hudson Bay. However, because climate-induced habitat change is the ultimate threat to polar bear viability, managing human harvest of polar bears can only mitigate declining polar bear numbers.

Currently, the Western Hudson Bay subpopulation is declining even without accounting for harvest. Regehr et al. (2007), in their assessment of the negative growth rate, called for innovative harvest management. The standard method of managing polar bears, which is incorporated into management plans, has been to maintain stability in a population of a target size, by harvesting at a rate equal to the population growth rate. Following this protocol, harvest in the Western Hudson Bay population should be eliminated. However, in this subpopulation there have also been localized areas of high densities and polar bears in poor condition, likely because of changing ice conditions. Such situations have and will increase human–bear interactions. In 2008, the annual polar bear quota in the Nunavut portion of

Western Hudson Bay was reduced from 57 to 8. Since this time, much of the harvest has been defence kills, and kills have exceeded sex-selective quotas in both 2008 and 2009 (Government of Nunavut data). No longer are polar bears harvested or intercepted on their way up to Nunavut as ice forms, but for the town of Arviat, kills mostly occur in conflict situations. According to this interception hypothesis (S. Medill, Government of Nunavut unpublished data, 2009), the drastic decrease in quota has contributed to increased human–bear interactions.

Innovative management requires the balancing of the legal rights and cultural traditions of harvesting polar bears, human safety and polar bear conservation. As a result of genuine efforts to maintain the sustainability of harvest, there has been public dissatisfaction with reductions in polar bear quotas (Nunavut Wildlife Management Board Public Hearings 2007, 2008, 2009; see Henri et al., this volume). We suggest such political situations, and the potential for dramatically increased and unregulated harvest, can be avoided using more effective management that aims to allow for a small managed aboriginal harvest, which may not or cannot be strictly sustainable. A small managed harvest would have limited demographic or genetic effects. The harvest could be designed to maintain polar bear abundance at reduced levels such that population health, i.e., body condition and productivity parameters, be maintained in the short term. We advocate a paradigm shift, from the concept of sustainability to one of a managed harvest, in light of declining abundance.

International concern about polar bears and the effects of climate change has caused confusion in the public about the role of harvesting and some have advocated a ban on harvest and trade in polar bear skins. Though harvest reductions may be necessary, the subpopulations of Southern Hudson Bay and Foxe Basin can currently sustain a harvest. The harvest in Western Hudson Bay, though theoretically unsustainable, can be managed to maintain the subpopulation in balance with safety concerns of the Inuit and a new, yet constantly changing, ecological carrying capacity. Efforts to manage the subpopulations using local observations have been met with caution by the scientific and non-aboriginal public; however, it is clear that means of integrating local observations with scientific methods will improve management efficacy (Henri et al. this volume).

A particular challenge in Hudson Bay is the existence of a tourism industry in Manitoba, which has goals that differ from the harvesting jurisdictions. The viewing industry is best served by more bears and in particular, large males which dominate the coastal areas. The high harvest of males in the Western Hudson Bay sub population has reduced the abundance of large bears for tourist viewing. Integration of consumptive and non-consumptive uses of polar bears has not been considered in any management plan to date.

Inherent to innovative harvest management is increased implementation of deterrence programs to reduce human–bear conflict and defence kills. It has been long known that highest numbers of defence kills occur during the on-shore season in seasonal ice populations such as those in Hudson Bay (Stenhouse et al. 1988). Consequently, it has been predicted that polar bear defence kills will increase as polar bears spend more time on land, as the ice-free season in Hudson Bay lengthens (Stirling and Derocher 1993; Derocher et al. 2004). Indeed, the number of problem bears in Churchill significantly increased with earlier ice break-up date (Towns et al. 2009). Furthermore, during the ice season, landfast ice is declining in expanse (Gagnon and Gough 2005a, b; Sahanatien and Derocher 2007), and thus polar bears will occur at higher localized densities and closer to land and communities. These two phenomena are likely to increase polar bear interactions with humans even during the ice season. It directly follows that an increase in problem bears will result in increased kills, if proper deterrent programs are not put in place. New efforts aimed to reduce human–bear conflict in Nunavut including data coordination, public campaigns, financial compensation for damaged property, and for the technology and supplies to prevent damage (S. Medill, personal communication 2009), should be encouraged and appropriately funded.

Intermediate Protection of Sea Ice Habitat

In 1973, the five nations that signed the International Agreement agreed not only to regulate harvest, but also to protect the habitat of polar bears. Despite the ensuing 37 years, very little has been done to map, much less protect key marine regions for polar bears. We regard four reasons for the lack of research and action for the explicit protection of sea ice habitat over the last 4 decades: (1) sea ice habitat is spatially and temporally dynamic, rendering identification of key areas difficult; (2) polar bears have occupied their entire historic range, minimizing focus on habitat protection; (3) human harvest has until now been the chief conservation concern and focus; and finally (4) sea ice as a habitat has been considered stable in quantity and quality. As reviewed in this chapter, these conditions are changing: sea ice habitat is declining, especially in Hudson Bay, and becoming more variable throughout the circumpolar region; development in the Arctic is increasing; and human harvest is now well-regulated and will become less singular in its importance.

Increased terrestrial and aquatic development in Hudson Bay and in the Arctic generally, will be associated with increased commercial shipping during both ice free and ice seasons, oil, gas, and mineral exploration, and concomitantly newly navigable waterways will increase human development (Arctic Council 2009). As an example, the proposed iron ore mine at Mary's River on Baffin Island will involve shipping every 32 h throughout the year, generating a 10-km-wide open water path stretching north through Foxe Basin from Hudson Strait to Steensby Inlet (Baffinland Iron Mines Corporation 2008; Nunavut Impact Review Board 2009). Shipping to the Port of Churchill is also expected to increase. Though it is unclear how ice-breaking, in particular, will affect polar bears and their prey, it is clear that polar bears rely on specific quantities of particular types of sea ice (Ferguson et al. 1998, 2000a, b, 2001; Mauritzen et al. 2001, 2003) and that anthropogenic changes in sea ice habitat will have some consequence to polar bears. The mechanism may be via direct avoidance of shipping channels by polar bears and/or their prey, or simply reduction in available suitable habitat. The threat of increased discharge of oil-based pollutants has not been examined, though it is well documented that polar bears are sensitive to oil pollution (Øritsland et al. 1981; Hurst and Øritsland 1982; Stirling 1990). Government regulators of terrestrial development and associated impacts on marine systems will need to comprehensively study, map, assess, and regulate impacts of industrial development on polar bears and their prey.

There are currently no federally administrated *Marine Protected Areas* in Hudson Bay or Arctic Canada; however one has been proposed by the Department of Fisheries and Oceans in northern Foxe Basin near Fury and Hecla Strait and is anticipated to be established in 2012. A *Marine Protected Area* is protected from gas and oil development, but allows for fishing, hunting and shipping. Fury and Hecla Strait is an important migration area for bowhead whales, seals and one of many important areas of springtime concentration for polar bears and walrus in greater Hudson Bay. We recommend that other regions in Hudson Bay be assessed and identified for their importance to polar bears in the crucial springtime foraging period. In addition, it is imperative to understand how particular expanses may change in importance for polar bears, as ice habitat will change in quantity and quality. Protected areas should incorporate spatial and temporal dynamic nature of sea ice habitat of polar bears (Prowse et al. 2009).

Protection of Terrestrial Habitat

Exploration and development in the Hudson Bay region includes: uranium mining (Kivalliq region of Nunavut), diamond mining (south of Wager Bay, Nunavut; west of Attawapiskat, Ontario), iron ore mining (Baffin Island, Nunavut), and oil and gas development in southern Hudson Bay off the Ontario and Manitoba coasts. Disturbance of denning areas has been known to cause polar bear den abandonment (Amstrup 1993; Lunn et al. 2004); disturbance can occur from direct development of denning habitat, and noise and seismic disturbance. Currently, Polar Bear Provincial Park (Ontario) and Wapusk (Manitoba) and Ukkusiksalik (Nunavut) National Parks provide protection to some denning areas in Hudson Bay. Because there has been no systematic mapping of polar bear denning in Foxe Basin, there is a need for mapping here to inform environmental assessment of development. It is also unknown how denning locations will change with climate change. For example, Gough and Leung (2002) estimated at least a 50% reduction in permafrost in the Hudson Bay lowlands by 2100 if temperature increases as predicted. Loss of permafrost will have an impact on maternity denning habitat selected by female polar bears in both Ontario and Manitoba, as palsas are permafrost structures. Potential impacts of loss of preferred maternity-den habitat on reproductive success are not well understood, but are expected to be negative. Because of climate change and increased terrestrial development, we emphasize the need for on-going monitoring of denning success and denning regions.

Conclusion

Polar bears are a highly evolved species with a highly specialized diet. They evolved to exploit an energy rich habitat: the sea ice. Though polar bears are still relatively abundant in Hudson Bay, the future looks uncertain. Earlier sea ice break-up and later ice formation will push the abilities of the bears to persist and thrive in the ecosystem. Because the entire food web upon which they depend is also reliant on the existence of sea ice, the future of the whole ecosystem, as well as the traditional and economic use of polar bears, as we now know it is at risk.

References

Amstrup, S.C. 1993. Human disturbances of denning polar bears in Alaska. Arctic 46: 246–250.

Arctic Council. 2009. Arctic Marine Shipping Assessment 2009 Report.

Ayotte, N. 1977. Polar bear denning survey: Wager Bay. NWT Wildlife Service.

Baffinland Iron Mines Corporation. 2008. Development proposal for the Mary River Project, p. 89.

Bentzen, T.W., E.H. Follmann, S.C. Amstrup, G.S. York, M.J. Wooller, and T.M. O'Hara. 2007. Variation in winter diet of southern Beaufort Sea polar bears inferred from stable isotope analysis. Canadian Journal of Zoology 85: 596–608.

Clark, D.A. and I. Stirling. 1998. Habitat preferences of polar bears in the Hudson Bay Lowlands during late summer and fall. Ursus 10: 243–250.

Clark, D.A., I. Stirling, and W. Calvert. 1997. Distribution, characteristics, and use of earth dens and related excavations by polar bears on the western Hudson Bay Lowlands. Arctic 50: 158–166.

Coltman, D.W., P. O'Donoghue, J.T. Jorgenson, J.T. Hogg, C. Strobeck, and M. Festa-Bianchet. 2003. Undesirable evolutionary consequences of trophy hunting. Nature 426: 655–658.

Crête, M., D. Vandal, L.P. Rivest, and F. Potvin. 1991. Double counts in aerial surveys to estimate polar bear numbers during the ice-free period. Arctic 44: 275–278.

Crompton, A.E., M.E. Obbard, S.D. Petersen, and P.J. Wilson. 2008. Population genetic structure in polar bears (Ursus maritimus) from Hudson Bay, Canada: Implications of future climate change. Biological Conservation 141: 2528–2539.

Derocher, A.E., N.J. Lunn, and I. Stirling. 2004. Polar bears in a warming climate. Integrative and Comparative Biology 44: 163–176.

Derocher, A.E. and I. Stirling. 1990. Distribution of polar bears (Ursus maritimus) during the ice-free period in western Hudson Bay. Canadian Journal of Zoology 68: 1395–1403.

Derocher, A.E., and I. Stirling. 1995. Estimation of polar bear population-size and survival in Western Hudson-Bay. Journal of Wildlife Management 59: 215–221.

Derocher, A.E., I. Stirling, and D. Andriashek. 1992. Pregnancy rates and serum progesterone levels of polar bears in western Hudson Bay. Canadian Journal of Zoology 70: 561–566.

Derocher, A.E., I. Stirling, and W. Calvert. 1997. Male-biased harvesting of polar bears in western Hudson Bay. Journal of Wildlife Management 61: 1075–1082.

Doutt, J.K. 1967. Polar bear dens on the Twin Islands, James Bay, Canada. Journal of Mammalogy 48: 468–471.

Dyck, M.G. 2006. Characteristics of polar bears killed in defence of life and property in Nunavut, Canada, 1970–2000. Ursus 17: 52–62.

Dyck, M.G. and R.K. Baydack. 2004. Vigilance behaviour of polar bears (Ursus maritimus) in the context of wildlife-viewing activities at Churchill, Manitoba, Canada. Biological Conservation 116: 343–350.

Dyck, M.G. and E. Kebreab. 2009. Estimating the energetic contribution of polar bear (Ursus maritimus) summer diets to the total energy budget. Journal of Mammalogy 90: 585–593.

Dyck, M.G., W. Soon, R.K. Baydack, D.R. Legates, S. Baliunas, T.F. Ball, and L.O. Hancock. 2007. Polar bears of western Hudson Bay and climate change: Are warming spring air temperatures the "ultimate" survival control factor? Ecological Complexity 4: 73–84.

Dyck, M.G., W. Soon, R.K. Baydack, D.R. Legates, S. Baliunas, T.F. Ball, and L.O. Hancock. 2008. Reply to response to Dyck et al. (2007) on polar bears and climate change in western Hudson Bay by Stirling et al. (2008). Ecological Complexity 5: 289–302.

Fast, H. and F. Berkes. 1994. Native land use, traditional knowledge, and the subsistence economy in the Hudson Bay bioregion. The Hudson Bay Programme, Canadian Arctic Resources Committee, p. 33.

Ferguson, S.H., M.K. Taylor, E.W. Born, and F. Messier. 1998. Fractals, sea-ice landscape and spatial patterns of polar bears. Journal of Biogeography 25: 1081–1092.

Ferguson, S.H., M.K. Taylor, E.W. Born, A. Rosing-Asvid, and F. Messier. 2001. Activity and movement patterns of polar bears inhabiting consolidated versus active pack ice. Arctic 54: 49–54.

Ferguson, S.H., M.K. Taylor, and F. Messier. 2000a. Influence of sea ice dynamics on habitat selection by polar bears. Ecology 81: 761–772.

Ferguson, S.H., M.K. Taylor, A. Rosing-Asvid, E.W. Born, and F. Messier. 2000b. Relationships between denning of polar bears and conditions of sea ice. Journal of Mammalogy 81: 1118–1127.

Freeman, M.M.R. and L. Foote. 2009. Inuit, polar bears and sustainable use: Local, national and international perspectives. CCI Press, University of Alberta, Edmonton, AB, p. 252.

Gagnon, A.S. and W.A. Gough. 2005a. Climate change scenarios for the Hudson Bay region: An intermodel comparison. Climatic Change 69: 269–297.

Gagnon, A.S. and W.A. Gough. 2005b. Trends in the dates of ice freeze-up and breakup over Hudson Bay, Canada. Arctic 58: 370–382.

Gough, W.A., A.R. Cornwell, and L.J.S. Tsuji. 2004. Trends in seasonal sea ice duration in southwestern Hudson Bay. Arctic 57: 299–305.

Gough, W.A. and A. Leung. 2002. Nature and fate of Hudson Bay permafrost. Regional Environmental Change 2: 177–184.

Herrero, J. and S. Herrero. 1997. Visitor safety in polar bear viewing activiites in the Churchill region of Manitoba, Canada.

Higdon, J.W. and S.H. Ferguson. 2009. Loss of Arctic sea ice causing punctuated change in sightings of killer whales (*Orcinus orca*) over the past century. Ecological Applications 19: 1365–1375.

Hobson, K.A., I. Stirling, and D.S. Andriashek. 2009. Isotopic homogeneity of breath CO_2 from fasting and berry-eating polar bears: Implications for tracing reliance on terrestrial foods in a changing Arctic. Canadian Journal of Zoology 87: 50–55.

Hurst, R.J. and N.A. Øritsland. 1982. Polar bear thermoregulation: Effect of oil on the insulative properties of fur. Journal of Thermal Biology 7: 201–208.

Iverson, S.J., I. Stirling, and S.L.C. Lang. 2006. Spatial and temporal variation in the diets of polar bears across the Canadian arctic: indicators of changes in prey populations and environment. Symposium of the Zoological Society of London, pp. 98–117.

Nunavut Tunngavik Incorporated. 2007. Nunavut Tunngavik Incorporated Preliminary western Hudson Bay Inuit Qaujimajatugangit workshop summary.

Jonkel, C., P. Smith, I. Stirling, and G.B. Kolenosky. 1976. The present status of the polar bear in the James Bay and Belcher Islands area. Canadian Wildlife Service Occasional Paper No. 26, p. 42.

Kolenosky, G.B. and J.P. Prevett. 1983. Productivity and maternity denning of polar bears in Ontario. International Conference on Bear Research and Management 5: 238–245.

Lunn, N.J., M.K. Gray, M.K. Taylor, and J. Lee. 1987. Foxe Basin polar bear research program: 1985 and 1986 field reports. Government of the Northwest Territories.

Lunn, N.J., I. Stirling, and S.N. Nowicki. 1997a. Distribution and abundance of ringed (*Phoca hispida*) and bearded seals (*Erignathus barbatus*) in western Hudson Bay. Canadian Journal of Fisheries and Aquatic Science 54: 914–921.

Lunn, N.J., I. Stirling, D. Andriashek, and G.B. Kolenosky. 1997b. Re-estimating the size of the polar bear population in Western Hudson Bay. Arctic 50: 234–240.

Lunn, N.J., I. Stirling, D. Andriashek, and E. Richardson. 2004. Selection of maternity dens by female polar bears in western Hudson Bay, Canada and the effects of human disturbance. Polar Biology 27: 350–356.

Mauritzen, M., S.E. Belikov, A.N. Boltunov, A.E. Derocher, E. Hansen, R.A. Ims, Ø. Wiig, and N. Yoccoz. 2003. Functional responses in polar bear habitat selection. Oikos 100: 112–124.

Mauritzen, M., A.E. Derocher, and Ø. Wiig. 2001. Space-use strategies of female polar bears in a dynamic sea ice habitat. Canadian Journal of Zoology 79: 1704–1713.

McKinney, M.A., E. Peacock, and R.J. Letcher. 2009. Sea ice-associated diet change increases the levels of chlorinated and brominated contaminants in Polar Bears. Environmental Science & Technology 43: 4334–4339.

Mills, L.S. and F.W. Allendorf. 1996. The one-migrant-per-generation rule in conservation and management. Conservation Biology 10: 1509–1518.

Molnár, P.K. 2009. Modelling future impacts of climate change and harvest on the reproductive success of female polar bears (*Ursus maritimus*). Ph.D. dissertation, University of Alberta, Edmonton, Alberta.

Molnár, P.K., A.E. Derocher, M.A. Lewis, and M.K. Taylor. 2008. Modelling the mating system of polar bears: a mechanistic approach to the Allee effect. Proceedings of the Royal Society B-Biological Sciences 275: 217–226.

Nunavut Impact Review Board. 2009. Draft guidlines for the preparation of an Environmental Impact Statement for Baffinland Iron Mine's Corporation's Mary River Project, p. 129. Cambridge Bay, NU.

Obbard, M.E., M.R.L. Cattet, T. Moody, L.R. Walton, D. Potter, J. Inglis, and C. Chenier. 2006. Temporal trends in the body condition of Southern Hudson Bay polar bears. Climate change research information note, No. 3. Applied Research and Development Branch, Ontario Ministry of Natural Resource, Sault Ste. Marie, ON, p. 8.

Obbard, M.E., T.L. McDonald, E.J. Howe, E.V. Regehr, and E. Richardson. 2007. Polar bear population status in southern Hudson Bay, Canada. US Geological Survey Administrative Report, p. 36.

Obbard, M.E. and L.R. Walton. 2004. The importance of polar bear provincial park to the Southern Hudson Bay polar bear population in the context of future climate change. In C.K. Rehbein, J.G. Nelson, T.J. Beechey, and R.J. Payne (eds.), Parks and protected areas research in Ontario, 2004: planning northern parks and protected areas. Proceedings of the Parks Research Forum of Ontario annual general meeting, Thunder Bay, ON, May 4–6, 2004.

Ontario Ministry of Natural Resources. 1980. Policies for the management of polar bears in Ontario. Wildlife Branch, Toronto, ON, p. 13.

Øritsland, N.A., F.R. Englehardt, F.A. Juck, R.J. Hurst, and P.D. Watts. 1981. Effect of crude oil on polar bears. Department of Indian and Northern Affairs, Ottawa, ON. Environmental Studies, No. 24, p. 268.

Oswalt, W.H. 1999. Eskimos and explorers, 2nd Edition. University of Nebraska Press, Lincoln, NB, pp. 341.

Paetkau, D., S.C. Amstrup, E.W. Born, W. Calvert, A.E. Derocher, G.W. Garner, F. Messier, I. Stirling, M.K. Taylor, Ø. Wiig, and C. Strobeck. 1999. Genetic structure of the world's polar bear populations. Molecular Ecology 8: 1571–1584.

Paetkau, D., W. Calvert, I. Stirling, and C. Strobeck. 1995. Microsatellite analysis of population structure in Canadian polar bears. Molecular Ecology 4: 347–354.

Peacock, E., A. Orlando, V. Sahanatien, S. Stapleton, A.E. Derocher, and D.L. Garshelis. 2008. Foxe Basin Polar Bear Project, Interim Report 2008, p. 55.

Peacock, E., N.F. Lunn, M.L. Branigan, R. Maraj, L. Carpenter, D. Larsen, A. Derocher, B. Ford, D. Hedman, J. Justus, S. Lefort, G. Mouland, G. Nirlungayuk, M.E. Obbard, and F. Pokiak. 2009a. 2009 Report of the status of Polar Bear in Canada. Provincial/Territorial and Federal Polar Bear Technical Committee of Canada, Whitehorse, YK.

Peacock, E., V. Sahanatien, and S. Stapleton. 2009b. Foxe Basin Polar Bear Project: 2009 Interim Report. Department of Environment, Government of Nunavut, Igloolik, NU.

Prestrud, P. and I. Stirling. 1994. The international Polar Bear agreement and the current status of polar bear conservation. Aquatic Mammals 20:113–124.

Prevett, J.P. and G.B. Kolenosky. 1982. The status of polar bears in Ontario. Naturaliste Canadien 109: 933–939.

Prowse, T.D., C. Furgal, F.J. Wrona, and J.D. Reist. 2009. Implications of climate change for Northern Canada: freshwater, marine, and terrestrial ecosystems. Ambio 38: 282–289.

Ramsay, M.A. and K.A. Hobson. 1991. Polar bears make little use of terrestrial food webs: evidence from stable-carbon isotope analysis. Oecologia 86: 598–600.

Ragen, T.J., H.P. Huntington, and G.K. Hovelsrud. 2008. Conservation of Arctic marine mammals faced with climate change. Ecological Applications 18: S166–S174.

Regehr, E.V., N.J. Lunn, S.C. Amstrup, and I. Stirling. 2007. Effects of earlier sea ice breakup on survival and populaiton size of polar bears in western Hudson Bay. Journal of Wildlife Management 71: 2673–2683.

Richardson, E., I. Stirling, and D.S. Hik. 2005. Polar bear (Ursus maritimus) maternity denning habitat in western Hudson Bay: A bottom-up approach to resource selection functions. Canadian Journal of Zoology 83: 860–870.

Rockwell, R.F. and L.J. Gormezano. 2009. The early bear gets the goose: climate change, polar bears and lesser snow geese in western Hudson Bay. Polar Biology 32: 539–547.

Rode, K.D., J.D. Reist, E., Peacock, and I. Stirling. In press. Comments in response to "Estimating the energetic contribution of polar bear (Ursus maritimus) summer diets to the total energy budget". Journal of Mammalogy.

Russell, R.H. 1975. The food habits of polar bears of James Bay and Southwest Hudson Bay in summer and autumn. Arctic 28: 117–129.

Sahanatien, V. and A. Derocher. 2007. Changing sea ice-scapes and polar bear habitat in Foxe Basin, Nunavut Territory, Canada (1979–2004). Oral presentation at International Conference on Bear Research and Management, Monterrey, Mexico.

Smith, T.G. 1975. Ringed seals in James Bay and Hudson Bay: Population estimates and catch statistics. Arctic 28: 170–182.

Smith, A.E. and M.R.J. Hill. 1996. Polar bear, Ursus maritimus, depredation of Canada Goose, Branta canadensis, nests. Canadian Field-Naturalist 110: 339–340.

Stenhouse, G.B., L.J. Lee, and K.G. Poole. 1988. Some characteristics of polar bears killed during conflicts with humans in the northwest territoires, 1976–86. Arctic 41: 275–278.

Stenhouse, G.B. and N.J. Lunn. 1987. Foxe basin polar bear research program: 1984 Field Report. Department of Renewable Resources, Government of the Northwest Territories, Rankin Inlet, NWT.

Stirling, I. 1990. Polar bears and oil: ecological perspectives. In J.R. Geraci and D.J. St. Aubin (eds.) Sea mammals and oil: confronting the risks. Academic Press, San Diego, pp. 223–234.

Stirling, I. and A.E. Derocher. 1993. Possible impacts of climatic warming on polar bears. Arctic 46: 240–245.

Stirling, I., A.E. Derocher, W.A. Gough, and K. Rode. 2008. Response to Dyck et al. (2007) on polar bears and climate change in western Hudson Bay. Ecological Complexity 5: 193–201.

Stirling, I., C. Jonkel, P. Smith, R. Robertson, and D. Cross. 1977. The ecology of the polar bear (Ursus maritimus) along the western coast of Hudson Bay. Canadian Wildlife Service, Occasional Paper 33, p. 64.

Stirling, I., N.J. Lunn, and J. Iacozza. 1999. Long-term trends in the population ecology of polar bears in western Hudson Bay in relation to climatic change. Arctic 52: 294–306.

Stirling, I., N.J. Lunn, J. Iacozza, C. Elliott, and M. Obbard. 2004. Polar bear distribution and abundance on the southwestern Hudson Bay coast during open water season, in relation to population trends and annual ice patterns. Arctic 57: 15–26.

Stirling, I. and C.L. Parkinson. 2006. Possible effects of climate warming on selected populations of polar bears (Ursus maritimus) in the Canadian Arctic. Arctic 59: 261–275.

Sturtevant, W.C. 1984. Handbook of North American Indians. Smithsonian Institute, Washington, DC.

Taylor, M. and J. Lee. 1995. Distribution and abundance of Canadian polar bear populations – a management perspective. Arctic 48: 147–154.

Taylor, M.K., D.P. DeMaster, F.L. Bunnell, and R.E. Schweinsburg. 1987. Modeling the sustainable harvest of polar bears. Journal of Wildlife Management 51: 811–820.

Taylor, M.K., J. Lee, and L. Gray. 1990a. Foxe basin Polar Bear research program 1987 Field Report.

Taylor, M.K., J. Lee, J. Laake, and P.D. McLoughlin. 2006. Estimating population size of polar bears in Foxe Basin, Nunavut using tetracycline biomarkers. File Report, Department of Environment, Government of Nunavut, p. 13.

Taylor, M.K., J. Lee, and R. Mulders. 1990b. Foxe Basin Polar Bear Research Program: 1988 Field Report. Department of Renewable Resources, Government of the Northwest Territries, Yellowknife, NWT.

Thiemann, G.W., S.J. Iverson, and I. Stirling. 2008. Polar bear diets and arctic marine food webs: insights from fatty acid analysis. Ecological Monographs 78: 591–613.

Towns, L., A.E. Derocher, I. Stirling, and N.J. Lunn. 2010. Changes in land distribution of Polar Bears in Western Hudson Bay. Arctic 63: In press.

Towns, L., A.E. Derocher, I. Stirling, N.J. Lunn, and D. Hedman. 2009. Spatial and temporal patterns of problem polar bears in Churchill, Manitoba. Polar Biology 32: 1529–1537.

Tyrrell, M. 2007. More bears, less bears: Inuit and scientific perceptions of polar bear populations on the west coast of Hudson Bay. Journal of Inuit Studies 30: 191–208.

Urquhart, D.R. and R.E. Schweinsburg. 1984. Polar Bear: Life history and known distribution of polar bear in the Northwest Territoires up to 1981. Department of Renewable Resources, Yellowknife, Northwest Territories.

The Rise of Killer Whales as a Major Arctic Predator

S.H. Ferguson, J.W. Higdon, and E.G. Chmelnitsky

Abstract Anecdotal evidence, sighting reports, Inuit traditional knowledge, and photographic identification indicate that killer whale (*Orcinus orca*) occurrence in Hudson Bay is increasing. Killer whales were not known to be present in the region prior to the mid-1900s but have since shown an exponential increase in sightings. More sightings from Foxe Basin, Nunavut in the north to Churchill, Manitoba in the south appear to be related to a decrease in summer sea ice in Hudson Strait. Killer whale activity during the open water season has been concentrated in the northwest Hudson Bay region that includes the Repulse Bay and northern Foxe Basin areas. Here, prey items are diverse and abundant. Killer whales are reported in western Hudson Bay on an annual basis with sighting reports and anecdotal evidence suggesting they are first observed heading through Hudson Strait in July and returning to the northwest Atlantic in September. However, arrival, occupancy, and departure times are likely related to yearly ice conditions and prey availability.

Killer whales have been observed preying on a number of marine mammal species in Hudson Bay. Of particular concern is predation on bowhead whales in Foxe Basin, narwhal in northwest Hudson Bay, and beluga in southwest Hudson Bay. The impact of killer whale predation on marine mammal species is unknown without long-term studies and direct observation of killer whale hunting behaviour. However, by defining population energetic requirement and considering population demography of prey, we can begin to assess the basic requirements of predator–prey dynamics in Hudson Bay marine ecosystem. To estimate predation impact we used a simple mass-balanced marine mammal model that includes age structure, population size, and predation rate inputs. Estimates of killer whale population size were variable, with the majority of information suggesting that at least 25 whales use the area each

S.H. Ferguson (✉)
Fisheries and Oceans Canada, Central and Arctic Region, Winnipeg, MB, Canada
and
Department of Environment and Geography, University of Manitoba, Winnipeg, MB, Canada
and
Department of Biological Sciences, University of Manitoba, Winnipeg, MB, Canada
e-mail: steve.ferguson@dfo-mpo.gc.ca

S.H. Ferguson et al. (eds.), *A Little Less Arctic: Top Predators in the World's Largest Northern Inland Sea, Hudson Bay*, DOI 10.1007/978-90-481-9121-5_6,
© Springer Science+Business Media B.V. 2010

summer. Results suggest that the Northern Hudson Bay narwhal population may be negatively impacted by continued killer whale predation. We conclude that conservation of marine mammals in Hudson Bay should consider killer whale effects since they have the potential to regulate population growth of prey populations.

Keywords Beluga • Bioacoustics • Bowhead • Inuit observations • Movements • Narwhal • Photo-identification • Rake marks

Introduction

The killer whale (*Orcinus orca*) exists in all oceans of the world but occurs at highest densities in productive temperate waters (Baird 1999). With climate change resulting in warming oceans and loss of sea ice, a major redistribution of whale species is underway with a predicted increase in killer whale abundance in Arctic waters (Moore and Huntington 2008). Historically, the presence of killer whales in Arctic waters has been limited by the presence of pack ice in winter (Reeves and Mitchell 1988; Dyke et al. 1996). The eastern and western Canadian Arctic is becoming more accessible to killer whales as the concentration of sea ice in choke points decreases with climate change (Higdon and Ferguson 2009). Here, we summarize research results from sighting reports, Inuit traditional ecological knowledge, acoustics, and photo-identification to better understand the ecology of killer whales in the Hudson Bay region.

Killer whales pose a potential dilemma for the Arctic ecosystem as they can exert significant regulatory effects to prey populations which may cause declines (Estes et al. 1998). Information on killer whale behaviour, group size, social structure, geographic movements, morphological characteristics, genetics, vocalizations, acoustic, and foraging behaviour is needed to understand changes occurring in the Arctic. Our summary of the history of use in this region by killer whales, seasonal movement patterns, feeding behaviour, photographic identification, bioacoustics, and predation by killer whales in the Hudson Bay region provides context in understanding potential ecosystem shifts and predation consequences for marine mammals.

History of Killer Whale Use of Greater Hudson Bay Region

Killer whales have received little directed study in the eastern Canadian Arctic, with the exception of two comprehensive reviews of available information (Reeves and Mitchell 1988; Higdon 2007). Killer whales were historically (1800s) present in Baffin Bay and Davis Strait and were often reported in commercial bowhead (*Balaena mysticetus*) whaling logbooks (Reeves and Mitchell 1988). However, all lines of evidence suggest that they are a recent addition to the marine mammal community in Hudson Bay (Higdon and Ferguson 2009). Inuit knowledge also

Fig. 1 Four of seven to nine killer whales observed in Western Hudson Bay (offshore of Rankin Inlet, Nunavut) by a "Students on Ice" cruise on August 5, 2007 (Photo by Trevor Lush)

suggests killer whales were not present prior to the mid-1900s but are now observed on a regular basis (Gonzalez 2001).

Higdon and Ferguson (2009) summarized killer whale sighting records in Hudson Strait, Hudson Bay, James Bay and Foxe Basin from 1900 to 2006 and demonstrated an exponential increase in sightings per decade (Fig. 1). The first killer whale sighting within Hudson Bay occurred in the 1940s. Most sightings in Hudson Bay have occurred since the 1960s, with the majority along the western coast. The majority of sighting reports are coastal, although the lack of offshore sightings is likely due to human effort being concentrated coastally. There is one offshore record provided by a wildlife observer program for oil and gas development (Milani 1986), and increased effort in the offshore region would likely increase the number of sightings.

Killer whales appear to be rare in James Bay and the south and east coasts of Hudson Bay (Higdon and Ferguson 2009). There is less effective reporting of killer whales along the Ontario coast, and our research efforts have focussed on Nunavut. However local knowledge does suggest that killer whales are rare along the southern coast. Cree residents report seeing killer whales "very occasionally" and note that they are more common further west (i.e., towards Churchill, Manitoba) (Higdon and Ferguson 2009). Ontario Cree do not have a word for killer whale (Johnston 1961), which suggests that their sporadic observations are a recent phenomenon. Higdon and Ferguson (2009) also found few sightings for the Nunavik region of northern Quebec (eastern Hudson Bay and the south coast of Hudson Strait). Nunavik researchers have heard anecdotal reports of killer whales sightings but there is no systematic data collection. It is possible that killer whales make relatively rapid movements through Hudson Strait on their way to western Hudson Bay and this would reduce the likelihood of observations. In recent years killer whales have been observed attacking beluga in Hudson Strait, and Inuit hunters report that killer whale numbers are increasing (J. Peters, Makivik Corp., personal communication 2008).

Chronological reporting by authors illustrates a progression in killer whale use of the Hudson Bay region, from no evidence (Degerbøl and Freuchen 1935; Peterson 1966; Banfield 1974), to sporadic occurrence (Davis et al. 1980), to occasional and possibly annual use (Reeves and Mitchell 1988), to an exponential increase (Higdon and Ferguson 2009), so that now killer whales occur in Hudson Bay on an annual basis. Higdon and Ferguson (2009) examined correlations between sighting frequency and a long-term sea ice dataset (Rayner et al. 2003) and found that the increase in killer whale sightings was significantly correlated to a decline in sea ice in Hudson Strait. Tynan and Demaster (1997) suggested that marine mammal responses to sea ice declines would first be shown with changes in distribution, and the analyses in Higdon and Ferguson (2009) provides a real-world example.

There are inherent biases in using sighting reports to infer changes in species abundance and distribution such as the clusters of where observers live. Despite this, there is reasonably good evidence that killer whales were historically absent in Hudson Bay. There is a long history of European exploration in the region, with over 600 voyages undertaken between 1610 and the early 1900s (Cooke and Holland 1978). A large volume of literature exists, but we know of no mention of killer whales previous to the twentieth century. Sailors were familiar with killer whales, and published accounts have included observations made in the North Atlantic while in transit (Chappell 1817), but no observations were recorded in Hudson Bay. In addition, all of the typical marine mammal species found in Hudson Bay were known well prior to the nineteenth century (Pinkerton and Doyle 1805). To our knowledge no Hudson Bay whaling logbooks mentioned killer whales, despite extensive effort there from 1860 to 1915 and several thorough reviews of the logs (Mitchell and Reeves 1982; Reeves et al. 1983; Reeves and Cosens 2003; Reeves and Mitchell 1988; Ross 1974; Stewart 2008). Hudson Bay whalers were also familiar with killer whales, and several logbooks mentioned trying to hunt them in the North Atlantic while travelling to the Hudson Bay whaling grounds (B. Stewart, Arctic Biological Consultants, Winnipeg, MB, personal communication 2009). It seems likely that the logbooks would have contained killer whale sightings had they occurred in the Hudson Bay region.

Degerbøl and Freuchen (1935) traveled extensively in the area from 1921 to 1924, conducting surveys and having many discussions with local Inuit. They heard no evidence of killer whales in Hudson Bay, and were told by Inuit that the whales were turned around in Hudson Strait by the presence of walrus (*Odobenus rosmarus*). Degerbøl and Freuchen (1935) felt that sea ice represented a more likely barrier to movement, and the analyses of Higdon and Ferguson (2009) supports this, suggesting that declining sea ice conditions are the most reasonable explanation for recent killer whale colonization of the Hudson Bay region. A change in ice conditions in central Hudson Strait in the 1930s opened up a former choke point (c.f. Wilson et al. 2004) and allowed a punctuated advancement of killer whales in the region.

Seasonality and General Movement Patterns

Killer whales are first seen in Hudson Strait in July, and reports in Hudson Bay peak in August (Higdon and Ferguson 2009). This was based on analyses of 80 records, and we have since updated the database with additional sightings (18) in the Hudson Bay region. Many of these new sightings (105) have come from an extensive effort to collect Inuit traditional knowledge (or *Inuit Qaujimajatuqangit*) in a number of Nunavut communities. Inuit observations have been collected and compiled from six communities in Hudson Strait (Kimmirut, 5 interviewees), Hudson Bay (Arviat [5], Rankin Inlet [10], and Repulse Bay [17]) and Foxe Basin (Hall Beach [7] and Igloolik [16]) (Westdal 2009). The most recent version of the killer whale sightings database (used in Higdon et al. in review) contains 203 records for the four Hudson Bay ecoregions (as identified by Stewart and Lockhart 2005).

Over half of the 207 records (n = 107) provide the month of the sighting, and an additional 31 include the season (two each in "spring" and "fall" and 27 in "summer") (Fig. 2). The updated database now contains records of sightings in June (n = 6). Most of these reports are from late June; including Inuit observations from Repulse Bay of occasional sightings of killer whales at the floe edge (Westdal 2009). More of the July sightings are in Hudson Strait not Hudson Bay and the peak

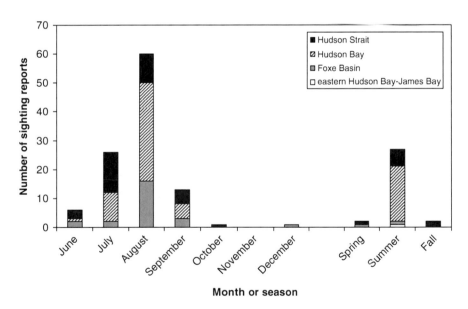

Fig. 2 Seasonality of killer whales in the Hudson Bay region, separated into the four ecoregions identified by Stewart and Lockhart WL (2005) (and used by Higdon and Ferguson 2009). Data on sighting records from the DFO database as summarized by Higdon et al. (in review) (n = 107 with month of sighting and 31 with season only). The one record from the eastern Hudson Bay–James Bay ecoregion is of a dead killer whale that washed up on shore (Reeves and Mitchell 1988)

in Hudson Bay records still occurs in August. In September there is a significant decline in the number of reports throughout the region, and the single October report occurred in Hudson Strait. There are no reports from November, and the one record from December is of an ice entrapment that occurred in Foxe Basin in the 1950s (Reeves and Mitchell 1988; Irngaut 1990; Kappianaq 2000; Higdon 2009). These reports indicate that killer whales generally travel west through Hudson Strait in July occurring most often in Hudson Bay and Foxe Basin in August, and typically depart in September.

Inuit knowledge of movement patterns is summarized in Fig. 3. Killer whales migrate in and out of Hudson Strait and use Frozen Strait to enter Repulse Bay in the summer. Many Inuit respondents in Repulse Bay felt that the same killer whales seen in Repulse Bay go south towards Arviat and north to Hall Beach (Westdal 2009). Some hunters suggest that killer whales head south through Roes Welcome Sound towards Arviat and Rankin Inlet, then return to the Repulse Bay area before travelling back through Frozen Strait and out into Hudson Strait (Higdon 2009; Westdal 2009). Inuit often report killer whale sightings via radio communication between coastal communities providing a chronology of seasonal sightings. For example, when killer whales are seen in Foxe Basin or in Rankin Inlet, they are generally seen near Repulse Bay about 5 days later (Westdal 2009). This suggests that the same groups of whales make north-south movements along the coast.

Inuit suggest that killer whales generally arrive in Foxe Basin from southern areas (around Cape Dorset and from Repulse Bay) when most of the ice has left (Higdon 2009). It is thought that they travel through the centre of Foxe Basin, following their prey (especially bowhead whales), and avoiding the coast. Some return south after coming north into Foxe Basin, but they have also been observed travelling through Fury and Hecla Strait in both eastern and western directions, following bowhead whales west, and smaller numbers following narwhal east in the fall (Higdon 2009). Interviewees in Kimmirut noted that killer whales are rarely seen in their area and do not stay for long, and they also did not provide any information on killer whale migration patterns. This supports the suggestion by Higdon and Ferguson (2009) that killer whales make rapid movements through Hudson Strait possibly to access the prey-rich Foxe Basin and western Hudson Bay coast.

Inuit observers have also noted several concentration areas in the Hudson Bay region (Fig. 3), specifically Repulse Bay (also see Higdon and Ferguson 2009), Lyon Inlet, and the area north of Igloolik. The first two areas are important summer concentration areas for narwhal, and the area in northern Foxe Basin contains large numbers of bowhead whales. Inuit in Igloolik note a direct link between increasing bowhead numbers and increasing presence of killer whales (Higdon 2009). Interviewees in Hall Beach noted that killer whales migrate quickly past their community, following bowhead to the region north of Igloolik. Many Inuit respondents in Hudson Bay and Foxe Basin indicate that, while numbers are increasing, killer whales are not as abundant as they are in other areas of Nunavut. For example, they identified Admiralty Inlet as an important concentration area (Fig. 3).

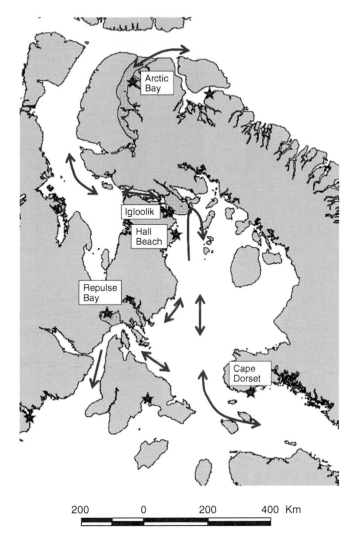

Fig. 3 Movements of killer whales into the Hudson Bay region and important concentration areas (Admiralty Inlet near Arctic Bay, Foxe Basin near Igloolik, and Repulse Bay near Repulse Bay) as identified by Inuit observations (Higdon 2009)

Killer Whale Prey Items and Evidence for Different Ecotypes

Killer whales eat a wide variety of prey items. In the North Pacific and Antarctica, there are sympatric ecotypes that consume specific (non-overlapping) prey. In the eastern North Pacific three ecotypes have been identified: (1) transients, which prey exclusively on marine mammals; (2) residents, which are exclusively fish eaters (mostly salmon); and (3) offshores, which have been little studied but appear to prey on a variety of fish species (Ford 2002; Ford et al. 2000; Dahlheim et al. 2008). These ecotypes also differ in vocal behaviour, genetics, morphology and group size and structure. Three different types also occur in Antarctic waters, referred to as Types A, B, and C, which preferentially consume Antarctic minke whales (*Balaenoptera bonaerensis*), seals and occasionally baleen whales, and a single fish species (Antarctic toothfish, *Dissostichus mawsoni*), respectively (Pitman and Ensor 2003). Conversely, some killer whale populations worldwide exhibit little prey specialization. In New Zealand, Visser (2000) identified four main prey types (rays, sharks, fin-fish and cetaceans) consumed by three proposed killer whale sub-populations. In the Northeast Atlantic, some killer whale populations and some killer whales within populations display a broad niche width that, for example, includes seals and herring (*Clupea harengus*) as prey (Foote et al. 2009).

Killer whale ecotypes have not been identified in the Northwest Atlantic including the Canadian eastern Arctic. Therefore, the degree of foraging specialization, if any, is not known. Killer whales have been observed feeding on a variety of cetacean and pinniped species, and there are occasional reports of predation on fish in the North Atlantic and West Greenland. Higdon (2007) summarized 132 recorded attempted or successful predation events in the Canadian Arctic, West Greenland, and northern Labrador. Marine mammal predation events dominated, and there were no recorded fish predation events in the Canadian Arctic, although several reports are available from West Greenland and Davis Strait.

Stomach contents of whales harvested in West Greenland provide evidence that at least some Arctic killer whales eat fish (Heide-Jørgensen 1988 (species not provided); Laidre et al. 2006 (lumpsucker fish, *Cyclopterus lumpus*)). Degerbøl and Freuchen (1935) also stated that killer whales in Davis Strait preyed on Greenland halibut (*Reinhardtius hippoglossoides*), and Vibe (1980a, in Jensen and Christensen 2003) stated that Greenland killer whales prey on cephalopods "a large degree". Killer whales have also been observed scavenging around longline fishing vessels of northern Newfoundland, coastal Labrador and southern Davis Strait (Sergeant and Fisher 1957; Mitchell and Reeves 1988; Lawson et al. 2007) and eating herring (*Clupea harengus*) off eastern Newfoundland (Steiner et al. 1979). Bluefin tuna (*Thunnus thynnus*) have been found in the stomachs of two stranded whales in Newfoundland (Lawson et al. 2007). Thus, it is apparent that at least some killer whales in the Northwest Atlantic and eastern Arctic (West Greenland/Davis Strait) consume fish.

Using the available published information, the vast majority of killer whales in the Canadian Arctic and the Hudson Bay region preferentially, if not exclusively, consume marine mammals (versus fish; Higdon 2007; Higdon et al. in review). Killer whales have been observed preying on all the typical Arctic marine mammal species. The most current version of the sightings database (as analysed by Higdon et al. in review) includes 111 reported predation events throughout the Canadian Arctic. Predation on monodontids (narwhal and beluga) was reported significantly more often than other prey groups, followed by predation on bowhead whales and phocid seals. The database includes 24 predation records from Hudson Bay, 20 from Foxe Basin, and 13 from Hudson Strait. Predation on monodontids, bowheads and

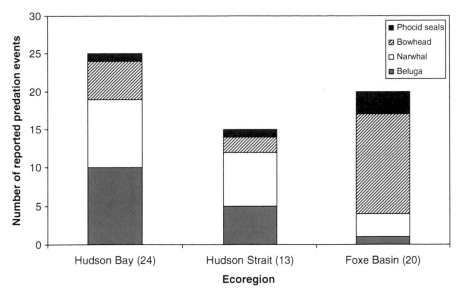

Fig. 4 Marine mammal prey items of killer whales reported in the Hudson Bay region (Modified from data in Higdon et al. in review). Data are shown by ecoregions; with no predation records reported for eastern Hudson Bay–James Bay. Values in parentheses are number of predation reports; the total is higher than the number of reports because some records included multiple species groups (particularly records from Inuit knowledge interviews)

seals was reported in Foxe Basin and Hudson Bay, and on monodontids and bowheads in Hudson Strait. Monodontid predation was most often reported in Hudson Bay and Hudson Strait, and bowhead predation most often in Foxe Basin (Fig. 4).

A number of published sources provide accounts of killer whale predation on marine mammals in the eastern Canadian Arctic (reviewed by Higdon 2007). However much of the information on predation summarized by Higdon et al. (in review) has been gathered through semi-directed interviews with Inuit hunters (Westdal 2009). Local Inuit are extremely knowledgeable on their local environment, and the wildlife species occurring there, and interviews have provided significant information on killer whale movements, distribution, and ecology. No interviewees in Arviat, Kimmirut, or Repulse Bay noted fish as a prey item, although one respondent in Rankin Inlet suggested that killer whales possibly ate shrimp and capelin (*Mallotus villosus*) (Westdal 2009). Foxe Basin interviewees were unsure whether or not killer whales ate fish, although several hunters thought they might. Ultimately, none of the interviewees provided a direct observation of killer whales preying on fish, and the results suggest that marine mammals are the primary, if not only, prey for killer whales in this area.

In certain cases Inuit observations can provide important information that scientific research approaches have difficulty addressing. The impact of killer whales on bowhead whales is one example. Several recent studies have used photo-identification

Fig. 5 Tooth rake marks on the flukes of an adult bowhead whale caused by killer whales (Photo credit Jeff Higdon: Foxe Basin floe edge, summer 2007)

methods (analyses of rake marks; Fig. 5) to conclude that killer whale attacks on large whales are rare (Mehta et al. 2007; Steiger et al. 2008). However photo-identification methods only identify survivors, as successful kills are not available. Observations of large whale attacks are rare, but they have occurred and been documented (reviewed by Reeves et al. 2006; Springer et al. 2008). Inuit hunters spend a considerable amount of time on the ocean hunting and fishing throughout Nunavut and have observed attacks on bowhead whales (NWMB 2000; Westdal 2009). Inuit regard killer whale predation as one of the major threats to bowhead recovery (Moshenko et al. 2003).

Foxe Basin is an important nursery area for bowhead cow-calf pairs (Cosens and Blouw 2003; NWMB 2000). Killer whales preferentially select baleen whale calves (Naessig and Lanyon 2004; Mehta et al. 2007; Steiger et al. 2008), and Foxe Basin should thus represent an important area for attempted predation. Inuit interviewees have confirmed this and have regularly seen attacks and found dead bowheads. The majority of Foxe Basin interviewees (91%, 15/16 in Igloolik and 6/7 in Hall Beach) identified bowhead whales as an important prey of killer whales. Many interviewees (eight in Igloolik and one in Hall Beach) noted a direct cause and effect relationship between bowhead population growth and increased killer whale presence (Higdon 2009; also see Piugaattuk 1994).

Eleven interviewees recounted either direct observation of killer whale attacks on bowheads (n = 7) or second-hand stories passed on by others, for a total of 13 independent predation events witnessed (Higdon 2009). These observations provided extensive descriptions of the different cooperative hunting techniques that killer whales use on bowheads. Seven interviewees noted that killer whales will go on top of the bowhead to cover the blowhole and suffocate it. Five hunters noted that killer whales will often kill the bowhead by ramming it from below and tearing chunks out of the belly, and six interviewees reported that killer whales will bite the bowhead's flippers (see Westdal 2009 for similar information from other communities). Inuit observations agree with scientific research and note that smaller bowheads are usually attacked, although several hunters did note that killer whales are capable of killing larger whales, particularly in other areas (e.g., Admiralty Inlet) where they tend to occur in larger groups. Fifteen interviewees reported observations of dead bowhead whales that had been killed by killer whales determined by the condition of the carcass (e.g., bite marks, chunks removed, broken ribs). A total of 32 observations of dead bowheads were reported, and after correcting for overlapping reports, a minimum of 22 different kills were documented (most observed in 1999).

Interviewees were asked to estimate how many bowhead whales are typically killed in Foxe Basin each year, and eight provided an opinion. The minimum estimated kill every summer was three to four whales, with a maximum of ten. Overall the responses indicate that on average about five bowhead whales are killed each year in Foxe Basin but in some years many more are killed (Higdon 2009). For example, in 1999 ice conditions were less extensive than usual, and killer whales were more abundant than usual (also see Cosens and Blouw 2003). In that year at least eight dead bowhead whales were discovered, including one fresh enough for Inuit to utilize. However, many hunters (n = 12) felt that current predation rates were not enough to negatively impact the growing bowhead population, although loss of sea ice could result in increased predation pressure on bowhead whales in this area.

Photo-Identification

In the 1970s, a technique was developed that uses photographs of the dorsal fin and gray saddle patch at the base of the killer whale fin to uniquely identify individual whales (Bigg 1982). The pigmentation, shape and scar pattern on these areas are unique for each whale. Scars in the saddle patch behind the dorsal fin show as clear permanent white or black marks and some whales also have saddle patches with a unique, easily recognizable shape. Nicks, cuts, and tears on the dorsal fin of killer whales also leave semi-permanent marks on the fin. The photo-identification method identified individuals photographed in some of the Hudson Bay sightings and provided inference on movement patterns.

Photo-identification studies indicate that there are at least 21 distinct killer whales within the western portion of Hudson Bay (Young et al. in review). There is no evidence of killer whale movements between Hudson Bay and the Baffin Bay, although there have been resightings of the same individuals within a region. Furthermore, the discovery curve, representing the number of new individuals identified each year, has increased linearly from 2004 to 2009 with no clear asymptote, which suggests that the entire killer whale population visiting the Hudson Bay region has not yet been identified. Continued and expanded photo-identification studies will likely refine estimates of the potential killer whale population numbers and distributions in the Canadian Arctic as well as movements between regions.

Bioacoustics of Killer Whales and Their Prey

Acoustic monitoring studies are become increasingly important in marine mammal science. Bioacoustics can answer many questions such as species and population distribution, abundance and behaviour (Marques et al. 2009; Burtenshaw et al. 2004). For killer whales, acoustic monitoring can detect presence and predation. Killer whale ecotypes in the North Pacific exhibit different acoustic behaviour while hunting; marine mammal eating killer whales reduce communication during prey searching to avoid detection and celebrate successful kills acoustically (Barrett-Lennard et al. 1996; Deecke et al. 2005).

Three areas in the Hudson Bay region were chosen for acoustic monitoring during the ice-free season (July–September) from 2006 to 2009 (Fig. 6).

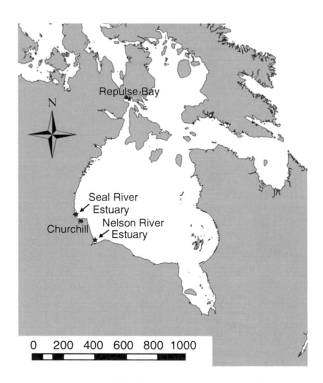

Fig. 6 Location of deployment of AURAL-M2 autonomous recording devices used to record whale acoustic calls

The areas were chosen based on past killer whale sightings and large aggregations of potential prey, including narwhal, beluga, bowhead, and seals. From 2006 to 2009, an AURAL-M2 (Multi-Électronique Inc., Rimouski, Quebec) was deployed in the Seal River Estuary near Churchill, Manitoba. In 2006, an AURAL-M2 was deployed near Repulse Bay, Nunavut and in 2008 and 2009 one was deployed in the Nelson River Estuary, Manitoba. An AURAL-M2 is an autonomous recording device that detects sounds within a frequency range up to 16 kHz, which includes most calls made by killer whales, belugas, bowhead, narwhal, and seals. In 2009 in addition to the AURAL-M2, a C-POD (Chelonia Limited, Cornwall, United Kingdom) was attached to the mooring at the Seal River and Nelson River Estuaries. C-PODs record echolocation clicks produced by toothed whales (killer whales, narwhal, and beluga) that are used for navigation and prey detection. Automated detection algorithms were used by JASCO Inc. to analyze the 2006 to 2008 recordings. Marine mammal calls were detected every day during all deployments however further analysis revealed that there were many false positive detections. Therefore, a more detailed analysis is planned to locate marine mammal calls.

Simple Predator–Prey Model

Large-bodied predators, such as killer whales, have high caloric demands that require a proportionally large intake of prey. The effect of predation on prey populations depends on predator and prey abundance and distributional overlap in time and space that influences kill rates (Lima and Dill 1990). If killer whales hunt in a focused area or on a particular prey then predation can have significant impact (Williams et al. 2004; Laidre et al. 2006). Anti-predator behaviour in response to killer whales has been observed in marine mammal species in the eastern Canadian Arctic by local residents and researchers. Narwhal and bowheads were observed grouping and swimming close to shore in the presence of killer whales (Finley 1990; Campbell et al. 1988). Narwhal also ceased vocalizing presumably to decrease probability of detection (Campbell et al. 1988). Richard (2005) found extreme aggregations of beluga whales near Churchill, Manitoba most likely in response to seven killer whales observed during the same aerial survey. Inuit have long-recognized fleeing to shallow nearshore waters as a response to killer whale presence ("*aarlirijuk*", or "fear of killer whales", in the south Baffin dialect of Inuktitut; NWMB 2000).

We apply a simple demographic model to predict the impact of killer whales on marine mammal prey populations in the greater Hudson Bay region. We chose this type of analysis because direct experimental analyses are logistically intractable due to low densities, rapid movements, and large seasonal ranges of these carnivorous marine predators. Also, killer whales can hunt underwater making direct assessment of foraging behaviour and predator–prey interactions challenging. The simple predator–prey model provides a scenario for possible trophic consequences of the current killer whale population residing in the Hudson Bay region during the ice-free period (Fig. 7).

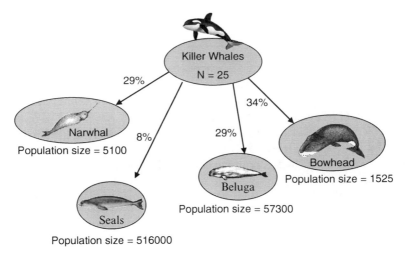

Fig. 7 Simple graphic showing prey selection by a group of killer whales in Hudson Bay used to model the possible effects of predation on marine mammal prey populations

We use the following assumptions. First, a group of 25 killer whales resides in Hudson Bay area for 3 months (1 July – 30 Sept.). We assume killer whales eat 3% of their body weight per day (2–4%; Kriete 1995; Hickie et al. 2007). The sex-age composition of the killer whale groups using the eastern Canadian Arctic region according to photo-identification work is 26% large males, 60% females or juvenile males, and 14% small young (Young et al. submitted). For 25 whales, this translates into 6.5 males, 15 females (and juvenile males), and 3.5 young killer whales. We assume large males weigh 5000 kg, females (or juvenile males) 2700 kg, and small young whales 1350 kg (Clark et al. 2000). We assume that the sightings database of predation events (Fig. 4) relates directly to amount of prey eaten. The biomass of prey eaten includes the likelihood that a number of beluga and narwhal are killed during an attack whereas only one seal or bowhead is killed during an attack. For the purposes of this scenario, we believe this assumption compensates for the inverse relationship between human observer ability to detect a predation event and prey size. Thus, we assume 29% of killer whale diet consisted of narwhal, 29% beluga, 34% bowhead whale, and 8% seals. The Hudson Bay regional sighting database estimates of predation events are comparable with results for the entire eastern Canadian Arctic (Higdon et al. in review). Further, we assume that during this summer period killer whales eat the majority (50% used in model) of their annual food intake as observed elsewhere for mammal-eating killer whales (Condy et al. 1978, Lopez and Lopez 1985, Baird and Dill 1995).

We used the following average weights of prey: for beluga 313 kg, 325 kg for narwhal (Trites and Pauly 1998; but see Heide-Jørgensen and Teilmann 1994 and Garde et al. 2007), 6,000 kg for a calf or juvenile bowhead (Węskawski et al. 2000) and 17,000 kg for a subadult (George et al. 2007), and 69 kg for ringed seals (Krafft et al. 2006). However, we consider that killer whales only eat 40% of a bowhead, 80% of beluga and narwhal, and 90% of seals due to selection of particular parts and loss of carcass at depths. For the killed and eaten bowhead, we assume killer whales select preferentially mostly (75%) calves and juveniles and some (25%) subadults. We make this assumption due to the novice behaviour and smaller size characteristic of juveniles resulting in reduced energetic costs, time associated with capture, and risk of injury to killer whales.

When a large whale is killed, killer whales typically consume only a small amount, leading Martinez and Klinghammer (1970) to report that they "feed very selectively". Observations of successful kills on a variety of baleen whale species indicate that usually only the tongue, throat area (possibly torn away for access to the tongue), skin, and occasionally a small portion of the blubber are consumed (Hancock 1965; Baldridge 1972; Silber et al. 1990; Jefferson et al. 1991; Ford et al. 2005). In longer feeding events (over several days) more substantial amounts of blubber and other flesh may be consumed (Ford et al. 2005). In many instances dead whales have open wounds on their abdomen with parts of the viscera extruding, although the organs are often uneaten (Hancock 1965; Silber et al. 1990). Observations of killer whales consuming grey whales (*Eschrichtius robustus*) in Alaskan waters indicate that substantially more of the whale is consumed if the mortality occurs in relatively shallow waters (C. Matkin personal communication 2009). Foxe Basin is relatively shallow and we assume a greater amount of killed bowhead may be consumed, and therefore we chose 40% consumption as a conservative estimate. Similarly, killer whales have been observed killing more monodontids than they eat and only eating part of the carcass (Laidre et al. 2006). Here, we assumed 80% of narwhal and beluga are actually consumed by killer whales. Similar considerations were used to estimate 90% of seals consumed.

The results of our simple predation model estimate that a group of 25 killer whales kill and eat 51 bowhead (45.4 calves and 5.3 subadults), 476 narwhal, 495 beluga, and 550 seals each year. Using summarized population estimates, this mortality represents 1% of the available beluga (57,300; Richard 2005) and less than 1% ringed seals (516,000; Stewart and Lockhart 2005), 9% of the narwhal (5,100; DFO 2009), and 3% of the available bowhead. For bowhead we used the population estimate provided by the International Whaling Commission Scientific Committee, who agreed on a fully corrected strip transect estimate of 1,525 (333–6,990) whales for Hudson Bay–Foxe Basin in 2004 (IWC 2009) using 0.24 correction according to Heide-Jørgensen et al. (2007). An additional consideration is that the bowheads in Foxe Basin and Hudson Bay are part of the larger East Canada-West Greenland bowhead population (COSEWIC 2009) of approximately 6,344 whales. Wade (1998) suggested using 4% as the intrinsic population growth rate for cetaceans which would suggest that killer whale predation in the Hudson Bay region has the potential to limit narwhal population growth. Altering the risk of predation by age classes changes these calculations considerably. If juvenile marine mammal prey is selected, as assumed for bowhead whales, then greater numbers of monodontid whales and seals are needed for killer whale consumption (due to smaller size of prey). To summarize, we consider the formulated predation model to be overly simplistic and based on numerous untested assumptions. The results should not be used as a realistic model of the Hudson Bay killer whale predation system. Instead, we consider these preliminary results necessary to foster interest in gathering the information required to appropriately assess possible impact of killer whale predation on prey populations.

Conclusion

Our research provides confirmation of the added value of combining Inuit Traditional Knowledge with western science in understanding the activities of a cryptic predator living at low densities and capable of moving rapidly across vast scales. Also, combining methods such as sighting reports, photographic identification, acoustic research, and oral knowledge can significantly improve our understanding of complex processes such as the ecosystem effects of killer whale activity in Hudson Bay.

The spatial and temporal occurrence of killer whales is related to predictable and abundant prey resources (Nichol and Shackleton 1996). Killer whale predation has been cited as a potential factor in the decline of several marine mammal populations (Guinet et al. 1992; Keith et al. 2001; Estes et al. 1998; Springer et al. 2003, 2008; but see DeMaster et al. 2006; Wade et al. 2007). We emphasize the need for long-term studies and direct observation of killer whale hunting behaviour to determine the factors that influence prey choice. However, by defining population energetic requirements and considering population demography of prey, we can begin to assess the basic requirements of predator–prey dynamics in Hudson Bay marine ecosystem. We conclude that conservation of marine mammals in Hudson Bay should consider killer whale predation in models of stock assessment for beluga, narwhal, and bowhead whales since killer whales may have the potential to regulate population growth of prey populations.

References

Baird, R.W. 2000. The killer whale – foraging specializations and group hunting. In: J. Mann, R.C. Connor, P.L. Tyack, and H. Whitehead (Eds.) in Cetacean societies: Field studies of dolphins and whales. University of Chicago Press, Chicago.

Baird, R.W., and L.M. Dill. 1995. Occurrence and behavior of transient killer whales -seasonal and pod-specific variability, foraging behavior, and prey handling. Can. J. Zool. 73: 1300–1311.

Baldridge, A. 1972. Killer whales attack and eat a gray whales. J. Mammal. 53: 898–900.

Banfield, A.W. 1974. The mammals of Canada. University of Toronto Press, Toronto, ON.

Barrett-Lennard, L.G., J.K.B. Ford, and K.A. Heise. 1996. The mixed blessing of echolocation: differences in sonar use by fish-eating and mammal-eating killer whales. Anim. Behav. 51: 553–565.

Bigg, MA. 1982. Assessment of killer whale (Orcinus orca) stocks off Vancouver Island, British Columbia. Rep. Intl. Whaling Comm. 32: 655–666.

Burtenshaw, J.C., E.M. Olsen, J.A. Hildebrand, M.A. McDonald, R.K. Andrew, B.M. Howe, and J.A. Mercer. 2004. Acoustically and satellite remote sensing of blue whale seasonality and habitat in the Northeast Pacific. Deep-Sea Res. II 51: 967–986.

Campbell, R.R., D.B. Yurick, and N.B. Snow. 1988. Predation on narwhals, *Monodon monoceros*, by killer whales, *Orcinus orca*, in the eastern Canadian Arctic. Can. Field Nat. 102: 689–696.

Chappell, E. 1817. Narrative of a voyage to Hudson's Bay in his majesty's ship Rosamond containing some account of the northeastern coast of America and of the tribes inhabiting that remote region. Printed for J. Mawman, Ludgate Street by R. Watts, Crown Court, Temple Bar. London.

Clark, S.T., D.K. Odell, and C.T. Lacinak. 2000. Aspects of growth in captive killer whales (*Oricnus orca*). Mar. Mamm. Sci. 16: 110–123.

Condy, P.R., R.J. van Aarde, and M.N. Bester. 1978. The seasonal occurrence and behaviour of killer whales *Orcinus orca*, at Marion Island. J. Zool. 184: 449–464.

Cooke, A., and C. Holland. 1978. The exploration of northern Canada 500 to 1920: a chronology. Arctic History Press, Toronto, ON, Canada.

Cosens, S.E., and A. Blouw. 2003. Size-and-age class segregation of Bowhead Whales summering in northern Foxe Basin: a photogrammetric analysis. Mar. Mamm. Sci. 19: 284–296.

COSEWIC. 2009. COSEWIC assessment and update status report on the Bowhead Whale *Balaena mysticetus*, Bering-Chukchi-Beaufort population and Eastern Canada-West Greenland population, in Canada. Committee on the Status of Endangered Wildlife in Canada. Ottawa, vii + 49 pp. (www.sararegistry.gc.ca/status/status_e.cfm).

Dahlheim, M.E., A. Schulman-Janiger, N. Black, R. Ternullo, D. Ellifrit, and K.C. Balcomb. 2008. Eastern temperate North Pacific offshore killer whales (*Orcinus orca*): occurrence, movements, and insights into feeding ecology. Mar. Mamm. Sci. 24: 719–729.

Davis, R.A., K.J. Finley, and W.J. Richardson. 1980. The present status and future management of arctic marine mammals in Canada. Science Advisory Board of the Northwest Territories. Yellowknife, Northwest Territories, Canada.

Deecke, V.B., J.K.B. Ford, and P.J.B. Slater. 2005. The vocal behaviour of marine mammal-eating killer whales: communicating with costly signals. Anim. Bchav. 69: 395–405.

Degerbøl, M., and P. Freuchen. 1935. Mammals, vol. II, no. 4–5, Report of the fifth Thule expedition 1921–24. Gyldendalske Boghandel, Nordisk Forlag, Copenhagen, Denmark.

DeMaster, D.P., A.W. Trites, P. Clapham, S. Mizroch, P. Wade, R.J. Small, and J. Ver Hoef. 2006. The sequential megafaunal collapse hypothesis: testing with existing data. Prog. Oceanogr. 68: 329–342.

DFO. 2008. Total allowable harvest recommendations for Nunavut narwhal and beluga populations. DFO Can. Sci. Advis. Sec. Sci. Advis. Report No. 2008/035.

Dyke, A.S., J. Hooper, and J.M. Savelle. 1996. A history of sea ice in the Canadian Arctic archipelago based on postglacial remains of the Bowhead Whale (*Balaena mysticetus)*. Arctic 49: 235–255.

Estes, J.A., M.T. Tinker, T.M. Williams, and D.F. Doak. 1998. Killer whale predation on sea otters linking oceanic and nearshore ecosystems. Science 282: 473–476.

Finley, K.J. 1990. Isabella Bay, Baffin Island: an important historical and present-day concentration area for the endangered bowhead whale (*Balaena mysticetus*) of the eastern Canadian Arctic. Arctic 43: 137–152.

Foote, A.D., J. Newton, S.B. Piertney, E. Willerslev, and M.T.P. Gilbert. 2009. Ecological, morphological and genetic divergence of sympatric North Atlantic killer whale populations. Mol. Ecol. 18: 5207–5217.

Ford, J.K.B. 2002. Killer Whale, *Orcinus orca*. In: W.F. Perrin, B. Wursig, and J.G.M. Thewissen (Eds.) Encyclopaedia of marine mammals pp. 669–676. Academic Press, San Diego, CA.

Ford, J.K.B., G.M. Ellis, and K.C. Balcomb. 2000. Killer whales: the natural history and genealogy of *Orcinus orca* in British Columbia and Washington. UBC Press, Vancouver, BC/ University ofWashington Press, Seattle, WA.

Ford, J.K.B., G.M. Ellis, D.R. Matkin, K.C. Balcomb, D. Briggs, and A.B. Morton. 2005. Killer whale attacks on minke whales: prey capture and antipredator tactics. Mar. Mamm. Sci. 21: 603–618.

Garde, E., M.P. Heide-Jorgensen, S.H. Hansen, G. Nachman, and M.C. Forchhammer. 2007. Age-specific growth and remarkable longevity in narwhals (*Monodon* monoceros) from west Greenland as estimated by Aspartic Acid Racemization. J. Mammal. 88: 49–58.

George, J.C., J.R. Bockstoce, A.E. Punt, and D.B. Botkin. 2007. Preliminary estimates of bow-head whale body mass and length from Yankee commercial oil yield records. Paper SC/59/BRG5 presented to the International Whaling Commission Scientific Committee.

Gonzalez, N. 2001. Inuit traditional ecological knowledge of the Hudson Bay narwhal (Tuugaalik) population. Department of Fisheries and Oceans, Iqaluit, Nunavut, Canada.

Guinet C., P. Jouventin, and H. Weimerskirch. 1992. Population changes, movements of southern elephant seals in Crozet and Kerguelen Archipelagos in the last decades. Polar Biol. 12: 349–356.

Hancock, D. 1965. Killer whales kill and eat a minke whale. J. Mammal. 46: 341–342.

Heide-Jørgensen, M.P. 1988. Occurrence and hunting of killer whales in Greenland. Rit Fiskideildar 11: 115–135.

Heide-Jørgensen, M.P, K. Laidre, D. Borchers, F. Samarra, and H. Stern. 2007. Increasing abundance of bowhead whales in West Greenland. Biol. Lett. 3: 577–580.

Heide-Jørgensen, M.P., and J. Teilmann. 1994. Growth, reproduction, age structure and feeding habits of white whales (*Delphinapterus leucas*) in West Greenland waters. Bioscience 39: 195–212.

Hickie, BE, P.S. Ross, R.W. Macdonald, and J.K.B. Ford. 2007. Killer whale (*Orcinus orca*) face protracted health risks associated with lifetime exposure to PCBs. Environ. Sci. Technol. 41: 6613–6619.

Higdon, J.W. 2007. Status of knowledge on killer whales (*Orcinus orca*) in the Canadian Arctic. Fisheries and Oceans Canada Canadian Science Advisory Secretariat Research Document 2007/048. Fisheries and Oceans Canada, Ottawa, ON, Canada.

Higdon, J.W. 2009. Inuit knowledge of killer whales in Foxe Basin, Nunavut. Unpublished report prepared for Steve Ferguson, Fisheries and Oceans Canada, Winnipeg, Manitoba. Available from Steve Ferguson.

Higdon, J.W., and S.H. Ferguson. 2009. Loss of Arctic sea ice causing punctuated change in sightings of killer whales (*Orcinus orca*) over the past century. Ecol. Appl. 19: 1365–1375.

Higdon, J.W., D.D.W. Hauser, and S.H. Ferguson. Submitted. Killer whales in the Canadian Arctic: distribution, prey items, group sizes, and seasonality. Arctic (in review).

Irngaut, I. 1990. IE-112, Archives of the Inullariit Society, Igloolik Research Centre, Igloolik, Nunavut.

IWC. 2009. Report of the Scientific Committee. J. Cetacean Res. Manag. 11(Suppl): 169–192.

Jefferson, T.A., P.F. Stacey, and R.W. Baird. 1991. A review of killer whale interactions with other marine mammals: predation to co-existence. Mamm. Rev. 21: 151–180.

Jensen, D.B., and K.D. Christensen (Eds.). 2003. The biodiversity of Greenland – a county study. Technical Report No. 55. Pinngortitalerifik, Grønlands Naturinstitut.

Johnston, D.H. 1961. Marine mammal survey Hudson Bay 1961. Ontario Department of Lands and Forests, Ottawa, ON, Canada.

Kappianaq, G. 2000. IE-456: Archives of the Inullariit Society. Igloolik Research Centre, Igloolik, Nunavut.

Keith, M., MN. Bester, P.A. Bartlett, D. Baker. 2001. Killer whales (*Orcinus orca*) at Marion Island, Southern Ocean. Afr. Zool. 36: 163–17.

Krafft, B.A., K.M. Kovacs, A.K. Frie, T. Haug, and C. Lydersen. 2006. Growth and population parameters of ringed seals (*Pusa hispida*) from Svalbard, Norway, 2002–2004. ICES J. Mar. Sci. 63: 1136–1144.

Kriete, B. 1995. Bioenergetics in the Killer Whale, *Orcinus orca*. Ph.D. dissertation, University of British Columbia, Vancouver.

Laidre, K.L., M.P. Heide-Jørgensen, and J. Orr. 2006. Reactions of Narwhals, *Monodon monoceros*, to killer whale, *Orcinus orca*, attacks in the eastern Canadian Arctic. Can. Field Nat. 120: 457–465.

Lawson, J., T. Stevens, and D. Snow. 2007. Killer whales of Atlantic Canada, with particular reference to the Newfoundland and Labrador Region. Fisheries and Oceans Canada Canadian

Science Advisory Secretariat Research Document 2007/062. Fisheries and Oceans Canada, Ottawa, ON, Canada.

Lima, S.L., and L.M. Dill. 1990. Behavioral decisions made under the risk of predation: a review and prospectus. Can. J. Zool. 68: 619–640.

Lopez, J.C., and D. Lopez. 1985. Killer whales (*Orcinus orca*) of Patagonia and their behaviour of intentional stranding while hunting nearshore. J. Mammal. 66: 181–183.

Marques, T.A., L. Thomas, J. Ward, and N. DiMarzio. 2009. Estimating cetacean population density using fixed passive acoustic sensors: an example with Blainville's beaked whales. J. Acoust. Soc. Am. 125: 1982–1994.

Martinez, D.R., and E. Klinghammer. 1970. The behavior of the whale *Orcinus orca*: a review of the literature. Zietschrift fur Tierpsychologie 27: 828–839.

Mehta, A.V., J.M. Allen, R. Constantine, C. Garrigue, B. Jann, C. Jenner, M.K. Marx, C.O. Matkin, D.K. Mattila, G. Minton, S.A. Mizroch, C. Olavarria, J. Robbins, K.G. Russell, R.E. Seton, G.H. Steiger, G.A. Vikingsson, P.R. Wade, B.H. Witteveen, and P.J. Clapham. 2007. Baleen whales are not important as prey for killer whales *Orcinus orca* in high latitude regions. Mar. Ecol. Prog. Ser. 348: 297–307.

Milani, D. 1986. Wildlife observation program, Hudson Bay 1985. Canterra Energy Ltd, Calgary, AL, Canada.

Mitchell, E.D., and R.R. Reeves. 1982. Factors affecting abundance of bowhead whales *Balaena mysticetus* in the eastern Arctic of North America, 1915–1980. Biol. Conserv. 22: 59–78.

Mitchell, E., and R.R. Reeves. 1988. Records of killer whales in the western North Atlantic, with emphasis on Canadian waters. Rit Fiskideildar 11: 161–193.

Moore, S.E., and H.P. Huntington. 2008. Arctic marine mammals and climate change: impacts and resilience. Ecol. Appl. 18(Suppl.)/Arctic Mar. Mamm.: S157–S165.

Moshenko, R.W., S.E. Cosens, and T.A. Thomas. 2003. Conservation Strategy for Bowhead Whales (*Balaena mysticetus*) in the Eastern Canadian Arctic. National Recovery Plan No. 24. Recovery of Nationally Endangered Wildlife (RENEW). Ottawa, ON, pp. 51.

Naessig, P.J., and J.M. Lanyon. 2004. Levels and probable origin of predatory scarring on humpback whale (*Megaptera novaeangliae*) in east Australian waters. Wildl. Res. 31: 163–170.

Nichol, L.M., and D.M. Shackleton. 1996. Seasonal movements and foraging behaviour of northern resident killer whales (*Orcinus orca*) in relation to the inshore distribution of salmon (*Oncorhynchus* spp.) in British Columbia. Can. J. Zool. 74: 983–991.

NWMB. 2000. Final report of the Inuit Bowhead Knowledge Study, Nunavut, Canada. Iqaluit, Nunavut: Nunavut Wildlife Management Board, p. 90.

Peterson, R.L. 1966. The mammals of eastern Canada. Oxford University Press, Toronto, Canada.

Pinkerton, J., and D. Doyle. 1805. Pinkerton's geography. Printed for Samuel F, Bradford, Philadelphia, Pennsylvania, USA.

Pitman, R.L., and P. Ensor. 2003. Three forms of killer whales in Antarctic waters. J. Cetacean Res. Manag. 5: 131–139.

Piugaattuk, N. 1994. IE-303: Archives of the Inullariit Society. Igloolik Research Centre, Igloolik, Nunavut.

Rayner, N.A., D.E. Parker, E.B. Horton, C.K. Folland, L.V. Alexander, D.P. Rowell, E.C. Kent, and A. Kaplan. 2003. Global analyses of sea surface temperature, sea ice, and night marine air temperature since the late nineteenth century. J. Geophys. Res. 108(D14): 4407.

Reeves, R.R., and S.E. Cosens. 2003. Historical population characteristics of bowhead whales (*Balaena mysticetus*) in Hudson Bay. Arctic 56: 283–292.

Reeves, R.R., and E. Mitchell. 1988. Distribution and seasonality of killer whales in the eastern Canadian Arctic. Rit Fiskideildar 11: 136–160.

Reeves, R.R., J. Berger, and P.J. Clapham. 2006. Killer whales as predators of large baleen whales and sperm whales. In: J.A. Estes, D.P. DeMaster, D.F. Doak, T.M. Williams, and R.L. Brownell, Jr. (Eds.) Whales, whaling and ocean ecosystems pp. 174–187. University of California Press, Berkeley, CA.

Reeves, R.R, E. Mitchell, A. Mansfield, and M. McLaughlin. 1983. Distribution and migration of the bowhead whale, *Balaena mysticetus*, in the eastern North American Arctic. Arctic 36: 5–64.

Richard, P.R. 2005. An estimation of the Western Hudson Bay beluga population size in 2004. DFO Can. Sci. Advis. Sec. Res. Doc. 2005/017.

Ross, W.G. 1974. Distribution, migration and depletion of bowhead whales in Hudson Bay, 1860 to 1915. Arctic Alpine Res. 6: 85–98.

Sergeant, D.E., and H.D. Fisher. 1957. The smaller cetacea of eastern Canadian waters. J. Fish. Res. Board Can. 14: 83–115.

Silber, G.K., M.W. Newcomer, and H. Perez-Cortez. 1990. Killer whales (*Orcinus orca*) attack and kill a Bryde's whale (*Balaenoptera edeni*). Can. J. Zool. 68: 1603–1606.

Springer, A.M., J.A. Estes, G.B. van Vliet, T.M. Williams, D.F. Doak, E.M. Danner, K.A. Forney, and B. Pfister. 2003. Sequential megafauna collapse in the North Pacific: an ongoing legacy of industrial whaling? Procs. Nat. Acad. Sci. USA 100: 12223–12228.

Springer, A.M., J.A. Estes, G.B. Van Vliet, T.M. Williams, D.F. Doak, E.M. Danner, and B. Pfister. 2008. Mammal eating killer whales, industrial whaling, and the sequential megafaunal collapse in the North Pacific Ocean: a reply to critics of Springer et al. 2003. Mar. Mamm. Sci. 24: 414–442.

Steiger, G.H., J. Calambokidis, J.M. Straley, L.M. Herman, S. Cerchio, D.R. Salden, J. Urbán-R., J.K. Jacobsen, O. von Ziegesar, K.C. Balcomb, C.M. Gabriele, M.E. Dahlheim, S. Uchida, J.K.B. Ford, P.L. de Guevara-P., M. Yamaguchi, and J. Barlow. 2008. Geographic variation in killer whale attacks on humpback whales in the North Pacific: implications for predation pressure. Endangered Species Res. 4: 247–256.

Steiner, W.W., J.H. Hain, H.E. Winn, and P.J. Perkins. 1979. Vocalizations and feeding behaviour of the killer whale (*Orcinus orca*). J. Mammal. 60: 823–827.

Stewart, D.B. 2008. Commercial and subsistence harvests of narwhals (*Monodon monoceros*) from the Canadian eastern Arctic. Arctic Biological Consultants, Winnipeg, MB for Canada Department of Fisheries and Oceans, Winnipeg, MB, ii + 97 p.

Stewart, D.B., and W.L. Lockhart. 2005. An overview of the Hudson Bay marine ecosystem. Canadian Technical Report of Fisheries and Aquatic Sciences no. 2856.

Trites, A.W., and D. Pauly. 1998. Estimating mean body masses of marine mammals from maximum body lengths. Canadian Journal of Zoology 76: 886–896.

Tynan, C.T., and D.P. DeMaster. 1997. Observations and predictions of Arctic climatic change: potential effects on marine mammals. Arctic 50: 308–322.

Visser, I.N. 2000. Orca (*Orcinus orca*) in New Zealand waters. Ph.D. dissertation, University of Auckland, Auckland, New Zealand.

Wade, P.R. 1998. Calculating limits to the human-caused mortality of cetaceans and pinnipeds. Mar. Mamm. Sci. 14: 1–37.

Wade, P.R., V.N. Burkanov, M.E. Dahlheim, N.A. Friday, L.W. Fritz, T.R. Loughlin, S.A. Mizroch, M.M. Muto, D.W. Rice, and L.G. Barrett-Lennard. 2007. Killer whales and marine mammal trends in the north Pacific – a re-examination of evidence for sequential megafauna collapse and the prey switching hypothesis. Mar. Mamm. Sci. 23: 766–802.

Węskawski, J.M., L. Hacquebord, L. Stempniewicz, and M. Malinga. 2000. Greenland whales and walruses in the Svalbard food web before and after exploitation. Oceanologia 42: 37–56.

Westdal, K. 2009. Inuit traditional ecological knowledge of killer whales: community responses at a glance. Unpublished report prepared for Steve Ferguson, Fisheries and Oceans Canada, Winnipeg, Manitoba. Available from Steve Ferguson.

Williams, T.M, J.A. Estes, D.F. Doak, and A.M. Springer. 2004. Killer appetites: assessing the role of predators in ecological communities. Ecology 85: 3373–3384.

Wilson, K.J., J. Falkingham, H. Melling, and R. De Abreu. 2004. Shipping in the Canadian Arctic: other possible climate change scenarios. Geoscience and Remote Sensing Symposium IGARSS '04. Proc. 2004 IEEE Int. 3: 1853–1856.

Young, B.G., E. Chmelnitsky, J.W. Higdon, and S.H. Ferguson. in review. Killer whale photo-identification in the eastern Canadian Arctic. Polar Research.

Hudson Bay Ringed Seal: Ecology in a Warming Climate

M. Chambellant

Abstract Ringed seals (*Phoca hispida*) have evolved to exploit snow covered sea-ice platforms for reproduction and survival and may face critical challenges with ongoing and predicted climate change. The Hudson Bay ecosystem is already showing signs of climate warming raising concerns for the ecological, economical, and culturally-significant ringed seals of Hudson Bay. This chapter summarizes the current knowledge on ringed seals in this subarctic region, including recent findings, and presents the data in regard to current climatic changes. In Hudson Bay, sandlance (*Ammodytes* sp.) is a major component of ringed seal fall diet, whereas Arctic cod (*Boreogadus saida*) representation in the diet is trivial, contrasting results from other Arctic locations. A comparison of density and demographic parameters between the 1990s and the 2000s suggests environmental conditions in the 1990s were not favourable for ringed seals, but improved in the 2000s. A decadal cycle in ringed seal numbers and reproductive performance may relate to variations in environmental conditions, particularly changes in the sea-ice regime. However, ringed seal pups are sensitive to snow cover and ice stability for survival. Thus, a long-term decline of ringed seal fitness in response to current and projected trends in Hudson Bay environmental variables may underlie the natural cycle. Long-term demographic studies of ringed seals at the southern limit of its range are needed to comprehend ringed seal population dynamics and its interaction with environmental variables.

Keywords Ringed seal • Sea ice • Sandlance • Arctic cod • Landfast ice • Diet • Polar bear • Euphausiid • Stable isotope • Capelin • Pregnancy rate • Demography • Range limit

M. Chambellant (✉)
Department of Biological Sciences, University of Manitoba,
Fisheries and Oceans Canada, 501 University Crescent, Winnipeg,
MB R3T 2N6, Canada
e-mail: mchambellant@yahoo.fr

S.H. Ferguson et al. (eds.), *A Little Less Arctic: Top Predators in the World's Largest Northern Inland Sea, Hudson Bay*, DOI 10.1007/978-90-481-9121-5_7,
© Springer Science+Business Media B.V. 2010

Introduction

Ringed seals (*Phoca hispida;* Order Carnivora; Family Phocidae; Picture 1) are among the smallest and more numerous seal species, and have a northern circum-polar distribution (Mansfield 1967; Frost and Lowry 1981). Sexually mature animals use primarily stable land-fast ice with sufficient snow cover to build sub-nivean birth lairs that are critical for pup survival (McLaren 1958; Smith and Stirling 1975; Hammill and Smith 1991). As a species adapted to exploit sea-ice habitat for reproduction and survival (Picture 2), the ringed seal may face critical challenges with predicted climate warming (IPCC [Intergovernmental Panel on Climate Change] 2007). Archaeological and historical studies have reported a decrease in ringed seals during warming/light-ice events (Woollett et al. 2000; see Laidre et al. 2008). More recently, the sensitivity of ringed seals to variations in their sea-ice habitat has also been reported. Early or late ice break-up, heavy/light ice conditions (Smith 1987; Kingsley and Byers 1998; Harwood et al. 2000) and unusual warm and/or rain events in the spring (Stirling and Smith 2004) can negatively affect

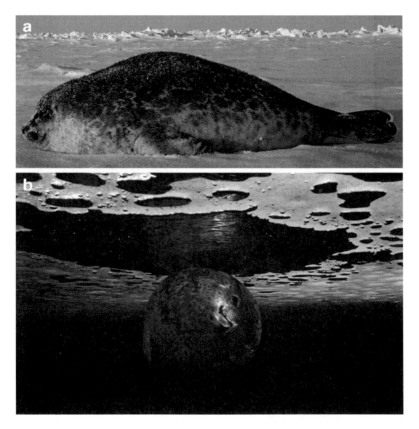

Picture 1 Ringed seal (**a**) on the ice (credit: John Moran) and (**b**) underwater (Credit: Jeremy Stewart-DFO)

Picture 2 Ringed seal winter habitat and breathing holes (*arrows*; Credit: Magaly Chambellant)

ringed seal reproduction and survival in the Arctic. In addition to the regular predation pressure by polar bears (*Ursus maritimus*) and Inuit, ringed seals in a warmer Arctic will likely suffer from habitat loss and shifts in prey availability and distribution. Expected shifts in species distribution could also bring new pressures on ringed seals such as predation by killer whales (*Orcinus orca*; Higdon and Ferguson 2009) or competition with invasive temperate seals such as harbour seals (*Phoca vitulina*; Stirling 2005). Effects of such changes on population health require long-term studies to assess and differentiate climate-induced changes from contemporaneous natural variations (Tynan and DeMaster 1997; Laidre et al. 2008).

In the Hudson Bay ecosystem, ringed seals are at the southern limit of the species' range (Frost and Lowry 1981), but, despite this fact, have seldom been studied in this region. In this chapter, I summarize current knowledge and recent findings on ringed seal population ecology in Hudson Bay and discuss their implications in the context of global warming.

Distribution and Density

Until recently, published information on density and distribution of ringed seals in Hudson Bay has been limited to estimates obtained by aerial surveys conducted in 1974 over the coastal areas of James Bay and southern and western sides of Hudson Bay (Smith 1975); 1978 over the coastal areas of southeastern Hudson Bay (Breton-Provencher 1979); and 1994–1995 over western Hudson Bay (Lunn et al.

1997). In 2007 and 2008, Fisheries and Oceans Canada (DFO) conducted aerial surveys over western Hudson Bay. The resulting data in combination with that collected by Environment Canada (EC) over the period 1995–2000, provide the first time-series for ringed seal density in Hudson Bay.

Both the DFO and EC studies were designed as systematic, strip-transect surveys following Lunn et al. (1997). Ten east-west transects (transects 7–16 of Lunn et al. 1997) were flown at intervals of 15' of north latitude between Churchill, MB (58°47' N; 94°10' W) in the south and Arviat, NU (61°6' N; 94°4' W) in the north, the west coast of Hudson Bay in the west, and 89°W longitude in the east (Fig. 1). Surveys were flown in late spring (26 May–6 June) during the annual moult when ringed seals haul out on the ice to bask in the sun (McLaren 1958; Smith 1973; Picture 3).

In western Hudson Bay, ringed seals favoured the land-fast ice as a prime habitat to haul-out in late spring (density: 1.3–3.4 seals/km^2). When hauled out on pack ice, ringed seals were found in lower densities (0.2–1.8 seals/km^2) and preferred regions of high ice coverage with small cracks. The inclination of ringed seals to use stable and high-ice covered, cracking platforms has already been observed in Hudson Bay (Smith 1975; Lunn et al. 1997) and across the Arctic (Stirling et al. 1977; Kingsley et al. 1985; Kingsley and Stirling 1991).

Ringed seal density estimates varied considerably from year-to-year (0.5–1.6 seals/km^2; Fig. 2) but were in the range of previous estimates (Smith 1975;

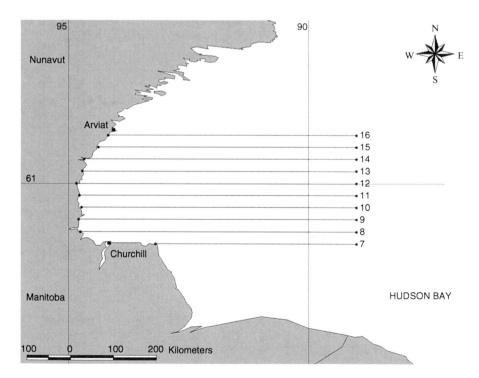

Fig. 1 Study area and transects flown during aerial surveys in western Hudson Bay, 1995–2008. Transect numbers (7–16) refer to Lunn et al. (1997)

Picture 3 Ringed seals hauled-out on the ice (**a**) along a crack and (**b**) around a hole during the spring moult (Credit: Blair Dunn-DFO)

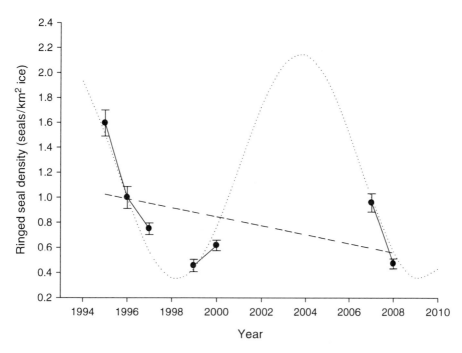

Fig. 2 Ringed seal density in western Hudson Bay estimated by aerial strip survey over the period 1995–2008. Seal density estimates ± standard error. A linear regression (*dashed line*) and a sine function (*dotted line*) were fitted to the data as y = −0.036x + 72.41 and y = 0.9 * sin((2π/11) * (x − 1990)) + 1.25, respectively

Breton-Provencher 1979; Lunn et al. 1997). The overall apparent declining trend was not statistically significant (Fig. 2). However, the decline in estimated density observed from 1995 to 1999 was supported by observations of low pregnancy rates and low percentages of pups in the harvest in the 1990s from the same area (Holst et al. 1999; Stirling 2005). Ferguson et al. (2005) confirmed low pup survival in the 1990s compared to the 1980s, and suggested the existence of a decadal pattern in ringed seal recruitment that reflected environmental cycles. In this study, an 11 year period sine function was a better fit to the density estimates than the linear regression (Residual Sum of Squares $(RSS)_{sine}$ = 0.08 versus RSS_{linear} = 0.75), suggesting that the number of ringed seals in western Hudson Bay may have followed a decadal cycle that could mirrored environmental changes, rather than a linear trend (Fig. 2). The age structure of ringed seals collected in the 2000s (Fig. 3), showing a high proportion of pups and juveniles, is typical of a growing population and would support the increasing number of ringed seals in the early 2000s suggested by the cycle. However, our ability to detect a trend in ringed seal density estimates is limited by the absence of survey data over a 6-year period (2001–2006) and strong interannual variations. Long time series are required to confirm, or not, the decadal cycle and/or the declining trend hypotheses.

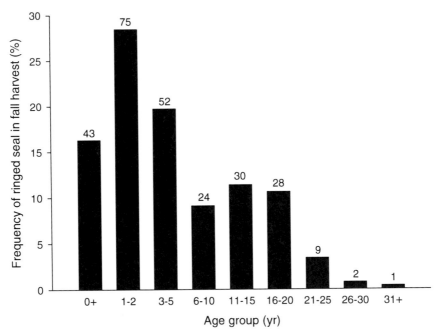

Fig. 3 Age distribution of ringed seals collected in western Hudson Bay from 2003 to 2006. Sample size for each age-class is presented on top of the bars

Growth, Body Condition, and Sexual Maturity

As part of a community-based monitoring program established by DFO Winnipeg, in several Hudson Bay communities, ringed seal samples and measurements were collected by Inuit hunters during their fall subsistence harvest in Arviat, Nunavut, from 2003 to 2006. Ringed seals were aged by reading growth layer groups in the cementum of decalcified, stained, thin canine tooth sections (Stirling et al. 1977). Ovaries and testes were frozen upon collection and later thawed to be weighed to the nearest 0.1 g.

Of the 262 seals examined over the 4-year study, 49.6% were females and 50.4% were males. The sex ratio was not different from unity as reported previously in Hudson Bay (Breton-Provencher 1979; Holst et al. 1999) and elsewhere in the Arctic (Smith 1973; Lydersen and Gjertz 1987; Smith 1987). Mean age of females was higher than that of males (Table 1), a result different from Holst et al. (1999). Maximum ages for females and males were 35 and 27 years, respectively. The relatively low mean population age could be explained by the fact that adults ($\geq$6 years old) represented only 36% of the collection (48% juveniles and 16% pups; Fig. 3). This is a typical age class distribution for ringed seals during the open-water/fall period, when the mating season is over and juveniles are not excluded by adults (McLaren 1958; Smith 1987; Holst et al. 1999). The mean age of ringed seal females collected in the 2000s was not different from females collected in the early 1990s at the same location (Holst et al. 1999), but males were significantly younger in the 2000s ($t_{176} = 3.27$, $P = 0.0013$).

Growth in length (Fig. 4) and mass (Fig. 5) of male and female ringed seals were estimated by Gompertz growth curves with three parameters. There was no sexual dimorphism in length, mass or body condition (fat depth) in western Hudson Bay ringed seals (Table 1). But, since the females were significantly older, the males may be slightly larger at a given age (Breton-Provencher 1979; McLaren 1993). Compared with other locations in the Arctic, ringed seals in Hudson Bay were

Table 1 Age of ringed seals, and standard length, mass and fat depth of adult ($\geq$6 years) male and female ringed seals collected in western Hudson Bay from 2003 to 2006. Mean $\pm$ SD (n) [range]

	Male	Female	P
Age (year)	5.4 $\pm$ 6.5 (128)	7.6 $\pm$ 7.9 (127)	0.027[a]
	[0 – 27]	[0 – 35]	
Standard length (cm)	120.2 $\pm$ 12.5 (40)	120.4 $\pm$ 13 (49)	0.94[b]
	[86 – 144.8]	[86.4 – 147.3]	
Mass (kg)	46.7 $\pm$ 17.5 (15)	48.8 $\pm$ 12.6 (24)	0.67[b]
	[22 – 75]	[21 – 72]	
Fat depth (cm)	5.6 $\pm$ 1.2 (38)	5.4 $\pm$ 1.5 (50)	0.57[b]
	[2.5 – 9.3]	[2 – 8]	

[a] P value from Mann–Whitney test.

[b] P value from 2-tail t-test.

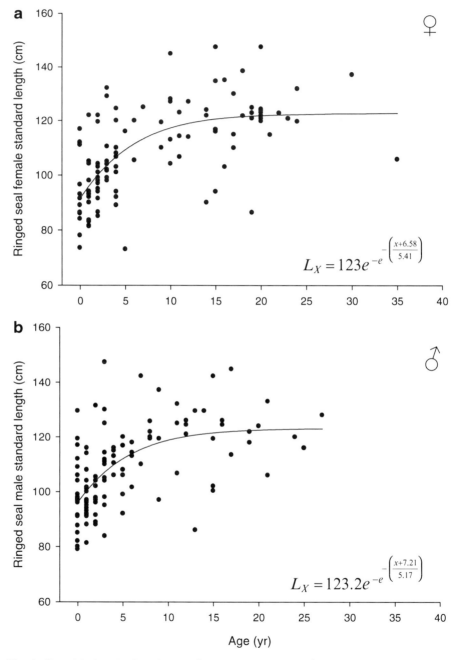

Fig. 4 Growth in length of (**a**) female (r^2 = 0.46) and (**b**) male (r^2 = 0.39) ringed seals collected in western Hudson Bay from 2003 to 2006

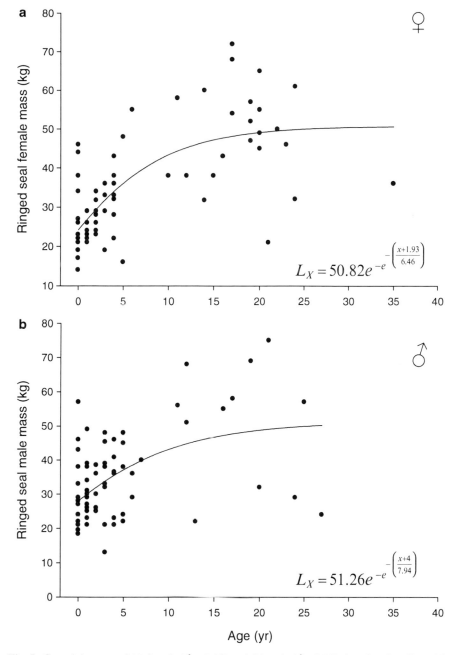

Fig. 5 Growth in mass of (**a**) female ($r^2 = 0.49$) and (**b**) male ($r^2 = 0.26$) ringed seals collected in western Hudson Bay from 2003 to 2006

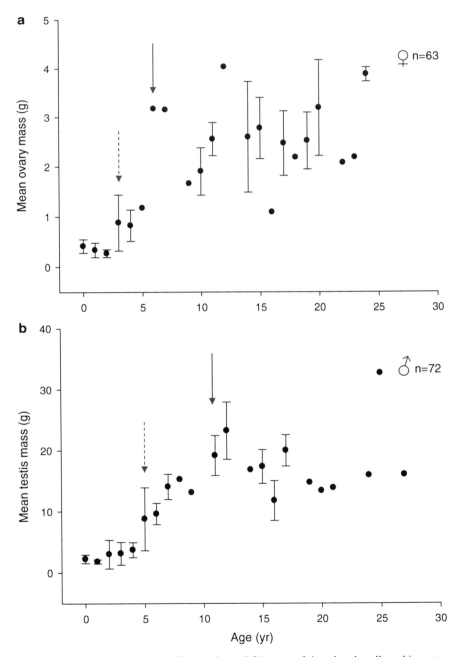

Fig. 6 Change over time in mass of (**a**) ovaries and (**b**) testes of ringed seals collected in western Hudson Bay from 2003 to 2006. Mean ± SD. The dashed-line arrow indicates physiological sexual maturity and the solid-line arrow indicates the behavioural sexual maturity

smaller both in length and mass supporting the hypothesis of latitudinal size differences (McLaren 1993; Holst and Stirling 2002; Krafft et al. 2006).

Body condition, measured by fat depth (cm), of ringed seals collected from 2003 to 2006 (5.5 ± 1.4 cm, $n = 88$) was similar to that reported in the 1990s from the same location (Stirling 2005).

In Hudson Bay, ovary mass increased dramatically around the age of 6 years but, increases could be detected as early as 3 years of age (Fig. 6a). Male ringed seals seemed to reach sexual maturity around 5 years of age, but testes mass continued to increase until 10–11 years of age (Fig. 6b). Both sexes appear to mature physically before they reach behavioural sexual maturity. The ages at maturity are in accordance with those of other ringed seals in Hudson Bay (Breton-Provencher 1979; Holst et al. 1999) and in the rest of the Arctic (Kingsley and Byers 1998; Holst and Stirling 2002; Krafft et al. 2006), albeit at the lower end of the spectrum.

Reproduction and Survival

The reproductive cycle of ringed seals in Hudson Bay has yet to be described but is likely similar to other Arctic locations. Pups are born on land-fast or stable pack ice in sub-nivean lairs that require a snow depth on the ground of 20 cm or more (Smith and Stirling 1975; Furgal et al. 1996; Ferguson et al. 2005) to provide sufficient protection against Arctic weather and predators (Smith and Stirling 1975; Hammill and Smith 1991; Smith et al. 1991; Picture 4). Nominal birth date for pups is set on the 1st of April but the pupping period could be spread over several weeks (McLaren 1958; Smith 1973, 1987). In Hudson Bay,

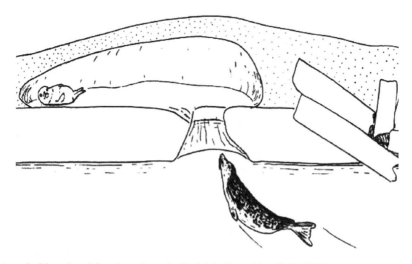

Picture 4   Ringed seal female and pup in birth lair (from Mansfield 1967)

traditional knowledge and recent data converge to an earlier pupping period, starting in February and peaking around mid-March (McDonald et al. 1997; Cleator 2001; M. Chambellant 2009). This supports the hypothesis of a latitudinal gradient of pupping suggested by McLaren (1958) and Smith (1987). Pups are weaned before break-up, after nursing for 5–7 weeks (Hammill et al. 1991; Lydersen and Hammill 1993).

Mating is thought to take place underwater around the time of weaning (Smith 1987). Breton-Provencher (1979) reported a peak of male sexual activity from February to April. Ringed seal gestation lasts around 10.5 months, with a period of suspended development during the first 2–3 months (McLaren 1958; Smith 1973, 1987; M. Chambellant 2009). In late spring, ringed seals undertake their annual moult and require an ice platform to haul-out (McLaren 1958).

Body condition of ringed seals is poorest in early summer after fasting during the breeding and moulting periods (McLaren 1958; Breton-Provencher 1979; Ryg et al. 1990). During the open water period all age-classes are mixed and feed intensively (McLaren 1958; Smith 1987).

When the ice starts to form in late fall, adults gather close to shore to establish territories (McLaren 1958; Smith 1987). During this period, juveniles are actively excluded from these habitats (McLaren 1958; Smith 1987; Holst et al. 1999; Krafft et al. 2007). Adult ringed seals show signs of site fidelity during the winter months (McLaren 1958; Smith and Hammill 1981; Boveng et al. 2009; S.P. Luque 2009), and may have a weakly polygynous, resource-defence mating system (Smith and Hammill 1981; Krafft et al. 2007; D. Yurkowski 2009).

Ovulation and pregnancy rates of adult females in western Hudson Bay in 2003, 2005 and 2006 were 100% ($n = 36$). This result contrasts with values obtained in the 1990s of 92% and 55% ($n = 100$) for ovulation and pregnancy rates, respectively (Holst et al. 1999; Stirling 2005). The low pregnancy rate in the 1990s could be an artefact of the 1992 collection, when no pregnant females were found (Holst et al. 1999; Stirling 2005). The percentage of pups in the harvest was low in the 1990s (9.6%) compared to the 2000s (16.3%; Fig. 3). More juveniles, especially 1 and 2 year olds, were also collected in the 2000s than in the early 1990s (Holst et al. 1999).

These results could suggest that environmental conditions in the 1990s were not favourable resulting in low pup production, low young survival and/or emigration of young animals toward more suitable regions. In the 2000s, this trend seemed to have reversed with better pup production and survival, and/or immigration of young animals from other areas. Ferguson et al. (2005) confirmed that recruitment was generally low in the 1990s compared to the 1980s, and hypothesized a decadal cycle in recruitment, consistent with data from the 2000s presented here. Change in marine productivity, prey distribution and/or availability resulting from a longer open water season, the detrimental effect on unweaned or freshly weaned pups of the loss of the ice platform earlier in the season, and the lack of sufficient snow depth to build strong and protective lairs, have been suggested as factors involved in the low pup production and/or survival in western Hudson Bay in the 1990s (Ferguson et al. 2005; Stirling 2005).

Feeding Habits

Information on ringed seal feeding habits in Hudson Bay is scarce. As in other regions of the Arctic, ringed seals in Hudson Bay are thought to feed year-round, but with intensive feeding in late summer and fall, as shown by the increase in fat depth measurements in the fall (McLaren 1958; Breton-Provencher 1979; Ryg et al. 1990). During the open water period, ringed seals of all age classes mix together and form large feeding aggregations (McLaren 1958; Smith 1987; Harwood and Stirling 1992). Diet composition varies greatly with geographical location, season and life-stage, but Arctic cod *(Boreogadus saida)* and invertebrates such as mysids (Mysida), amphipods (Amphipoda) and euphasiids (Euphausiacea) are common prey (McLaren 1958; Breton-Provencher 1979; Lowry et al. 1980; Bradstreet and Finley 1983; Gjertz and Lydersen 1986; Smith 1987; Weslawski et al. 1994; Siegstad et al. 1998; Wathne et al. 2000; Holst et al. 2001; Labansen et al. 2007).

In southeastern Hudson Bay, the hyperiid amphipod *Parathemisto libellula.* euphasiids and the pelagic fish sandlance (*Ammodytes* sp.) were major prey of ringed seals, but Arctic cod were absent from the 218 stomach contents analyzed (Breton-Provencher 1979; Picture 5). In western Hudson Bay, 93% of the otoliths found in the stomach contents of ringed seals collected from 1998 to 2000 were from sandlance and 6% from Arctic cod (Stirling 2005).

Stomach content analyses provide qualitative and quantitative information on diet although differences in the digestion rates of large and small, hard and soft prey can bias diet estimation (Iverson et al. 2004). Stomach contents also represent only prey ingested shortly before death. Consequently, indirect methods, like stable isotope (SI) analysis, have been developed to determine diets.

Stable isotope analysis is based on the natural occurrence of different isotopes of the same element and their differential fractionation during biological processes (Kendall et al. 1995; Kendall and Caldwell 1998). The differential fractionation of carbon (C) SI (^{13}C and ^{12}C) during photosynthesis confers phytoplankton with a unique carbon-isotopic signature that is passed on almost directly to consumers. In diet studies of marine mammals, carbon SI ratio ($^{13}C/^{12}C$) gives indication of foraging behaviour (e.g., benthic versus pelagic) and locations (if feeding occurs in isotopically different water masses; Hirons et al. 2001; Lesage et al. 2001; Kurle and Worthy 2002). The differential fractionation of the nitrogen (N) isotopes, ^{15}N and ^{14}N, occurs in consumers that preferentially integrate the heavier isotope to their tissues, resulting in an enrichment of 2–5‰ from diet to consumer through the food chain (Hobson and Welch 1992; Hobson et al. 1996; Lesage et al. 2002). In diet studies of marine ecosystems, nitrogen SI ratio ($^{15}N/^{14}N$) provides information on the relative place a given organism occupies in the food chain (Hobson and Welch 1992). Due to the specific metabolic rate of each tissue, looking at SI ratios in different tissues of the same animal provides dietary information of food assimilated over a range of time scales (Kurle and Worthy 2002; Lesage et al. 2002). Typically, tissues with a high metabolic rate represent a time scale of days or weeks (liver, kidney, serum), whereas tissues with a low metabolic rate represent food ingested months before (muscle, red blood cells).

Picture 5 Ringed seal stomach contents showing (**a**) Sand Lance (*Ammodytes* sp.), (**b**) hyperiid amphipod (*Parathemisto libellula*) and (**c**) euphausiids (*Tysanoessa* sp.) (Credit: DFO)

In southeastern Hudson Bay, stomach contents of ringed seals collected during the Inuit fall and early winter (October–January) subsistence harvest from 2003 to 2005, confirmed the importance of amphipods and sandlance in the diet of ringed seals around the Belcher Islands (Fig. 7). However, capelin (*Mallotus villosus*) and mysids represented important prey as well, and Arctic cod were present in more than 20% of the stomachs, which contrasts with results from the late 1970s (Breton-Provencher 1979). Most (95%) of the energy acquired came from fish, including 54% from capelin (M. Chambellant 2009).

Results from stomach contents of ringed seals collected during the Inuit fall (October) subsistence harvest in western Hudson Bay from 2003 to 2005 confirmed that sandlance is the main food resource of ringed seals there at that time of the year (Fig. 8). The relative contribution of invertebrates was small and that of Arctic cod insignificant (Fig. 8). In a traditional knowledge study conducted in western Hudson Bay, nine of ten Inuk hunters reported finding amphipods in ringed seal stomachs (Cleator 2001). Capelin was mentioned by six hunters and sandlance and

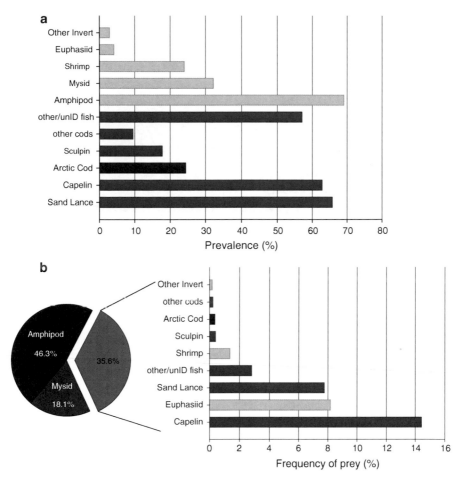

Fig. 7 Diet of ringed seals expressed as (**a**) prevalence and (**b**) frequency of prey found in stomachs collected in southeastern Hudson Bay from 2003 to 2005. Prevalence was defined as the number of stomachs containing a given prey item divided by the total number of stomach examined

Arctic cod by five. Other prey mentioned were Greenland cod (*Gadus ogac*), sculpin (Scorpaenifromes) and Arctic char (*Salvelinus alpinus*). Mysids, euphausiids, shrimps (Decapoda) and snailfish (*Liparis* sp.) were not identified by the hunters as prey of ringed seals. Contribution of invertebrates to the energy acquisition by ringed seals represented only 0.04% while sandlance contributed 65% (M. Chambellant 2009).

The mean carbon and nitrogen SI ratios observed in muscle and liver collected in western Hudson Bay in 2004 and 2006 provided some insights into ringed seal diet over a longer time frame (Fig. 9). The carbon SI signature of ringed seal tissues could indicate that they fed on a mixed diet of pelagic and benthic/epi-benthic organisms. The nitrogen SI signature indicated that cods in general, and Arctic cod

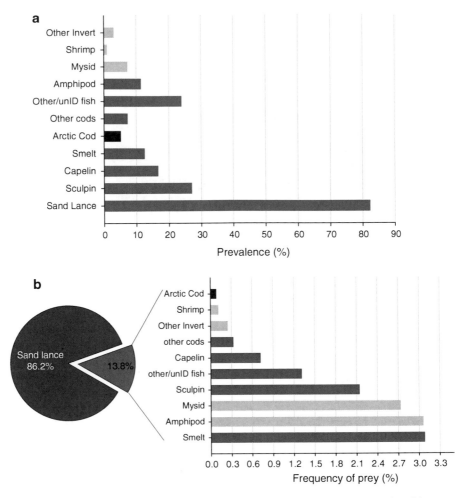

Fig. 8 Diet of ringed seals expressed as (**a**) prevalence and (**b**) frequency of prey found in stomachs collected in western Hudson Bay from 2003 to 2005. Prevalence was defined as the number of stomachs containing a given prey item divided by the total number of stomach examined

in particular, did not seem to be part of the diet. The muscle nitrogen SI signature suggested that ringed seals diet in early summer may be based on amphipods whereas in the fall (liver signature), the SI analyses supported the importance of sandlance in this region.

Previous and present results indicated that sandlance is, and has been, a major component of ringed seal fall diet in Hudson Bay at least since the 1980s (Breton-Provencher 1979; Stirling 2005; this study). Arctic cod consumption however seems to be trivial in Hudson Bay compared to other Arctic locations, where more Arctic cod than any other fish were found in ringed seals stomachs (Northern Foxe Basin and southwestern Baffin Island: McLaren 1958; western Canadian Arctic: Smith 1987;

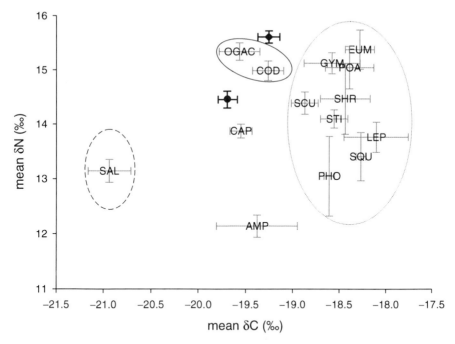

Fig. 9 Mean carbon (C) and nitrogen (N) stable isotope ratios (dC and dN) in ringed seal muscle (*dot*) and liver (*diamond*) collected in western Hudson Bay in 2004 and 2006 and potential prey species (*grey error bars*) collected in Hudson Bay in 2006 and 2007. Circles indicate groupings of animals from the same ecological background (pelagic: *dashed line*, benthic: *dotted line* and epibenthic/cods: *solid line*). SAL: sandlance (*Ammodytes sp.*); AMP: amphipods (*Parathemisto libellula*); CAP: capelin (*Mallotus villosus*); PHO: banded gunnel (*Pholis fasciata*); SQU: squids; LEP: daubbed shanny (*Leptoclinus maculatus*); STI: Arctic shanny (*Stichaeus punctatus*); SHR: shrimps; SCU: sculpins (*Triglops sp.*); POA: Atlantic poacher (*Leptagonus decagonus*); EUM: fourline snakeblenny (*Eumesogrammus praecisus*); GYM: fish doctor (eelpout; *Gymnelus viridis*); COD: Arctic cod (*Boreogadus saida*); OGAC: Greenland cod (*Gadus ogac*)

high Canadian Arctic: Bradstreet and Finley 1983; Barents Sea: Wathne et al. 2000; Greenland: Siegstad et al. 1998). Stomach content analysis from ringed seal inhabiting Ungava Bay and the northern coast of Labrador revealed that sandlance was the most abundant fish preyed upon, and that Arctic cod number in stomachs was anecdotal (McLaren 1958). These dietary differences support the existence of a latitudinal gradient in ringed seal feeding habits, based on preference and/or availability of fish, as has been suggested by McLaren (1958) and Siegstad et al. (1998).

Conclusion

Changes in the Hudson Bay ecosystem have been occurring over the last 3 decades. Surface air temperatures in spring (Skinner et al. 1998) and the length of the ice-free period (Gagnon and Gough 2005b) have increased significantly, whereas sea-ice extent (Parkinson and Cavalieri 2008) and snow depth (Ferguson et al. 2005) have decreased. Sea ice break-up in western Hudson Bay is now occurring ~3 weeks earlier than in the 1970s (Gagnon and Gough 2005b; Stirling and Parkinson 2006) and climate change scenarios for the Hudson Bay region predict that trends observed in recent years will continue (Gough and Wolfe 2001; Gagnon and Gough 2005a).

The consequences of a shorter period of ice cover on marine, ice-associated species have been observed in Hudson Bay. At the end of the 1990s, the reduction in the mid-July sea-ice cover in northern Hudson Bay was correlated with a regime transition, wherein the prey thick-billed murres (*Uria lomvia*) brought back to their chicks shifted away from species typical of Arctic waters (e.g., Arctic cod) and toward those typical of subarctic waters (e.g., capelin; Gaston et al. 2003). As well, the body condition, reproduction rate, cub survival and abundance of polar bears in western Hudson Bay have declined over the last 25 years, and this trend has been correlated to earlier break-up of sea ice (Stirling et al. 1999; Regehr et al. 2007). Ringed seals in Hudson Bay have received less research and only recently have been the focus of dedicated studies, and concerns in the context of the ongoing climate warming have been raised (Ferguson et al. 2005; Stirling 2005).

In this chapter, I reviewed our current knowledge of ringed seals in Hudson Bay. A decline in ringed seal density estimates in western Hudson Bay occurred from 1995 to 1999 but was not statistically significant when the 2000s survey results were included. As Ferguson et al. (2005) suggested for recruitment, a decadal cycle in ringed seal abundance has been hypothesized, with low number of seals in western Hudson Bay in the late 1990s and a peak in the mid-2000s. This result suggested environmental conditions were not favourable for ringed seals in the 1990s but changed positively in the 2000s. The positive change in pregnancy rates and pup recruitment that occurred at the beginning of the 2000s supported this hypothesis.

In the Arctic environment, decadal fluctuations, particularly in the ice regime through atmospheric forcing (e.g., North Atlantic Oscillation) have been observed (Mysak and Manak 1989; Hurrell 1995; Mysak et al. 1996). Variations in life-history parameters of several Arctic species have been linked to climatic variations (Skinner et al. 1998; Ottersen et al. 2001; Post and Forchhammer 2002; Stirling 2002; Derocher 2005; Ferguson et al. 2005; Regehr et al. 2007). In western Hudson Bay, snow depth has been identified as a key factor in ringed seal recruitment but so far, correlations with environmental variables remained unclear and/or statistically not significant (Ferguson et al. 2005; M. Chambellant 2009). However, exceptionally cold and heavy ice conditions were recorded over the eastern Arctic at the beginning of the 1990s (McCormick et al. 1995; Mysak et al. 1996; Gough et al. 2004) and could have triggered the decline in number and demographic parameters of ringed seals in Hudson Bay, as occurred in the western Arctic in the 1970s and 1980s (Smith 1987; Kingsley and Byers 1998; Stirling 2002).

Ringed seals are sensitive to specific environmental factors, like snow depth and ice stability, for reproduction and survival. Thus, if current environmental trends continue in Hudson Bay, as projected, a long term decline of ringed seals might be underlying the natural decadal cycle. In fact, while the percentage of pups in the harvest in western Hudson Bay has almost doubled in the 2000s relative to the 1990s, the absolute number (16%) is still low when compared to other Arctic locations with similar ovulation/pregnancy rates (Smith 1973; Breton-Provencher 1979; Smith 1987). Supporting a possible decline in ringed seal numbers in western Hudson Bay is the declines in population parameters of polar bears over the past 25 years (Stirling et al. 1999; Regehr et al. 2007). Indeed, polar bear diet relies heavily on ringed seals, and especially on ringed seal pups, for reproductive success and survival and a positive linear relationship was found between ringed seal and polar bear population estimates (Stirling and Oritsland 1995).

The impacts of rapid and unidirectional climatic changes on ringed seals are not yet explicit. In order to better comprehend ringed seal population dynamic and its interaction with environmental variables, and be able to assess and predict effects of current environmental trends, long time-series of density, demographic and dietary data are needed. Intense research efforts are particularly critical in Hudson Bay, a subarctic region where climate change is expected to occur first and faster (Walsh 2008) and, where ringed seals occur at the southern limit of their range.

References

Boveng, P.L., London, J.M., Cameron, M.F., and Kelly, B.P. 2009. Long-term satellite telemetry records provide new insights about ice-associated seal habitat preference and breeding site fidelity. Poster *in* 18th Biennial Conference on the Biology of Marine Mammals, Quebec City, QC, 12–16 Oct. 2009.

Bradstreet, M.S.W., and Finley, K.J. 1983. Diet of ringed seals (*Phoca hispida*) in the Canadian high Arctic. LGL Limited, Environmental Research Associates, 44 Eglington Ave. W., Toronto, ON, M4R 1A1, Calgary, Alberta, 36 pp.

Breton-Provencher, M. 1979. Etude de la population de phoques annelés (*Phoca hispida*) et des autres pinnipèdes de la région de Poste-de-la-baleine (Nouveau-Québec). GIROQ, Rapport à l'hydro-Québec, Projet Grande-Baleine (Mandat d'avant-projet préliminaire OGB/76-1), 148 pp.

Cleator, H. 2001. Traditional knowledge study of ringed seals: a transcript of interviews with hunters from Chesterfield Inlet, Nunavut. Fisheries and Oceans Canada, Central and Arctic Region, Winnipeg, MB, R3T 2N6, 108 pp.

Derocher, A.E. 2005. Population ecology of polar bears at Svalbard, Norway. Popul. Ecol. 47: 267–275.

Ferguson, S.H., Stirling, I., and McLoughlin, P. 2005. Climate change and ringed seal (*Phoca hispida*) recruitment in western Hudson Bay. Mar. Mamm. Sci. 21(1): 121–135.

Frost, K.J., and Lowry, L.F. 1981. Ringed, Baikal and Caspian seals *Phoca hispida* Schreber, 1775; *Phoca sibirica* Gmelin, 1788 and *Phoca caspica* Gmelin, 1788. In S.H. Ridgway and R.J. Harrison (Eds.) Handbook of marine mammals, pp. 29–53. F.R.S. Academic Press, London/New York/Toronto/Sydney/San Francisco.

Furgal, C.M., Innes, S., and Kovacs, K.M. 1996. Characteristics of ringed seal, *Phoca hispida*, subnivean structures and breeding habitat and their effect on predation. Can. J. Zool. 74: 858–874.

Gagnon, A.S., and Gough, W.A. 2005a. Climate change scenarios for the Hudson Bay region: an intermodel comparison. Climat. Change 69: 269–297.

Gagnon, A.S., and Gough, W.A. 2005b. Trends in the dates of ice freeze-up and break-up over Hudson Bay, Canada. Arctic 58(4): 370–382.

Gaston, A.J., Woo, K., and Hipfner, J.M. 2003. Trends in forage fish populations in Northern Hudson Bay since 1981, as determined from the diet of nestling thick-billed Murres, *Uria lomvia*. Arctic 56(3): 227–233.

Gjertz, I., and Lydersen, C. 1986. The ringed seal (*Phoca hispida*) spring diet in northwestern Spitsbergen, Svalbard. Polar Res. 4: 53–56.

Gough, W.A., Cornwell, A.R., and Tsuji, L.J.S. 2004. Trends in seasonal sea ice duration in southwestern Hudson Bay. Arctic 57(3): 299–305.

Gough, W.A., and Wolfe, E. 2001. Climate change scenarios for Hudson Bay, Canada, from general circulation models. Arctic 54(2): 142–148.

Hammill, M.O., Lydersen, C., Ryg, M., and Smith, T.G. 1991. Lactation in the ringed seal (*Phoca hispida*). Can. J. Fish. Aquat. Sci. 48: 2471–2476.

Hammill, M.O., and Smith, T.G. 1991. The role of predation in the ecology of ringed seal in Barrow strait, Northwest Territories, Canada. Mar. Mamm. Sci. 7(2): 123–135.

Harwood, L.A., Smith, T.G., and Melling, H. 2000. Variation in reproduction and body condition of the ringed seal (*Phoca hipsida*) in the Western Prince Albert sound, NT, Canada, as assessed through a harvest-based sampling program. Arctic 53(4): 422–431.

Harwood, L.A., and Stirling, I. 1992. Distribution of ringed seals in the southeastern Beaufort Sea during late summer. Can. J. Zool. 70: 891–900.

Higdon, J.W., and Ferguson, S.H. 2009. Loss of Arctic sea ice causing punctuated change in sightings of killer whales (*Orcinus orca*) over the past century. Ecol. Appl. 19(5): 1365–1375.

Hirons, A.C., Schell, D.M., and Finney, B.P. 2001. Temporal records of $d^{13}C$ and $d^{15}N$ in North Pacific pinnipeds: inferences regarding environmental change and diet. Oecologia 129: 591–601.

Hobson, K.A., Schell, D.M., Renouf, D., and Noseworthy, E. 1996. Stable carbon and nitrogen isotopic fractionation between diet and tissues of captive seals: implications for dietary reconstructions involving marine mammals. Can. J. Fish. Aquat. Sci. 53: 528–533.

Hobson, K.A., and Welch, H.E. 1992. Determination of trophic relationships within a high Arctic marine food web using $d^{13}C$ and $d^{15}N$ analysis. Mar. Ecol. Prog. Ser. 84: 9–18.

Holst, M., and Stirling, I. 2002. A comparison of ringed seal (*Phoca hispida*) biology on the east and west sides of the North water Polynya, Baffin Bay. Aquat. Mammal. 28(3): 221–230.

Holst, M., Stirling, I., and Calvert, W. 1999. Age structure and reproductive rates of ringed seals (*Phoca hispida*) on the Northwestern coast of Hudson Bay in 1991 and 1992. Mar. Mamm. Sci. 15(4): 1357–1364.

Holst, M., Stirling, I., and Hobson, K.A. 2001. Diet of ringed seals (*Phoca hispida*) on the east and west sides of the north water polynya, Northern Baffin Bay. Mar. Mamm. Sci. 17(4): 888–908.

Hurrell, J.W. 1995. Decadal trends in the North Atlantic oscillation: regional temperatures and precipitation. Science 269: 676–679.

IPCC (Intergovernmental Panel on Climate Change). 2007. Climate Change 2007: Synthesis Report, p. 104, R.K. Pachauri and A. Reisinger (Eds.). http://www.ipcc.ch/ipccreports/ar4-syr.htm

Iverson, S.J., Field, C., Bowen, W.D., and Blanchard, W. 2004. Quantitative fatty acid signature analysis: a new method of estimating predator diets. Ecol. Monogr. 72(2): 211–235.

Kendall, C., and Caldwell, E.A. 1998. Fundamentals of isotope geochemistry. In C. Kendall and J.J. McDonnell (Eds.) Isotope tracers in catchment hydrology, pp. 51–86. Elsevier Science, Amsterdam.

Kendall, C., Sklash, M.G., and Bullen, T.D. 1995. Isotopes tracers of water and solute sources in catchments. In S.T. Trudgill (Ed.) Solute modelling in catchment systems, pp. 261–303. Wiley, New York.

Kingsley, M.C.S., and Byers, T.J. 1998. Failure in reproduction of ringed seals (*Phoca hispida*) in Amunsden Gulf, Nortwest Territories in 1984–1987. In M.P. Heide-Jorgensen and C. Lydersen (Eds.) Ringed seals in the North Atlantic, pp. 197–210. The North Atlantic Marine Mammal Commission, Tromso.

Kingsley, M.C.S., and Stirling, I. 1991. Haul-out behaviour of ringed and bearded seals in relation to defence against surface predators. Can. J. Zool. 69: 1857–1861.

Kingsley, M.C.S., Stirling, I., and Calvert, W. 1985. The distribution and abundance of seals in the Canadian high Arctic, 1980–82. Can. J. Fish. Aquat. Sci. 42: 1189–1210.

Krafft, B.A., Kovacs, K.M., Frie, A.K., Haug, T., and Lydersen, C. 2006. Growth and population parameters of ringed seals (*Pusa hispida*) from Svalbard, Norway, 2002–2004. ICES J. Mar. Sci. 63: 1136–1144.

Krafft, B.A., Kovacs, K.M., and Lydersen, C. 2007. Distribution of sex and age groups of ringed seals *Pusa hispida* in the fast-ice breeding habitat of Kongsfjorden, Svalbard. Mar. Ecol. Prog. Ser. 335: 199–206.

Kurle, C.M., and Worthy, G.A.J. 2002. Stable nitrogen and carbon isotope ratios in multiple tissues of the northern fur seal *Callorhinus ursinus*: implications for dietary and migratory reconstructions. Mar. Ecol. Prog. Ser. 236: 289–300.

Labansen, A.L., Lydersen, C., Haug, T., and Kovacs, K.M. 2007. Spring diet of ringed seals (*Phoca hispida*) from northwestern Spitsbergen, Norway. ICES J. Mar. Sci. 64(6): 1246–1256.

Laidre, K.L., Stirling, I., Lowry, L.F., Wiig, O., Heide-Jorgensen, M.P., and Ferguson, S.H. 2008. Quantifying the sensitivity of Arctic marine mammals to climate-induced habitat change. Ecol. Appl. 18(Suppl. 2): S97–S125.

Lesage, V., Hammill, M.O., and Kovacs, K.M. 2001. Marine mammals and the community structure of the estuary and Gulf of St Lawrence, Canada: evidence from stable isotope analysis. Mar. Ecol. Prog. Ser. 210: 203–221.

Lesage, V., Hammill, M.O., and Kovacs, K.M. 2002. Diet-tissue fractionation of stable carbon and stable isotopes in phocid seals. Mar. Mamm. Sci. 18(1): 182–193.

Lowry, L.F.K., Frost, J., and Burns, J.J. 1980. Variability in the diet of ringed seals, *Phoca hispida*, in Alaska. Can. J. Fish. Aquat. Sci. 37: 2254–2261.

Lunn, N.J., Stirling, I., and Nowicki, S.N. 1997. Distribution and abundance of ringed (*Phoca hispida*) and bearded seals (*Erignathus barbatus*) in western Hudson Bay. Can. J. Fish. Aquat. Sci. 54: 914–921.

Lydersen, C., and Gjertz, I. 1987. Population parameters of ringed seals (*Phoca hispida* Schreber, 1775) in the Svalbard area. Can. J. Zool. 65: 1021–1027.

Lydersen, C., and Hammill, M.O. 1993. Diving in ringed seal (*Phoca hispida*) pups during the nursing period. Can. J. Zool. 71: 991–996.

Mansfield, A.W. 1967. Seals of arctic and eastern Canada. Fisheries Research Board of Canada, Bulletin 137, p. 35.

McCormick, M.P., Thomason, L.W., and Trepte, C.R. 1995. Atmospheric effects of the Mt Pinatubo eruption. Nature 373: 399–404.

McDonald, M., Arragutainaq, L., and Novalinga, Z. 1997. Voices from the Bay: traditional ecological knowledge of Inuit and Cree in the Hudson Bay bioregion. Canadian Arctic Resources Committee, Environmental Committee of the municipality of Sanikiluaq, 97 pp.

McLaren, I.A. 1958. The biology of the ringed seal (*Phoca hispida* Schreber) in the eastern Canadian Arctic. Fisheries Research Board of Canada, Bulletin 118, 97 pp.

McLaren, I.A. 1993. Growth in pinnipeds. Biol. Rev. 68: 1–79.

Mysak, L.A., Ingram, R.G., Wang, J., and Van der Baaren, A. 1996. The anomalous sea-ice extent in Hudson Bay, Baffin Bay and the Labrador sea during three simultaneous NAO and ENSO episodes. Atmos. Ocean 34(2): 313–343.

Mysak, L.A., and Manak, D.K. 1989. Arctic sea-ice extent and anomalies, 1953–1984. Atmos. Ocean 27(2): 376–405.

Ottersen, G., Planque, B., Belgrano, A., Post, E., Reid, P.C., and Stenseth, N.C. 2001. Ecological effects of the North Atlantic Oscillation. Oecologia 128: 1–14.

Parkinson, C.L., and Cavalieri, D.J. 2008. Arctic sea ice variability and trends, 1979–2006. J. Geophys. Res. 113: C07003.

Post, E., and Forchhammer, M.C. 2002. Synchronization of animal population dynamics by large-scale climate. Nature 420: 168–171.

Regehr, E.V., Lunn, N.J., Amstrup, S.C., and Stirling, I. 2007. Effects of earlier sea ice break-up on survival and population size of polar bears in Western Hudson Bay. J. Wildl. Manage. 71(8): 2673–2683.

Ryg, M., Smith, T.G., and Oritsland, N.A. 1990. Seasonal changes in body mass and composition of ringed seals (*Phoca hispida*) on Svalbard. Can. J. Zool. 68: 470–475.

Siegstad, H., Neve, P.B., Heide-Jorgensen, M.P., and Harkonen, T. 1998. Diet of the ringed seal (*Phoca hispida*) in Greenland. In M.P. Heide-Jorgensen and C. Lydersen (Eds.) Ringed seals in the North Atlantic, pp. 229–241. The North Atlantic Marine Mammal Commission, Tromso.

Skinner, W.R., Jefferies, R.L., Carleton, T.J., Rockwell, R.F., and Abraham, K.F. 1998. Prediction of reproductive success and failure in lesser geese based on early season climatic variables. Glob. Change Biol. 4: 3–16.

Smith, T.G. 1973. Population dynamics of the ringed seal in the Canadian Eastern Arctic. Fisheries Research Board of Canada, Bulletin 181, 55 pp.

Smith, T.G. 1975. Ringed seals in James Bay and Hudson Bay: population estimates and catch statistics. Arctic 28: 170–182.

Smith, T.G. 1987. The ringed seal, *Phoca hispida*, of the Canadian western Arctic. Canad. Bull. Fish. Aquat. Sci. 216: 81.

Smith, T.G., and Hammill, M.O. 1981. Ecology of the ringed seal, *Phoca hispida*, in its fast ice breeding habitat. Can. J. Zool. 59: 966–981.

Smith, T.G., Hammill, M.O., and Taugbol, G. 1991. A review of the developmental, behavioral and physiological adaptations of the ringed seal, *Phoca hispida*, to life in the Arctic winter. Arctic 44(2): 124–131.

Smith, T.G., and Stirling, I. 1975. The breeding habitat of the ringed seal (*Phoca hispida*). The birth lair and associated structures. Can. J. Zool. 53: 1297–1305.

Stirling, I. 2002. Polar bears and seals in the eastern Beaufort Sea and Amundsen Gulf: a synthesis of population trends and ecological relationships over three decades. Arctic 55(Suppl. 1): 59–76.

Stirling, I. 2005. Reproductive rates of ringed seals and survival of pups in Northwestern Hudson Bay, Canada, 1991–2000. Polar Biol. 28: 381–387.

Stirling, I., Archibald, W.R., and DeMaster, D. 1977. Distribution and abundance of seals in the Eastern Beaufort Sea. Fish. Res. Board Can. 34: 976–988.

Stirling, I., Lunn, N.J., and Lacozza, J. 1999. Long-term trends in the population ecology of polar bears in western Hudson Bay in relation to climatic change. Arctic 52(3): 294–306.

Stirling, I., and Oritsland, N.A. 1995. Relationships between estimates of ringed seal (*Phoca hispida*) and polar bear (*Ursus maritimus*) populations in the Canadian Arctic. Can. J. Fish. Aquat. Sci. 52: 2594–2612.

Stirling, I., and Parkinson, C.L. 2006. Possible effects of climate warming on selected populations of polar bears (*Ursus maritimus*) in the Canadian Arctic. Arctic 59(3): 261–275.

Stirling, I., and Smith, T.G. 2004. Implications of warm temperatures and an unusual rain event for the survival of ringed seals on the coast of Southeastern Baffin Island. Arctic 57(1): 59–67.

Tynan, C.T., and DeMaster, D.P. 1997. Observations and predictions of arctic climatic change: potential effects on marine mammals. Arctic 50(4): 308–322.

Walsh, J.E. 2008. Climate of the Arctic marine environment. Ecol. Appl. 18(Suppl. 2): S3–S22.

Wathne, J.A., Haug, T., and Lydersen, C. 2000. Prey preference and niche overlap of ringed seals *Phoca hispida* and harp seals *Pagophilus groenlandica* in the Barents Sea. Mar. Ecol. Prog. Ser. 194: 233–239.

Weslawski, J.M., Ryg, M., Smith, T.G., and Oritsland, N.A. 1994. Diet of ringed seals (*Phoca hispida*) in a Fjord of West Svalbard. Arctic 47(2): 109–114.

Woollett, J.M., Henshaw, A.S., and Wake, C.P. 2000. Palaeoecological implications of Archaeological seal bone assemblages: case studies from Labrador and Baffin Island. Arctic 53(4): 395–413.

Past, Present, and Future for Bowhead Whales (*Balaena mysticetus*) in Northwest Hudson Bay

J.W. Higdon and S.H. Ferguson

Abstract The bowhead whale (*Balaena mysticetus*) is the largest Arctic cetacean and the only baleen whale to live year-round in high-latitude waters. In this chapter we discuss the ecology and shifts in ecosystem structure and energy transfer among trophic levels due to changes in bowhead population size in the Hudson Bay region. Eastern Canadian Arctic bowheads comprise a single large, wide-ranging population that is shared between Canada and West Greenland, with considerable age- and sex-based segregation. In the Hudson Bay region important areas of aggregation include the spring nursery in northern Foxe Basin, summer locations in northwest Hudson Bay, and wintering habitat in Hudson Strait. Bowhead whales have been important for Inuit subsistence for millennia, and were commercially hunted in northwest Hudson Bay from 1860 to 1915. By the time whaling ended bowhead whales had been hunted to low numbers. However the bowhead population has grown in recent decades. We speculate on how rapid changes in northwest Hudson Bay population size, from pristine to depleted and now expanding, may have affected the marine ecosystem. First, bowhead removal would have resulted in a "freeing up" of zooplankton biomass, with potential cascading effects throughout the food web. Second, with population increases, more bowhead whales are now consuming more zooplankton, creating another ecosystem shift. Third, the region is rapidly losing sea ice due to warming likely resulting in future ecosystem changes. Population growth and warming are now occurring together necessitating a better understanding of bowhead ecology and ecosystem role to inform resource managers and policy makers. With better ecological data and explicit assumptions, detailed ecosystem models could be used to examine changes in ecosystem structure over time and into the future.

J.W. Higdon (✉)
Department of Environment and Geography, University of Manitoba and Fisheries and Oceans Canada, Central and Arctic Region, Winnipeg, MB, Canada
e-mail: jeff.higdon@gmail.com; Jeff.Higdon@dfo-mpo.gc.ca

S.H. Ferguson et al. (eds.), *A Little Less Arctic: Top Predators in the World's Largest Northern Inland Sea, Hudson Bay*, DOI 10.1007/978-90-481-9121-5_8,
© Springer Science+Business Media B.V. 2010

Keywords Commercial whaling • Ecosystem shifts • Population growth • Killer whale predation • Zooplankton • Food habits • Food web • Trophic structure • Subsistence hunts • Inuit traditional ecological knowledge

Introduction

The bowhead whale (*Balaena mysticetus* Linnaeus 1758, *Arvik* or *Arviq* in Inuktitut) (Fig. 1) is a member of the right whale family (Balaenidae). It is the largest Arctic cetacean and the only baleen whale to remain at high latitudes year-round, occurring in conditions ranging from open water to thick, extensive (unconsolidated) pack ice (COSEWIC 2009). They are well-adapted to ice-covered waters (Fig. 2), able to break thick ice (over 20 cm) to breathe and navigate under extensive ice fields. Bowheads have a nearly circumpolar distribution in the Northern Hemisphere in both the North Pacific and North Atlantic (Moore and Reeves 1993). Physical characteristics are described by Haldiman and Tarpley (1993). Adults can be >18 m long, and it is one of the stockiest whales, with a round body and a very large head comprising ca. 30% of the total length. The upper jaw (with ca. 330 baleen plates up to 4 + m long in each side) is bowed sharply upward, giving the bowhead whale its common name. The body is black with white (nonpigmented) regions including the chin and caudal peduncle (tail stock), with no dorsal fin or hump.

Two eastern populations (Hudson Bay-Foxe Basin [HB-FB] and Davis Strait-Baffin Bay [DS-BB]) were provisionally separated on the assumption of Fury and Hecla Strait representing a barrier to intermingling (Reeves et al. 1983; Reeves and Mitchell 1990). Bowheads in eastern Canada received little study for decades, due to their low numbers after centuries of commercial whaling, but there have been

Fig. 1 An adult bowhead whale surfacing in Foxe Basin, Nunavut (Photo by J.W. Higdon/DFO)

Fig. 2 A bowhead whale resting amongst loose pack ice, typical early summer habitat (Photo by J.W. Higdon/DFO)

significant discoveries in recent decades. Satellite tagging and genetic research have shown that bowhead whales in the eastern Canadian Arctic and West Greenland represent a single wide-ranging population (the Eastern Canada-West Greenland [EC-WG] population, COSEWIC 2009) with long seasonal migrations and significant age and sex segregation (Dueck et al. 2006; Heide-Jorgensen et al. 2003, 2006; Postma et al. 2006; IWC 2008).

Hudson Strait is an important wintering ground (Koski et al. 2006), and Inuit in Foxe Basin and northwest Hudson Bay occasionally observe bowheads at the floe edge in some winters (NWMB 2000). In April and May, some whales move west through Hudson Strait to summer aggregation areas in northwest Hudson Bay (Reeves and Mitchell 1990) and northern Foxe Basin (NWMB 2000). Northwest Hudson Bay was also a focal area for commercial whalers in the late 1800s and early 1900s (Ross 1979; Reeves and Cosens 2003). The floe edge in northern Foxe Basin is currently used by cow-calf pairs as a nursery area (Cosens and Blouw 2003).

Bowhead whales have long been important to Inuit subsistence and were pivotal in the movements and migrations of early Thule people, who were active bowhead whalers (McCartney and Savelle 1985). A long history of commercial whaling reduced numbers to low levels throughout the eastern Arctic (Higdon 2008, in press). In recent decades numbers have increased significantly, and both Inuit knowledge and scientific research indicate a growing bowhead population (COSEWIC 2009). Changes in the COSEWIC status reflect this: they were classed as "Endangered" in 1980, downgraded to "Threatened" in 2005, and further downgraded to "Special Concern" in 2009 based on an updated status report (COSEWIC 2009).

The nineteenth century commercial whaling grounds in northwest Hudson Bay covered 60,000 km², extending from Marble Island north through Roes Welcome Sound to Lyon Inlet and into Fisher Strait (Ross 1974; Reeves and Cosens 2003). Much of this area is still used by bowheads (NWMB 2000), but at lower densities than during historical times (Cosens and Innes 2000). In Foxe Basin, animals congregate north of Igloolik Island. Satellite tracking has shown that bowheads move through Fury and Hecla Strait and into the Gulf of Boothia (and hence cross the "dividing line" that had provisionally separated the two stocks; Dueck et al. 2006). In northern Foxe Basin bowhead whales aggregate along the land-fast ice edge in June and July before ice break-up, where they are observed socializing and feeding (Thomas 1999).

Rapid climatic changes are occurring in Arctic ecosystems, and these changes are having an impact on ice-adapted marine mammals and seabirds. However another significant change in the Arctic marine system relates to the removal of bowhead whales by commercial hunting, and the current growth in the population would be expected to result in addition changes in ecosystem structure and function. Bowhead whales consume zooplankton, and large reductions in abundance would have resulted in a zooplankton surplus. We argue that this surplus provided benefits to other marine species that trickled upwards throughout the entire ecosystem (Węskawski et al. 2000). In this chapter we review the history of bowhead exploitation in northwest Hudson Bay, summarize population characteristics, and speculate on food web impacts that may have resulted from rapid declines in bowhead abundance followed by partial and continued recovery. Finally, we review current literature to discuss possible future scenarios for bowhead whales in this region. We provide a starting point for further inquiry into the role of major ecosystem players, like bowhead whales, and how past, present, and future changes in abundance can influence large-scale processes and trophic ecosystem linkages.

The Past – History of the Bowhead Whale in the Hudson Bay Region

Bowhead range in the Arctic has been closely tied with changes in sea ice over millennia (Dyke et al. 1996; Savelle et al. 2000), and aboriginal subsistence use has also varied with climatic changes. The Thule people, ancestors of the Inuit, started to migrate eastward from the Bering Sea ca. AD1000 and actively hunted bowheads (Savelle and McCartney 1990). Aboriginal whaling declined in the central Arctic after ca. AD1500 but survived in several areas, including western Hudson Bay, where open water conditions allowed continued hunting (Schledermann 1979; Stoker and Krupnik 1993). In more recent historic times, Inuit also traded blubber and baleen to whalers and traders (Barr 1994).

Commercial harvesting of the Eastern Canada-West Greenland (EA-WG) population started with Basque whalers off southern Labrador in the 1500s and continued until the early 1900s, with >70,000 whales taken (Higdon 2008, in press). Many different whaling grounds were exploited, and the Hudson Bay ground was the last of these, active from 1860 to 1915 (Ross 1974, 1979). Harvests were minor in relation to other whaling grounds, but by the time Scottish and American whalers entered this region, commercial harvests had been going on for ca. 350 years and the EA-WG bowhead population was already seriously reduced. Northern Foxe Basin is the only known summer aggregation area for this population where commercial whaling did not occur because it was inaccessible due to heavy ice conditions.

Ross (1974, 1979) estimated commercial bowhead harvests in northwest Hudson Bay from 1860 to 1915, which have been used by other authors to estimate pristine stock size. Ross (1979) used whalebone (baleen) and oil returns to estimate that 565 whales were killed by Scottish and American whalers from 1860 to 1915, not counting those struck and lost. Estimates of the pre-whaling population size in northwest Hudson Bay have ranged from 450 to 680 whales (Mitchell 1977; Woodby and Botkin 1993). These estimates assumed a closed population (now unsupported) and excluded Foxe Basin, where no commercial kills occurred.

Figure 3 summarizes the estimated commercial whaling take (landed whales only) on the northwest Hudson Bay ground (Ross 1979). The catch history shows extremely high catches in the early decades, a situation seen on each of the different whaling grounds, with most whales taken early in the fishery (Higdon 2008, in press). The majority of whales taken in Hudson Bay were killed in the first decade (n = 341, 60%), with three quarters of the total removal complete in just 20 years.

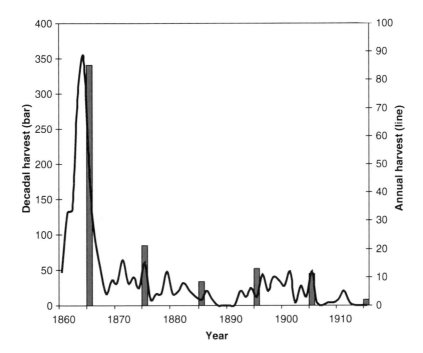

Fig. 3 Estimated catches of bowhead whales in northwest Hudson Bay by Scottish and American vessels, 1960–1915 (using data from Ross 1979). Numbers represent landed whales only, with no correction for struck and lost. *Bars* represent total catch per decade, *line* is yearly landed catch

By 1900 over 90% of the total take had been removed, and by 1915 bowhead whales were rare throughout their EC-WG range and whaling was no longer commercially viable.

The Present – Current Distribution and Estimates of Bowhead Population Size

In recent decades bowheads have shown significant recovery, and the species is once again common in many parts of its range (reviewed in COSEWIC 2009). Cosens and Innes (2000) conducted aerial surveys in August 1995 throughout the northwest Hudson Bay region where commercial kills occurred. They observed bowheads in Roes Welcome Sound, Frozen Strait and Repulse Bay, but did not see any on transect in the southern part of the study area, despite this being a major area for commercial harvests. This suggests that while the population is growing, full recovery has yet to occur (also see NWMB 2000). Cosens and Innes (2000) used strip-transect methods and estimated 75 whales. This is negatively biased because they did not account for diving animals (availability bias) or whales at the surface that were not observed (perception bias). The estimate can be corrected for availability bias using data from time-depth recorders on satellite tags. Using a proportion of time at surface value of 0.24 (Heide-Jørgensen et al. 2007) provides a simple correction of $75/0.24 = 313$ whales. This is not corrected for whales at the surface but not seen so it is still negatively biased. Similar surveys flown in northern Foxe Basin (Cosens et al. 1997) indicated higher abundance in that summer aggregation area.

Population changes between the start of commercial whaling and the present time are not known with any certainty, but it is apparent that large numbers of bowheads were taken throughout their range. The population is growing (NWMB 2000; Heide-Jørgensen et al. 2007; COSEWIC 2009), but the current abundance is not known with certainty (IWC 2009). Using a bowhead catch history (Higdon 2008, in press) with a population model to estimate pre-exploitation size (reviewed by Baker and Clapham 2004) would assist with gauging the level of recovery, setting recovery goals, and understanding ecosystem changes resulting from population decline and subsequent growth. Available information clearly indicates that bowhead numbers in northwest Hudson Bay were severely reduced over a short-time time (over 300 whales killed in the first 10 years). Numbers are now increasing, but the population has not yet fully recovered nor fully colonized their former range (Cosens and Innes 2000; NWMB 2000).

Population Characteristics – Historic and Current

There are no rigorous data available on the historic population structure of Hudson Bay bowhead whales, but Reeves and Cosens (2003) provide some information from commercial whaling logbooks. They examined logs covering 50 American voyages, representing a total of 92 "ship-seasons" (vessels would often overwinter and operate for two ship-seasons, Ross 1974). Their database (provided by the authors) contains 355 records of killed whales, and 104 of these have a production value, and/or the sex of the animal, and/or a qualitative estimate of whale age/size. In total, 48 have a category (calf, very small, small, large, etc.), 26 have baleen length, 14 have baleen weight, and 54 have oil yield (with some overlap between these categories – 7 have information for 3 different measures and 23 have data for 2 measures).

For whales with production data (baleen or oil), we used the regression equations of Lowry (1993) and Finley and Darling (1990) to estimate total length using data on baleen length and weight and oil yield. Whales were assigned to an age-class based on their estimated length (using mean values from both sets of regression equations). Calves were considered to be those whales with an estimated total length of 7.5 m or less, or baleen less than 70 cm in length, or producing less than 25 barrels of oil (Reeves and Cosens 2003; George and Suydam 2006). Bowheads do not begin to actively feed until they are 4 years old, when their baleen plates become developed enough (Schell and Saupe 1993). Based on baleen growth rates (George and Suydam 2006; Schell and Saupe 1993), a 4-year old whale would have baleen between 157.5 and 175 cm long and a total length of 10.2–10.8 m (equation from Lowry 1993). There is again significant variation in total length (George and Suydam 2006), and we classed whales estimated at 7.5–10 m long as juveniles (1–4 years old).

Typical length of sexual maturity for females (which tend to be larger than males) is ca. 13.5 m (Koski et al. 1993; George et al. 2004). Males exhibit significant testicular development at 12.5–13 m in length (O'Hara et al. 2002), agreeing with photogrammetric data (Koski et al. 1993) that sexual maturity generally occurs at 12–13 m length (estimated age of sexual maturity is 25, George et al. 1999). Therefore, all males >12.5 m were considered adults, along with females or whales of unknown sex >13 m. Subadult whales (>4 years old but not yet sexually mature) were 10–13 m for females and those of unknown sex and 10–12.5 m for males.

The historic age-class (and sex where available) distribution for 104 harvested whales from the Reeves and Cosens (2003) database (Fig. 4) indicates 40% calves or juveniles, 18% subadults, and 42% adults. Data on gender are available for few whales but indicate that adults and subadults of both sexes were present. All the different age/sex classes were widely distributed throughout the region, with large numbers of cow-calf pairs in the southern portion (Reeves and Cosens 2003). The authors suggested that the waters south of Wager Bay were possibly used as a summer nursery area (as northern Foxe Basin currently is). Before the 1870s logbooks tended to provide fewer details and the age/sex class profile is more applicable to the second half of the whaling period (1880s to early 1900s) than the first (1860s–1870s).

However limited (see discussion of biases by Reeves and Cosens 2003), these are the only available data on age and sex classes of the historic bowhead population. The logbook data are not necessarily representative of the historic population structure, and caution is required in using these data for estimating the age and sex

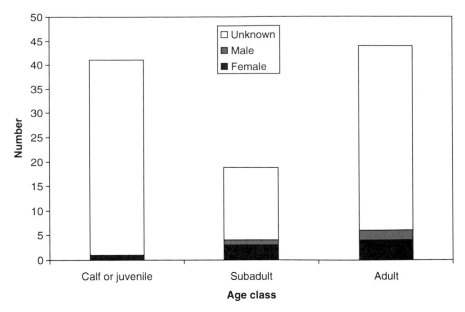

Fig. 4 Age-class distribution of the historic northwest Hudson Bay bowhead whale population (n = 104) based on information found in commercial whaling logbooks (Data from Reeves and Cosens 2003)

distribution of the historical commercial catch given the unsystematic record keeping. The data are also from American logbooks only, and no Scottish logbooks were included. The proportion of calves and juveniles in the harvest remained fairly constant over time, ranging from 33% to 45%. In contrast, the proportion of adults and subadults in the harvest varied considerably by decade (Fig. 5). A higher proportion of adults were taken during the early and later stages of the fishery, with a larger proportion of subadult whales taken in the 1870s. All different age and sex classes were present and taken by whalers over these years.

Foxe Basin is currently the primary nursery ground for cow/calf pairs occupying the Hudson Bay region. In northern Foxe Basin calves and juveniles comprise >80% of the summer population (Cosens and Blouw 2003). The contemporary Foxe Basin population therefore contains significantly fewer adult whales and more juvenile and subadult whales than the historic northwest Hudson Bay population (based on the commercial whaling data).

There is no published data on the contemporary age-class structure of northwest Hudson Bay bowhead whales. Cosens and Blouw (2003) suggested that adult males and non-calving females currently aggregate in northwest Hudson Bay. Photo-identification research (Higdon and Ferguson 2008–2009) indicates that most bowheads in Repulse Bay are large adults. However cow-calf pairs were observed in 2008 and 2009. We predict greater use of the northwest Hudson Bay area with continued population growth. Inuit also observe bowhead cows with

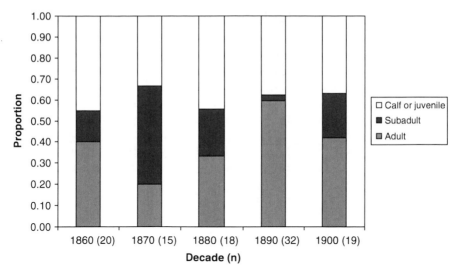

Fig. 5 Proportion of different age classes in American harvests of Hudson Bay bowhead whales, by decade (sample size, as number of whales from logbooks, in parentheses)

calves near Repulse Bay, and occasionally further south near Chesterfield Inlet. Bowheads are still rarely observed near Rankin Inlet, and those seen tend to be adults (NWMB 2000).

Bowhead Feeding Ecology and Zooplankton Consumption

Bowhead whales are specialized filter feeders that primarily eat pelagic crustacean zooplankton, particularly copepods (primarily *Calanus* spp.) and euphausiids (*Thysanoessa* spp.), in addition to epibenthic organisms (Lowry 1993; Lowry et al. 2004). Smaller whales (<10.5 m long) tend to consume more epibenthic organisms (Lowry 1993), and copepods appear more important in the diet of larger whales (Schell et al. 1987). Lowry (1993) suggested that bowheads rely on abundant food in late summer and fall to acquire the lipid reserves necessary to sustain them during the winter. This would suggest that northwest Hudson Bay is an important fall foraging area for a segment of the EC-WG population.

A large adult (>17 m) bowhead can weigh 60,000 kg, a 13 m adult ca. 34,000 kg, and a 10 m subadult ca. 17,000 kg (George et al. 2007), with a daily food requirement of 1–4% of the total body weight (Frost and Lowry 1984). Węskawski et al. (2000) used 1.5% as a conservative value, which equals a daily estimated consumption of 255–900 kg of zooplankton per whale for the weights noted above. In Disko Bay, West Greenland, it has been estimated that the current spring population of 250 whales (85% adults) consumes 223 t of zooplankton per day, or almost 27,000 t over a 4-month residency period (Laidre et al. 2007).

Ecosystem Impacts of Bowhead Population Change

The main impact of bowhead decline would be a "freeing up" of significant zooplankton biomass that would have otherwise been bowhead food. This would lead to increased availability of zooplankton for consumption by other trophic groups. In contrast, a currently increasing population results in a reduction of the zooplankton biomass available to other consumers. Such effects would then trickle up (and down) throughout the food web and likely affect species at all levels, ranging from phytoplankton to polar bears. However, changes are unlikely linear and the return of large whales does not necessarily mean the return of past marine ecosystems (e.g., Ballance et al. 2006).

Despite the uncertainty in pre-whaling and current abundance and changes over time, several basic trends can be identified. First, bowhead abundance declined rapidly, with 60% of the total estimated landed harvest (565 whales) being removed in the first 2 decades (Ross 1979). Second, bowhead whales were reduced to low numbers (commercial extinction) by the early 1900s. The reduction in the local whale population was therefore both extensive and rapid. Third, both Inuit knowledge and scientific research indicate a currently growing bowhead population. Detailed population modelling is required to fully address questions related to ecosystem changes, but several generalizations can be made. When bowhead whale populations were reduced, zooplankton consumption declined accordingly. This "freeing up" of zooplankton biomass likely had significant effects on energy transfer and ecosystem structuring in the marine ecosystem (e.g., Węskawski et al. 2000 for Svalbard).

We speculate on some possible effects, which may help inform ecosystem models (Hoover this volume) that are more useful in quantifying changes. We show three stages in a generalized and simplified marine food web for northwest Hudson Bay (Fig. 6). The historic food web (Fig. 6a) represents the relationships before commercial bowhead removal, and also before the occurrence of killer whales in Hudson Bay (Ferguson et al. this volume).

Figure 6b represents the system during and immediately following the commercial whaling era, and still prior to killer whale occurrence. With a decrease in bowhead abundance, Arctic cod (*Boreogadus saida*) was likely one of the first species to take advantage of the zooplankton surplus (Węskawski et al. 2000). This in turn would have benefited fish-eating seabirds such as thick-billed murres (*Uria lomvia*) and seals. Węskawski et al. (2000) suggested that bowhead reduction ultimately led to the growth of the large present-day colonies of piscivorous seabirds on Svalbard. The large colonies of thick-billed murres in northern Hudson Bay and Hudson Strait may have directly benefited from increased Arctic cod production. Ringed seals are the primary prey for polar bears throughout most of their range (e.g., Stirling 2002), and any increase in seals would likely be beneficial. Polar bears are also opportunistic scavengers, and commercial whalers provided an abundant potential food source (e.g., Miller et al. 2004; Bentzen et al. 2007). Opportunities for scavenging during commercial whaling may have had long-term consequences for bear populations (Whitehead and Reeves 2005).

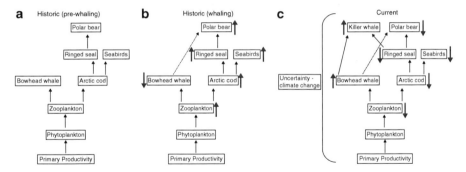

Fig. 6 A simplified and generalized food web for the northwest Hudson Bay marine ecosystem: (**a**) the historic food web, pre-commercial whaling and also before killer whales started to occur in the area, (**b**) the commercial whaling era, when a reduction in bowhead numbered "freed-up" zooplankton, benefiting Arctic cod, seabirds, ringed seals, and polar bears (*dotted line* represents polar bear scavenging on bowhead carcasses left by whalers), and still prior to killer whale expansion, and (**c**) the current ecosystem, with an increasing bowhead population, increasing killer whale presence, and declines in other species (polar bears scavenging bowhead from natural strandings, killer whale predation and Inuit subsistence harvests). Impacts of changing climate represent a major source of uncertainty for all species and trophic levels

A number of recent studies have identified negative impacts of a changing climate on various marine species groups in Hudson Bay. In the current system (Fig. 6c) the bowhead population is growing, but speculation on top-down and bottom-up effects is made difficult with uncertainty due to climate change. Declines in Hudson Bay sea ice extent and duration have been implicated in changes in fish communities (Gaston et al. 2003) and subsequent negative effects on local murre colonies (Mallory et al. this volume). Trophic effects from increased bowhead consumption may also be playing a concurrent role. Disentangling warming effects, such as loss of sea ice, and increasing bowhead zooplankton consumption to understand trophic changes will be difficult since both effects have similar trophic consequences (e.g., decreased seabird colony size).

Earlier spring break-up of sea ice and reduced snow cover (for birth lairs) has been implicated in reduced recruitment of ringed seals in western Hudson Bay (Ferguson et al. 2005). Declines in sea ice and snow cover (or increases in springtime rain events, Stirling and Smith 2004) will have negative effects on ice-adapted ringed seals. However, these effects should be explored in the context of ecosystem changes associated with increased bowhead abundance. Juvenile ringed seals in Hudson Bay feed on zooplankton prey (Young et al. 2010), and this life stage may be adversely affected by reduced euphasiids with an increase in consumption by bowhead whales.

Declines in ringed seal numbers would be expected to negatively affect polar bears. In recent years there have been declines in polar bear body condition and population size, which have been linked to climate warming (Peacock et al. this volume), particularly during the spring (earlier ice break-up). It is possible that access to bowhead carcasses during the commercial whaling era increased survival and reproduction and led to increased bear population sizes. Regardless of whether such effects did exist, changing ice conditions are clearly impacting bears in Hudson Bay today. However, a better understanding of the role of bowhead scavenging to polar bear ecology, both historic and current, would benefit the conservation of bears in relation to changing climatic conditions. Polar bears are also currently observed scavenging bowhead carcasses (natural mortality and Inuit subsistence harvests), which could provide a significant resource for at least some individual bears (Bentzen et al. 2007).

Killer whales have also now become established as a top predator in Hudson Bay, where they are preying on bowhead whales and ringed seals, in addition to narwhal and beluga (Ferguson et al. 2010). Declining sea ice in Hudson Strait has been a factor in this range increase (Higdon and Ferguson 2009), but local Inuit also note a link between an increasing bowhead population and increased killer whale presence.

The Future for Hudson Bay Bowhead Whales

Rapid climatic changes are occurring throughout the Arctic region, and in the Hudson Bay region both the extent and duration of sea ice has decreased in recent decades (Parkinson and Cavalieri 2002; Houser and Gough 2003; Gough et al. 2004; Stirling et al. 2004; Gagnon and Gough 2005). These changes have been implicated in negative effects on a number of species groups, and continued declines are expected to exacerbate these effects. However the impact of bowhead population decline and subsequent growth on zooplankton abundance and availability has also likely affected the marine system. Climate-related negative effects are therefore likely to be happening in concert with impacts from altered trophic interactions.

Changing climate will influence bowhead whales and other species through both direct and independent effects (Forchhammer and Post 2004). Accurate predictions are elusive (IPCC 2007) and responses to change are unlikely to be consistent across a species' range (Tynan and DeMaster 1997). Predicting impacts in the Arctic is more complex than in temperate regions due to increased intra- and interannual variability (Ferguson and Messier 1996).

Loss of Sea Ice

The effects of declining sea ice on marine mammals are likely to be first reflected in shifts in range and abundance (Tynan and DeMaster 1997). Bowhead ecology and movements are closely associated with ice (Dyke et al. 1996; Savelle et al. 2000; Ferguson et al. in review). Future changes in ice extent are expected to again influence distribution, and several recent reviews have predicted bowhead range contraction with climate warming (Learmonth et al. 2006; MacLeod 2009). There is also suggestion that bowhead whales are heat intolerant (IWC 1997; Bannister 2002) due to their extensive blubber layer, and an increase in water temperatures with declining ice and increased solar radiation may have negative effects.

Food

As ice retreats in the spring, open water is exposed to sunlight, which provides the conditions necessary for the "spring bloom" in production and transfer to higher trophic levels through zooplankton (Bluhm and Gradinger 2008; Hansen et al. 2002). As a zooplankton-specialist, bowheads may be one of the first species affected by any trophic decoupling associated with loss of sea ice (Laidre et al. 2008). How, and if, bowheads are able to respond to trophic mismatch between sea ice, sunlight, productivity and zooplankton remains to be seen (Laidre et al. 2007), and this is an important question to address in predictions of future impacts.

Decreases in summer ice extent, and the resulting increase in open water, could initially benefit whales by enhancing local prey populations, extending the foraging period, and/or extending the habitat available for foraging (Moore and Laidre 2006; George et al. 2005). The western Arctic bowhead population has been consistently growing since 1978 (George et al. 2004), a positive demographic change during a period of declining ice extent (Laidre et al. 2008). There is a similar trend in Hudson Bay, where the bowhead population has grown while the duration of the ice-free season has increased. However interpreting this as a demographic cause and effect relationship with climate change is difficult given that both populations are recovering from commercial overharvesting (Laidre et al. 2008). In the western Arctic decreases in ice extent have also been linked with an improvement in whale body condition, with body condition of landed whales being higher in years when summer ice concentration was lower (George et al. 2005). The authors hypothesized that local increases in primary production due to reduced ice cover provided improved feeding opportunities.

Predation

Killer whales were historically present in Baffin Bay and Davis Strait and were reported in whaling logbooks in the 1800s (Reeves and Mitchell 1988). We hypothesize that bowhead whales use Foxe Basin as a nursery area because it historically was not occupied by killer whales. Juvenile whales are more susceptible to predation (Ford and Reeves 2008) and use of habitat that precludes killer whales may be necessary for their protection. The area north of Igloolik Island provides an open-water refuge (polynya) along the land-fast ice floe edge, and in June and July it is separated from Hudson Strait by hundreds of kilometres of heavy (9/10 +) pack ice. This may prevent access by killer whales. However, with sea ice declines the region may become less useful for bowhead whales as a habitat that restricts killer whale access to calves.

Anthropogenic Stressors

An increase in the length of the open water season may bring an increase in shipping traffic. Bowhead whales are sensitive to ocean noise (Richardson 2006), and an increase in seismic and shipping activity could disrupt migration patterns and habitat use. In addition, large whale ship strikes are often fatal (Jensen and Silber 2003), and an increase in traffic would increase the probability of such events. An increase in fishing activity could also affect bowheads, as gear entanglement is a major mortality source for the closely-related North Atlantic right whale (Kraus 1990).

Future

Ultimately, predictions of the future for Hudson Bay bowhead whales, and the other species in the system, will be difficult. Laidre et al. (2008) concluded that bowhead whales were moderately sensitive to climate change – less so than pinnipeds and polar bears that require ice as a physical platform, more so than belugas, but less sensitive than narwhal. In the interim, we predict that the population will continue to grow and zooplankton consumption will increase accordingly. There was a strong Inuit tradition of bowhead whale hunting in Hudson Bay, and throughout the eastern Canadian Arctic, which declined after commercial whalers decimated the population. Inuit movements and survival have been closely tied with bowhead whales for millennia, and they are once again able to harvest them. The bowhead harvest will help to restore and maintain an important aspect of Inuit culture and a vital link between the marine system and those that depend on it.

Conclusion

Commercial whaling resulted in the removal of tens of thousands of bowhead whales from circumpolar Arctic waters, with significant ecosystem impacts (Węskawski et al. 2000). Catches from the Hudson Bay region were low in comparison to some areas (e.g., Baffin Bay, Ross 1979), but nonetheless relatively large numbers of whales were removed from a relatively small geographic area over an incredibly short time frame. We consider it likely that the removal of bowhead whales as a major zooplankton consumer had significant impacts on ecosystem structuring and trophic processes. Similarly, current growth and recovery of bowheads, with increased zooplankton consumption, will again result in changes. Current changes are also occurring in the face of sea ice declines and warming temperatures, and disentangling effects will be difficult. In this chapter we have speculated on possible impacts of bowhead population change. Bowhead whales are linked, via zooplankton, to the entire marine food chain, including many species that are currently experiencing climate-related consequences. A structured assessment of ecosystem impacts (Hoover this volume) is required, and we hope our discussion and speculation will provide a starting point for further inquiry.

Much has been learned about bowhead whales in recent years (see COSEWIC 2009); although many important questions remain unanswered. A better understanding of life-history parameters or population vital rates, such as calving intervals, calf survival, and age of maturity, is necessary to ensure effective hunt management and recovery planning. Photo-identification projects to address these questions have been started (Ferguson IPY-GWAMM project) but long-term studies will be required. Bowhead harvest numbers throughout eastern Canada and West Greenland have been compiled (Higdon 2008, in press) and these data can be used with a population model to estimate total pre-exploitation abundance. This information, coupled with current abundance estimates, will provide information to gauge the level of recovery and model possible future growth. This will also require comprehensive surveys to get an accurate estimate of current abundance as well as monitoring changes over time. Finally, a better understanding of bowhead foraging dynamics is needed, which can be coupled with past, present and projected future population estimates to better understand the impacts of decline and recovery on ecosystem structure and function. Scientists and northerners will need to work together on these issues to ensure that bowhead whales, and their Arctic habitat, are protected for future generations.

References

Baker, C.S., and P.J. Clapham. 2004. Modelling the past and future of whales and whaling. Trends in Ecology and Evolution 19: 365–371.

Ballance, L.T., R.L. Pitman, R. Hewitt, D. Siniff, W. Trivelpiece, P. Clapham, and R.L. Brownell, Jr. 2006. The removal of large whales from the Southern Ocean. Evidence for long-term ecosystem effects? In J.A. Estes, D.P. DeMaster, D.F. Doak, T.M. Williams, and R.L. Brownell, Jr., (Ed.) Whales, whaling, and ocean ecosystems, pp. 215–230. University of California Press, Berkeley, CA.

Bannister, J.L. 2002. Baleen whales. In W.F. Perrin, B. Würsig, and J.G.M. Thewissen (Eds.) Encyclopedia of marine mammals, pp. 62–72. Academic Press, San Diego, CA.

Barr, W. 1994. The eighteenth century trade between the ships of the Hudson's Bay Company and the Hudson Strait Inuit. Arctic 47: 236–246.

Bentzen, T.W., E.H. Follmann, S.C. Amstrup, G.S. York, M.J. Wooller, and T.M. O'Hara. 2007. Variation in winter diet of southern Beaufort Sea polar bears inferred from stable isotope analysis. Canadian Journal of Zoology 85: 596–608.

Bluhm, B.A., and R. Gradinger. 2008. Regional variability in food availability for Arctic marine mammals. Ecological Applications 18(Suppl.): S77–S96.

Cosens, S.E., and A. Blouw. 2003. Size- and age-class segregation of bowhead whales summering in northern Foxe Basin: a photogrammetric analysis. Marine Mammal Science 19: 284–296.

Cosens, S.E., and S. Innes. 2000. Distribution and numbers of bowhead whales (*Balaena mysticetus*) in northwestern Hudson Bay in August 1995. Arctic 53: 36–41.

Cosens, S.E., T. Qamukaq, B. Parker, L.P. Dueck, and B. Anardjuak. 1997. The distribution and numbers of bowhead whales, *Balaena mysticetus*, in northern Foxe Basin in 1994. Canadian Field Naturalist 111: 381–388.

COSEWIC. 2009. COSEWIC assessment and update status report on the Bowhead Whale *Balaena mysticetus*, Bering-Chukchi-Beaufort population and Eastern Canada-West Greenland population, in Canada. Committee on the Status of Endangered Wildlife in Canada. Ottawa. vii + 49 pp. (www.sararegistry.gc.ca/status/status_e.cfm).

Dueck, L.P., M.P. Hiede-Jørgensen, M.V. Jensen, and L.D. Postma. 2006. Update on investigations of bowhead whale (*Balaena mysticetus*) movements in the eastern Arctic, 2003–2005, based on satellite-linked telemetry. Fisheries and Oceans Canada Canadian Science Advisory Secretariat Research Document 2006/050.

Dyke, A.S., J. Hooper, and J.M. Savelle. 1996. A history of sea ice in the Canadian Arctic Archipelago based on postglacial remains of the bowhead whale (*Balaena mysticetus*). Arctic 49: 235–255.

Ferguson, S.H., and F. Messier. 1996. Ecological implications of a latitudinal gradient in interannual climatic variability: a test using fractal and chaos theories. Ecography 19: 382–392.

Ferguson, S.H., I. Stirling, and P. McLoughlin. 2005. Climate change and ringed seal (*Phoca hispida*) recruitment in western Hudson Bay. Marine Mammal Science 21: 121–135.

Ferguson, S.H., J.W. Higdon, and E.G. Chmelnitsky. 2010. The rise of killer whales as a major arctic predator. In S.H. Ferguson, L.L. Loseto, and M.L. Mallory (Eds.) A little less arctic: Top predators in the worlds largest northern inland sea, Hudson Bay, pp.117–136. Springer, New York.

Ferguson, S.H., L. Dueck, and L.L. Loseto. in review. Bowhead whale (*Balaena mysticetus*) seasonal selection of sea ice. Marine Ecology Progress Series.

Finley, K.J., and L.M. Darling. 1990. Historical data sources on the morphometry and oil yield of the bowhead whale. Arctic 43: 153–156.

Forchhammer, M.C., and E. Post. 2004. Using large-scale climate indices in climate change ecology studies. Population Ecology 46: 1–12.

Ford, J.K.B., and R.R. Reeves. 2008. Fight or flight: antipredator strategies of baleen whales. Mammal Review 38: 50–86.

Frost, K., and L.F. Lowry. 1984. Trophic relationships of vertebrate consumers in the Alaskan Beaufort Sea. In P.W. Barnes, D.M. Schell, and E. Reimnitz (Eds.) Ecosystem and environments, pp. 381–401. Academic Press, London.

Gagnon, A.S., and W.A. Gough. 2005. Trends in the dates of ice freeze-up and break-up over Hudson Bay, Canada. Arctic 58: 370–382.

Gaston, A.J., K. Woo, and J.M. Hipfner. 2003. Trends in forage fish populations in northern Hudson Bay since 1981, as determined from the diet of nestling Thick-billed Murres *Uria lomvia*. Arctic 56: 227–233.

George, J.C., and R. Suydam. 2006. Length estimates of bowhead whale (*Balaena mysticetus*) calves. Paper SC/58/BRG23 presented to the International Whaling Commission Scientific Committee.

George, J.C., J. Bada, J. Zeh, L. Scott, S.E. Brown, T.O'Hara, and R. Suydam. 1999. Age and growth estimates of Bowhead Whales (*Balaena mysticetus*) via aspartic acid racemization. Canadian Journal of Zoology 77: 571–580.

George, J.C., E. Follmann, J. Zeh, M. Sousa, R. Tarpley, and R. Suydam. 2004. Inferences from bowhead whale ovarian and pregnancy data: age estimates, length at sexual maturity and ovulation rates. Paper SC/56/BRG8 presented to the International Whaling Commission Scientific Committee.

George, J.C., C. Nicholson, S. Drobot, and J. Maslanik. 2005. Progress Report. Sea ice density in the Beaufort Sea and bowhead whale body condition. Paper SC/57/E13 presented to the IWC Scientific Committee.

George, J.C., J.R. Bockstoce, A.E. Punt, and D.B. Botkin. 2007. Preliminary estimates of bowhead whale body mass and length from Yankee commercial oil yield records. Paper SC/59/BRG5 presented to the International Whaling Commission Scientific Committee.

Gough, W.A., A.R. Cornwall, and L.J.S. Tsuji. 2004. Trends in seasonal sea ice duration in southwestern Hudson Bay. Arctic 57: 299–305.

Haldiman, J.T., and R.J. Tarpley. 1993. Anatomy and physiology. In J.J. Burns, J.J. Montague, and C.J. Cowles (Eds.) The bowhead Whale, pp. 71–156. Special Publication no. 2. Society for Marine Mammalogy, Lawrence, KS.

Hansen, A.S., T.G. Nielsen, H. Levinsen, S.D. Madsen, T.F. Thingstad, and B.W. Hansen. 2002. Impact of changing ice cover on pelagic productivity and food web structure in Disko Bay, West Greenland: a dynamic model approach. Deep-Sea Research I 50: 171–187.

Heide-Jorgensen, M.P., K.L. Laidre, O. Wiig, M.V. Jensen, L. Dueck, L.D. Maiers, H.C. Schmidt, and R.C. Hobbs. 2003. From Greenland to Canada in ten days: tracks of bowhead whales, *Balaena mysticetus*, across Baffin Bay. Arctic 56: 21–31.

Heide-Jørgensen, M.P., K.L. Laidre, M.V. Jensen, L. Dueck, and L.D. Postma. 2006. Dissolving stock discreteness with satellite tracking: bowhead whales in Baffin Bay. Marine Mammal Science 22: 34–45.

Heide-Jørgensen, M.P., K.L. Laidre, D. Borchers, F. Samarra, and H. Stern. 2007. Increasing abundance of Bowhead Whales in West Greenland. Biology Letters 3: 577–580.

Higdon, J.W. in press. Commercial and subsistence harvests of Bowhead Whales (*Balaena mysticetus*) in eastern Canada and West Greenland. Journal of Cetacean Research and Management.

Higdon, J. 2008. Commercial and subsistence harvests of Bowhead Whales (*Balaena mysticetus*) in eastern Canada and West Greenland. DFO Canadian Science Advisory Secretariat Research Document 2008/008.

Higdon, J.W., and S.H. Ferguson. 2009. Loss of Arctic sea ice causing punctuated change in sightings of killer whales (*Orcinus orca*) over the past century. Ecological Applications 19: 1365–1375.

Hoover, C. 2010. Hudson Bay ecosystem: past, present and future. In S.H. Ferguson, L.L. Loseto and M.L. Mallery (Eds.) A little less Arctic: Top predators in the worlds largest northen inland sea, Hudson Bay, Springer, New York 217–236.

Houser, C., and W.A. Gough. 2003. Variations in sea ice in the Hudson Strait: 1971–1999. Polar Geography 27: 1–14.

IPCC. 2007. Climate Change 2007: Impacts, Adaptation, and Vulnerability. In M.L. Parry, O.F. Canziani, J.P. Palutikof, P.J. van der Linden, and C.E. Hanson (Eds.) Contribution of working group II to the fourth assessment report of the intergovernmental panel on climate change, p. 1000. Cambridge University Press, Cambridge.

IWC. 1997. Report of the IWC workshop on climate change and cetaceans. Report of the International Whaling Commission 47: 293–331.

IWC. 2008. Report of the Scientific Committee. Journal of Cetacean Research and Management 10(Suppl.): 1–74.

IWC. 2009. Report of the Scientific Committee. Journal of Cetacean Research and Management 11(Suppl.): 169–192.

Jensen, A.S., and G.K. Silber. 2003. Large whale ship strike database. U.S. Department of Commerce, NOAA Technical Memorandum. NMFS-OPR-25, p. 37.

Koski, W.R., M.P. Heide-Jørgensen, and K.L. Laidre. 2006. Winter abundance of bowhead whales, *Balaena mysticetus*, in the Hudson Strait, March 1981. Journal of Cetacean Ressearch and Management 8: 139–144.

Koski, W.R., Davis, R.A., Miller, G.W., and Withrow, D.E. 1993. Reproduction. In J.J. Burns, J.J. Montague, and C.J. Cowles (Eds.) The bowhead Whale, pp. 239–269. Allen Press, Lawrence, KS.

Kraus, S.D. 1990. Rates and potential cause of mortality in North Atlantic right whales (*Eubalaena glacialis*). Marine Mammal Science 6: 278–291.

Laidre, K.L., M.P. Heide-Jørgensen, and T.G. Nielsen. 2007. Role of the Bowhead Whale as a predator in West Greenland. Marine Ecology Progress Series 346: 285–297.

Laidre, L.L., I. Stirling, L.F. Lowry, O. Wiig, M.P. Heide-Jørgensen, and S.H. Ferguson. 2008. Quantifying the sensitivity of Arctic marine mammals to climate induced habitat change. Ecological Applications 18(Suppl.): S297–S125.

Learmonth, J.A., C.D. MacLeod, M.B. Santos, G.J. Pierce, H.Q.P. Crick, and R.A. Robinson. 2006. Potential effects of climate change on marine mammals. Oceanography and Marine Biology: An Annual Review 44: 431–464.

Lowry, L.F. 1993. Foods and feeding ecology. In J.J. Burns, J.J. Montague, and C.J. Cowles (Eds.) The bowhead Whale, pp. 201–238. Special Publication no. 2. Society for Marine Mammalogy, Lawrence, KS.

Lowry, L.F., G. Sheffield, and J.C. George. 2004. Bowhead whale feeding in the Alaskan Beaufort Sea, based on stomach contents analyses. Journal of Cetacean Research and Management 6: 215–223.

MacLeod, C.D. 2009. Global climate change, range changes and potential implications for the conservation of marine cetaceans: a review and synthesis. Endangered Species Research 7: 125–136.

Mallory, M.L. A.J. Gaston, H.G. Gilchrist, G.J. Robertson and B.M. Braune. 2010. Effects of climate change, altered sea-ice distribution and seasondal phenology on marine birds. In S.H. Ferguson, L.L. Leseto and M.L. Mallory (Eds.) A little less Arctic: Top predators in the worlds largest northern inland sea, Hudson Bay, Springer, New York. 179–195.

McCartney, A.P., and J.M. Savelle. 1985. Thule Eskimo whaling in the central Canadian Arctic. Arctic Anthropology 22: 37–58.

Miller, S., K. Proffitt, and S. Schliebe. 2004. Demography and behaviour of polar bears feeding on marine mammal carcasses. OCS study, MMS 2004, U.S. Department of the Interior, Anchorage, Alaska.

Mitchell, E.D. 1977. Initial population size of bowhead whale (*Balaena mysticetus*) stocks: cumulative catch estimates. Paper SC/29/33 presented to the IWC Scientific Committee.

Moore, S.E., and K.L. Laidre. 2006. Analysis of sea ice trends scaled to habitats used by bowhead whales in the western Arctic. Ecological Applications 16: 932–944.

Moore, S.E., and R.R. Reeves. 1993. Distribution and movement. In J. J. Burns, J.J. Montague, and C.J. Cowles (Eds.) The bowhead Whale, pp. 313–386. Society for Marine Mammalogy Special Publication no. 2. Allen Press, Lawrence, KS.

NWMB. 2000. Final report of the Inuit Bowhead Knowledge Study, Nunavut, Canada. Nunavut Wildlife Management Board, Iqaluit, Nunavut, p. 90.

O'Hara, T.M., J.C. George, R.J. Tarpley, K. Burek, and R.S. Suydam. 2002. Sexual maturation in male bowhead whales (*Baleana mysticetus*) of the Bering-Chukchi-Beaufort Seas stock. Journal of Cetacean Research and Management 4: 143–148.

Parkinson, C.L., and D.J. Cavalieri. 2002. A 21 year record of Arctic sea ice extents and their regional, seasonal and monthly variability and trends. Annals of Glaciology 34: 441–446.

Peacock, E., A.E. Derocher, N.J. Lunn and M.E. Obbard. 2010. Polar bear ecology and management in Hudson Bay in the face of climate change. In S.H. Ferguson, L.L. Loseto, and M.L. Mallory (Eds.) A little less Arctic: Top predators in the worlds largest northern inland sea, Hudson Bay, 93–115. Springer, New York.

Postma, L.D., L.P. Dueck, M.P. Heide-Jørgensen and S.E. Cosens. 2006. Molecular genetic support for a single population of bowhead whales (*Balaena mysticetus*) in Eastern Canadian Arctic and Western Greenland waters. Fisheries and Oceans Canada Canadian Science Advisory Secretariat Research Document 2006/051.

Reeves, R.R., and S.E. Cosens. 2003. Historical population characteristics of Bowhead Whales (*Balaena mysticetus*) in Hudson Bay. Arctic 56: 283–292.

Reeves, R.R., and E. Mitchell. 1988. Distribution and seasonality of killer whales in the eastern Canadian Arctic. Rit Fiskideildar 11: 136–160.

Reeves, R.R., and E. Mitchell. 1990. Bowhead Whales in Hudson Bay, Hudson Strait and Foxe Basin: A review. Naturaliste Canadienne 117: 25–43.

Reeves, R., E. Mitchell, A. Mansfield, and M. McLaughlin. 1983. Distribution and migration of the bowhead whale, *Balaena mysticetus*, in the eastern North American Arctic. Arctic 36: 5–64.

Richardson, W.J. (Ed.) 2006. Monitoring of industrial sounds, seals, and bowheads whales, near BP's Northstar Oil Development, Alaskan Beaufort Sea, 1999–2004. [Updated Comprehensive Report, April 2006.]

Ross, W.G. 1974. Distribution, migration and depletion of bowhead whales in Hudson Bay, 1860 to 1915. Arctic and Alpine Research 6: 85–98.

Ross, W.G. 1979. The annual catch of Greenland (bowhead) whales in waters north of Canada 1719–1915: a preliminary compilation. Arctic 32: 91–121.

Savelle, J.M., and A.P. McCartney. 1990. Prehistoric Thule Eskimo whaling in the Canadian Arctic Islands: current knowledge and future research directions. In C.R. Harington, (Ed.) Canada's missing dimension: science and history in the Canadian Arctic islands, vol. II, pp. 695–723. Canadian Museum of Nature, Ottawa, ON.

Savelle, J.M., A.S. Dyke, and A.P. McCartney. 2000. Holocene bowhead whale (*Balaena mysticetus*) mortality patterns in the Canadian Arctic Archipelago. Arctic 53: 414–421.

Schell, D.M., and S.M. Saupe. 1993. Feeding and growth as indicated by stable isotopes. In J.J. Burns, J.J. Montague, and C.J. Cowles (Eds.) The bowhead whale, pp. 491–509. Special Publication no. 2. Society for Marine Mammalogy, Lawrence, KS.

Schell, D.M., S.M. Saupe, and N. Haubenstock. 1987. Bowhead Whale feeding: allocation of regional habitat importance based on stable isotope abundances. In W.J. Richardson (Ed.) Importance of the eastern Alaskan Beaufort Sea to feeding Bowhead Whales 1985–86, pp. 369–415. Report to U.S. Minerals Management Service by LGL Ecological Research Associates Inc., NTIS No. PB88-150271.

Schledermann, P. 1979. The 'baleen period' of the Arctic Whale Hunting Tradition. In A.P. McCartney (Ed.) Thule Eskimo Culture: an anthropological retrospective, pp. 134–148. National Museum of Man, Mercury Series, Archaeological Survey of Canada Paper No. 88.

Stirling, I. 2002. Polar bears and seals in the eastern Beaufort Sea and Amundsen Gulf: a synthesis of population trends and ecological relationships over three decades. Arctic 55(Suppl.): 59–76.

Stirling, I., and T.G. Smith. 2004. Implications of warm temperatures and an unusual rain event on the survival of ringed seals on the coast of southeastern Baffin Island. Arctic 57: 59–67.

Stirling, I., N.J. Lunn, J. Iacozza, C. Elliot, and M. Obbard. 2004. Polar bear distribution and abundance on the southwestern Hudson Bay coast during the open water season, in relation to population trends and annual ice patterns. Arctic 57:15–26.

Stoker, S.W., and I.I. Krupnik. 1993. Subsistence whaling. In J.J. Burns, J.J. Montague, and C.J. Cowles (Eds.) The bowhead whale, pp. 579–629. Special Publication no. 2. Society for Marine Mammalogy, Lawrence, KS.

Thomas, T.A. 1999. Behaviour and habitat selection of Bowhead Whales (*Balaena mysticetus*) in northern Foxe Basin, Nunavut. M.Sc. thesis, University of Manitoba, Winnipeg, MB, Canada, 107 p.

Tynan, C.T., and D.P. DeMaster. 1997. Observations and predictions of Arctic climatic change: potential effects on marine mammals. Arctic 50: 308–322.

Węskawski, J.M., L. Hacquebord, L. Stempniewicz, and M. Malinga. 2000. Greenland whales and walruses in the Svalbard food web before and after exploitation. Oceanologia 42: 37–56.

Whitehead, H., and R. Reeves. 2005. Killer whales and whaling: the scavenging hypothesis. Biology Letters 1: 415–418.

Woodby, D.A., and D.B. Botkin. 1993. Stock sizes prior to commercial whaling. In J.J. Burns, J.J. Montague, and C.J. Cowles (Eds.) The bowhead whale, pp. 387–407. Special Publication no. 2. Society for Marine Mammalogy, Lawrence, KS.

Young, B.G., L.L. Loseto, and S.H. Ferguson. 2010. Diet differences among age classes of Arctic seals: evidence from stable isotope and mercury biomarkers. Polar Biology. 33:153–162 DOI 10.1007/s00300-009-0693-3.

Effects of Climate Change, Altered Sea-Ice Distribution and Seasonal Phenology on Marine Birds

M.L. Mallory, A.J. Gaston, H.G. Gilchrist, G.J. Robertson, and B.M. Braune

Abstract The Hudson Bay region supports internationally significant populations of marine birds (>2,000,000 individuals), particularly thick-billed murres (*Uria lomvia*) and common eiders (*Somateria mollissima*). The breeding ecology of both of these species is inextricably linked to distribution of sea ice and the timing of its breakup and freeze-up, which determine the availability and distribution of open water in which to feed. For the piscivorous murres, earlier ice breakup is creating a mismatch between the timing of breeding and the peak in food availability, and the birds have not, to date, advanced their breeding phenology to keep up with the pace of environmental change. However, at the end of the season, delayed freeze-up is extending the period that birds can remain in the Bay. Earlier ice breakup may allow migratory eiders earlier access to the benthic mollusks they require to gather nutrient resources prior to breeding, and for the non-migratory eider population in southern Hudson Bay, warmer temperatures mean more open water (i.e., larger and more numerous polynyas and floe edges) needed to gather food supplies necessary for their overwinter survival. Thus, the effects of changes in sea ice cover vary according to each species' ecological needs. We anticipate considerable changes in populations of marine birds in the Hudson Bay region in future, due to direct and indirect effects of reduced sea ice cover.

Keywords Thick-billed murre • Common eider • Arctic cod • Capelin • Sea ice • Break-up • Freeze-up • Breeding ecology • Dietary shift • Energetics

Introduction

The vast inland sea that is Hudson Bay supports some of the most important habitats for Arctic birds in North America. At least 20 "key habitat sites" for migratory birds (i.e., sites where ≥1% of the Canadian population can be found at some

M.L. Mallory (✉)
Environment Canada, Canadian Wildlife Service, Box 1714, Iqaluit,
NU X0A 0H0, Canada
e-mail: mark.mallory@ec.gc.ca

S.H. Ferguson et al. (eds.), *A Little Less Arctic: Top Predators in the World's Largest Northern Inland Sea, Hudson Bay*, DOI 10.1007/978-90-481-9121-5_9,
© Springer Science+Business Media B.V. 2010

point during the year) have been identified in the greater Hudson Bay Region. To the public (bird hunters, bird watchers, television documentaries), the Bay is perhaps best known for its seemingly never-ending, flat coastal plains and marshes. Millions of snow geese (*Chen caerulescens*) breed in huge colonies along these coastlines, and hundreds of thousands of other waterfowl and shorebirds breed in or rely on coastal habitats as critical feeding areas during migration (Latour et al. 2008). Seven of these sites are national Migratory Bird Sanctuaries.

Perhaps less well known is the importance of the Hudson Bay Region for marine birds, that is, birds that live close to the sea and feed in the sea. This region is home to more than 2,000,000 marine birds. Arctic terns (*Sterna paradisaea*), various gulls (*Larus* spp., *Xema sabini, Rhodostethia rosea*), jaegers (*Stercorarius* spp.), guillemots (*Cepphus grylle*) and loons (*Gavia* spp.) rely on the waters of Hudson Bay for breeding and feeding annually. However, the most numerous marine birds in the Bay are thick-billed murres (*Uria lomvia*) and common eiders (*Somateria mollissima*). In fact, the two largest thick-billed murre colonies in Canada occur here, at Akpatok Island and at Digges Sound, which together support approximately 1,600,000 birds (Gaston and Hipfner 2000). Similarly, the largest, single common eider colony in Arctic Canada is found at East Bay, Southampton Island, where up to 9,000 birds breed in some years (Mallory and Fontaine 2004). Furthermore, small island archipelagoes that dot the eastern parts of the region contain widely distributed colonies that support tens of thousands of nesting eiders (Abraham and Finney 1986; Nakashima and Murray 1988).

Both murres and eiders have been the focus of conservation research in Hudson Bay for more than 20 years. Adults and eggs of both species are harvested by Inuit living in communities around Hudson Bay (Priest and Usher 2004), as well as being part of regionally important hunts in wintering areas along Newfoundland and Labrador, the northern Gulf of St. Lawrence in Quebec, and southwest Greenland (Chardine et al. 2008; Merkel and Christensen 2008). Here, we review some of the ecological needs of these species that have been revealed during these studies, with particular attention given to the relationship between sea ice and marine bird reproduction and survival.

A Tale of Two Marine Birds

It has been well-established that tracking the population size and reproductive success of marine birds serves as a strong and sensitive monitor of marine environmental conditions (Cairns 1987; Furness and Greenwood 1993, Parsons et al. 2008, Gaston et al. 2009). As well, some marine birds travel long distances to feed, with individuals sometimes feeding at different trophic levels, but they return to their colony during incubation or to rear chicks, so information integrating conditions over a geographically-large marine area can be gathered by studying birds at a single colony (Montevecchi 1993; Parsons et al. 2008).

Waters of the Arctic are covered with sea-ice for much of the year. As air-breathing predators, marine birds and mammals require open water in which to surface and breathe while feeding. The timing of ice breakup and the position of the ice edge is a critical factor influencing the ecology of seasonally ice-covered waters (Welch et al. 1992), for two reasons. First, sea-ice forms a physical barrier for predators to be able to access marine foods. Second, breakup of sea-ice releases a food web and nutrients that developed under the ice, generating a flush of herbivorous zooplankton, which provides abundant food for upper trophic levels, and effectively "sets the clock" for the whole marine ecosystem (e.g., Le Fouest et al. 2005; Wang et al. 2005). Moreover, variation in the schedule of these events can create year-to-year variation in primary production that can affect growth and reproduction at higher trophic levels (Arrigo and van Kijken 2004; Johnston et al. 2005).

The relationships between the timing of open water availability and population size and reproductive success have been studied for both murres and eiders in Hudson Bay. These two bird species occupy very different ecological niches, and thus it should not be surprising that each tells a different story about sea-ice, climate change and the marine environment of Hudson Bay.

Thick-Billed Murres (*Uria lomvia*)

Natural History

Thick-billed murres are robust, black and white, pursuit diving seabirds that are found in Arctic waters (Gaston and Hipfner 2000). They resemble penguins except that they can fly (Fig. 1). Murres are long-lived (at least 29 years), and a breeding pair tends to stay "partnered" for many years (i.e., socially monogamous). Murres return to the exact same nesting location on the cliff annually, where they lay a single egg that is incubated on bare rock. They are principally pelagic piscivores, although they also consume other marine organisms like crustaceans, and evidence suggests that birds may become "specialists" on certain prey (Woo et al. 2008). Murres usually feed within 170 km of their colony, diving down to 140 m (Gaston and Hipfner 2000; Elliott et al. 2007, 2008). Because they feed relatively high in the marine food web, murres bioaccumulate various contaminants, and these concentrations differ among colonies in the Canadian Arctic (Braune et al. 2002; Braune 2009a). Both the male and the female incubate the egg and help rear the chick. During chick-rearing, murres return to their colony usually with a single fish held in their bill, which makes it easy for researchers to quantify what adult murres are feeding to their offspring (Gaston 1985; Elliott et al. 2008).

Reproduction and Climate

Murres have been studied intensively for nearly 3 decades at the Coats Island colony (62°57'N, 82°00'W) in northern Hudson Bay (Gaston et al. 2009).

Fig. 1 Thick-billed murres (*Uria lomvia*) on the breeding ledges at the Coats Island colony (Photo by Jennifer Provencher)

This site supports approximately 30,000 breeding pairs of murres, making it much smaller than Digges Sound or Akpatok Island, but the physical configuration of the colony cliffs makes working there much safer. Murres from Coats Island spend the winter in the waters of Davis Strait, Labrador Sea and off Newfoundland, and then migrate through Hudson Strait to return to their colony each spring (Gaston and Hipfner 2000). In a "normal" year, murres arrive at the colony in mid- to late May, and lay their single egg in the second half of June (Gaston and Hipfner 1998). However, in years when the sea-ice is late breaking up in Hudson Bay, murres initiate their laying by as much as a week later. This delay is attributable to the need for murres to forage and acquire resources for egg production; in years of heavy ice, open water may be too distant from the colony, and birds must wait longer before they can initiate egg formation.

Interestingly, the relationship between nest initiation and ice cover shows a parallel relationship with diet. Arctic cod *Boreogadus saida*, an ice-associated, key species in Arctic marine food webs (Bradstreet and Cross 1982) was a major component of murre nestling diet at Coats Island up to the mid-1990s. During that period murres delivered a higher proportion of cod to their chicks in years when extensive ice cover persisted later into the season (Gaston and Hipfner 1998). This situation (i.e., later ice break-up associated with a cod-dominated diet) is thought to represent the historical norm for the Coats Island colony. However, that relationship has changed since the 1990s.

Over the past few decades, ice conditions in Hudson Bay have changed, with break-up occurring earlier and the consequent date of 50% clearing of ice cover in surrounding waters advancing approximately 17 days, especially since 1994 (Gaston et al. 2009; Fig. 2). Along with this physical change in the marine eco-system, monitoring of thick-billed murres at Coats Island has detected biological changes. Although murre chicks were fed mostly Arctic cod from 1981 to 1990, a period of dietary transition occurred between 1990 and 1996, and since 1996, capelin (*Mallotus villosus*) has dominated chick diets (Gaston et al. 2003). Capelin is a fish more typical of subarctic waters, while Arctic cod is common among High Arctic areas. Thus, over the 3 decades of monitoring chick diets, there has been a shift in the marine food web towards warmer water species (Fig. 3). During the 1990s this change in diet was associated with lower adult body mass during chick-rearing than was observed in the early years of this study and with lower chick growth rates (Gaston et al. 2005). One possible reason for lower mass or growth rates comes from nutritional changes. Arctic cod and capelin have similar energy densities, but capelin delivered by adult murres tend to be smaller than cod. Consequently, chicks now receive less energy for growth than they did in the past, and adults may be working as hard or harder to deliver this energy (Gaston et al. 2005; Elliott and Gaston 2008). There is also evidence that the change in diet may be altering the exposure of these birds to different contaminants (Braune 2009b). The extent to which this may be affecting the birds is currently being explored.

Interestingly, another auk, the more typically boreal-water breeding razorbill (*Alca torda*) began appearing at the Coats Island colony coincident with an increase in the delivery of sandlance to murre chicks at this site (Gaston and Woo 2008). Sandlance is an important constituent of razorbill diet more or less throughout the species' range. The razorbills appeared in 1997, following a sharp increase in the representation of sandlance in thick-billed murre chick diets and disappeared again after 2004 as sandlance became rarer. If a long-term shift in available prey near Coats Island leads to a greater abundance of sandlance, it is likely that razorbills will colonize this site and compete for food resources with murres.

There is another route by which global warming is affecting murres, which has nothing to do with sea-ice. Coats Island supports an irritatingly abundant mosquito population which attack the incubating murres on the cliff. Mosquitoes can be so thick on the exposed feet of incubating murres that the birds look as if they are wearing fur boots. In the 1980s and early 1990s mosquitoes generally became abundant only after mid-July. Although they affected the murres on warm, still days, they appeared to have had little effect on reproduction during those years. However, by the late 1990s high mosquito activity was occurring from late June onwards. When mosquito attacks were most intense, in a few particularly hot years, the combination of mosquito parasitism and heat stress caused some incubating murres to die at their nests through a process analogous to heat stroke in humans (Gaston et al. 2002). In addition, many more birds left their breeding sites to cool off on the sea, leaving their eggs unattended. Many of these eggs were removed by

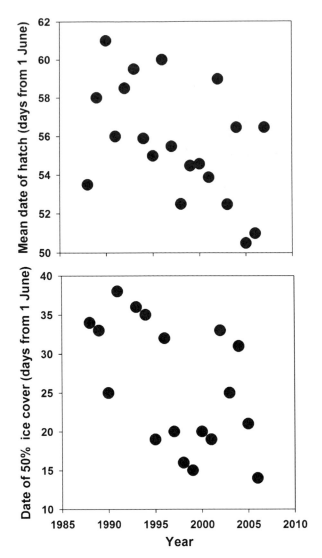

Fig. 2 Thick-billed murre hatch dates and the date at which 50% of the ice has cleared from Hudson Bay have both been advancing over the past 2 decades. In the top panel, earlier murre nesting is described by the regression: Date (days from 1 June) = −0.24*Year + 535, r^2 = 0.23, $F_{1, 18}$ = 5.3, P = 0.03. In the bottom panel, earlier ice clearing over the study period is described by the regression: Date (50% ice cover) = 1729 − 0.85(Year), r^2 = 0.26, $F_{1, 18}$ = 7.6, P = 0.01

glaucous gulls *Larus hyperboreus*, which patrol the colony constantly, on the look-out for exposed eggs or chicks, lowering the overall reproductive success of the colony. Because the first appearance and peak abundance of mosquitoes has become earlier at Coats Island over the past 3 decades (mirroring the earlier break up of sea-ice),

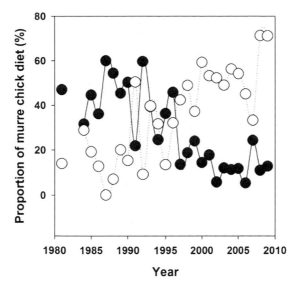

Fig. 3 The proportion of Arctic cod in the diet of thick-billed murre chicks has decreased since the 1980s (*filled circles*), while the proportion of capelin fed to chicks has increased (*unfilled circles*)

this is exposing the murre population to earlier and more prolonged mosquito attacks, creating increased mortality and lowered reproductive success.

Common Eiders (*Somateria mollissima sedentaria*)

Natural History

Common eiders are large, robust diving seaducks typical of cool waters (Fig. 4; Goudie et al. 2000). Like most Arctic marine birds, eiders are relatively long-lived, surviving up to 21 years, and they return to the same nesting colony in successive years, although not necessarily the same nest cup (Bustnes and Erikstad 1993). Arctic eiders lay clutches of three to six eggs in a nest cup lined with down feathers which the female plucks from her breast. Unlike murres, only the female incubates the eggs, and she departs the colony with the young when they are 1–2 days old. There is considerable variation in the size of eggs laid by eiders in different locations around Hudson Bay, presumably attributable to local differences in the genetic make-up of populations (Robertson et al. 2001; see below). Eiders dive to depths generally <20 m and feed on the ocean bottom, typically eating mollusks, sea urchins, sea stars, fish eggs and some crustaceans (Goudie et al. 2000). Like most other marine birds in the Arctic (Braune et al. 2002), common eiders exhibit

Fig. 4 A female common eider (*Somateria mollissima*) on her nest (Photo by Mark Mallory)

elevated levels of persistent organic pollutants (POPs) and certain toxic trace elements (Wayland et al. 2001; Mallory et al. 2004). In Hudson Bay, and notably in the Belcher Islands, eiders have relatively higher levels of certain POPs like DDT and mirex, and birds from western and northern Hudson Bay tended to have higher levels of mercury (Mallory et al. 2004).

In the Hudson Bay region, two races or subspecies of eiders inhabit coastal islands. Although the dividing line is not clear, it is thought that most eiders breeding around Hudson Strait west to Southampton Island and all areas north of that, belong to northern common eider race (*Somateria mollissima borealis*) (Abraham and Finney 1986). These eiders migrate from breeding colonies to overwinter along the west coast of Greenland, or along coastal Labrador, Newfoundland, and into the Gulf of St. Lawrence (Goudie et al. 2000; Merkel et al. 2006; Mosbech et al. 2006). Unlike murres, northern common eiders wintering in Greenland and Newfoundland do not rely on ice edges for foraging per se; they simply require open water areas over the shallow coastal mussel beds on which they feed. Sea ice has the potential to influence the timing of their spring migration, however. An earlier breakup of sea-ice in Hudson Strait, northern Hudson Bay, and Foxe Basin may initially benefit these populations by providing earlier access to food supplies as they approach their breeding colonies to nest.

Within Hudson Bay itself, from the Ottawa Islands south, most common eiders belong to the Hudson Bay common eider race (*Somateria mollissima sedentaria*); the largest race of eiders found in the world. These birds never leave Hudson Bay: they breed on coastal islands, and then aggregate in open water leads and polynyas in the Bay where they must survive the winter (Nakashima and Murray 1988; McDonald et al. 1997; Robertson and Gilchrist 1998; Gilchrist and Robertson 2000). It is these eiders that have been the focus of climate and ice research.

Overwintering and Climate

There are estimated to be more than 100,000 Hudson Bay common eiders, according to recent population surveys (Gilchrist et al. 2006). The vast majority of these birds converge on the Sleeper and Belcher Islands during the winter, to huddle together in the recurrent polynyas and the leads and floe edges that open and close with tide and currents (Nakashima and Murray 1988; Mallory et al. 2006; Fig. 5). However, the number of eiders found there in the winter fluctuates dramatically both seasonally and among years, in relation to climate and ice patterns (Nakashima and Murray 1988).

Part of this variation in numbers is attributable to climate-influenced reproductive success of nesting eiders the previous summer. For example, working in southwestern Hudson Bay, Robertson (1995a) showed that Hudson Bay eiders laid smaller eggs, delayed egg-laying and laid smaller clutches in a year following a winter of extreme ice (1991–1992), and when weather conditions were cooler prior to breeding (Fig. 6). Small eggs result in smaller chicks at hatch which typically have lower survival (Williams 1994). Thus, in particularly cold years, the combination of fewer eggs laid, and those eggs tending to be smaller means that we should expect fewer ducklings that survive to their first winter. This pattern has been observed in other Arctic waterbirds (Boyd 1996; Ganter and Boyd 2000).

Fig. 5 Common eiders huddled together in all available open water in a polynya, Belcher Islands (Photo by Grant Gilchrist)

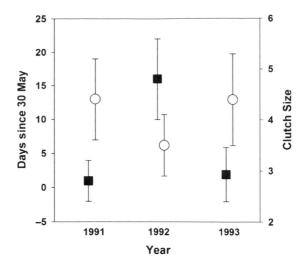

Fig. 6 Following a cold winter when sea ice lasted longer and was more extensive, eiders initiated nests later (*filled squares* ± SD; $F_{2,312} = 27.8$, $P < 0.001$), and laid smaller clutches (*unfilled circles* ± SD; $F_{2,312} = 342.3$, $P < 0.0001$)

Exacerbating this issue, cool spring weather in 1992 led to ice bridges remaining between islands and increased predation on clutches by arctic foxes (*Vulpes alopex*; Robertson 1995b).

However, what probably has a far larger and more lasting impact on the Hudson Bay eider population is the availability of open water during the winter, and how that constrains their access to food. This has been long-recognized by the Inuit in Sanikiluaq, as they have recorded aboriginal traditional knowledge of massive die-offs of starving eiders during some winters (McDonald et al. 1997), notably in the winter of 1991–1992 (see Fig. 7), the same winter that preceded Robertson's (1995a) observations of reduced reproductive effort by eiders.

During "normal" winters, polynyas and leads in the sea ice are numerous around the Belcher Islands (Mallory and Fontaine 2004). During the day, eiders forage in open water areas offshore or along coastlines, where they dive through leads in the ice to feed on mussel beds. Towards the end of the day, eiders return to recurring polynyas, often in great numbers, where they roost overnight (Fig. 8). In the morning, the pattern reverses itself and eiders commute back to offshore or coastal feeding areas (Gilchrist and Robertson 2000). Ice cover is constantly shifting in the winter in this region, so the number of eiders that return to roost in the predictable open water at polynyas varies with the dynamic sea ice patterns that occur over feeding areas (Gilchrist et al. 2006). However, in years when particularly cold conditions and calm winds combine to generate heavy ice cover, many of the coastal and offshore feeding areas freeze over entirely. Under these conditions, eiders have no choice but to take refuge in the few recurrent polynyas that are kept open by strong tidal currents. If heavy ice conditions persist, all of the overwintering eider

Fig. 7 Dead common eiders along the edge of a freezing polynya, Belcher Islands (Photo by Grant Gilchrist)

Fig. 8 Common eiders hauled out to digest food along the edge of a polynya, Belcher Islands (Photo by Grant Gilchrist)

population must feed and roost at these polynyas (Gilchrist et al. 2006). Such a situation occurred in 1991–1992, when Arctic temperatures were lowered and ice was abnormally heavy in Hudson Bay, following the Mount Pinatubo eruption in 1991 (Ganter and Boyd 2000).

When eiders are trying to feed in these polynyas, they are facing three major constraints. First, there are many other eiders also trying to feed on sessile benthic prey, so competition for food in available food patches is high. For this reason, during normal winters it appears to be principally juvenile, inexperienced eiders that remain to feed at the polynyas, while more experienced, adult birds commute to open water patches where food is more abundant (Gilchrist et al. 2006). Second, strong currents, limited periods of suitable light, and freezing temperatures (which cause rapid ice formation during slack tide) mean that the amount of time available for eiders to effectively forage is actually quite limited (Heath et al. 2007). This constraint is further exacerbated by the need for eiders to have appropriate time to digest prey items, which they do when hauled out on the edge of the polynya (Gilchrist et al. 2006; Heath et al. 2006, 2007). The combination of these two constraints means that in any "normal" year some eiders die of starvation, generally inexperienced birds that are ill-equipped to withstand the winter stress. However, a third constraint emerges during years of heavy ice as occurred in the winter of 1991–1992. In these winters, thousands of eiders are forced to feed in a few polynyas, which results in rapid depletion of available prey. Mussels get eaten, and eiders then need to search for more mobile, but perhaps less nutritious, prey like urchins that move in and out of the patch. The net result is that under heavy ice conditions, food supplies are quickly depleted at polynyas and eiders begin to starve en masse. If sea ice remains extensive, mass starvation and death ensues.

In 1996, Inuit of Sanikiluaq thought that their local eider population had declined considerably since surveys conducted in the 1980s (Nakashima and Murray 1988), and they contacted government scientists to see if new surveys of known nesting islands could be initiated. The next breeding season brought favorable summer conditions, and surveys were undertaken on 426 islands where 1,414 nesting eiders were counted. In the late 1980s, these same islands had supported 5,651 eiders on nests, therefore representing a 75% decline in the local breeding population (Robertson and Gilchrist 1998). Because eiders exhibit breeding philopatry to nesting islands or island clusters, it was highly unlikely that the entire breeding population had moved. Moreover, clutch sizes were generally large in 1997, suggesting that it was also unlikely that many birds simply skipped breeding (something eiders will do in poor years; Goudie et al. 2000). However, populations of long-lived marine birds are sensitive to reductions in adult survival, and Inuit local had observed that many females had died in 1992. Subsequent collaborative research between the community and government scientists has supported the interpretation that the huge decline in the eider population detected in 1997 was a lingering result of the large die-off of the breeding population from 5 years earlier.

Collectively, research on eiders in Hudson Bay indicates that annual sea-ice conditions play a major role in regulating population size and, through occasional severe winters, structuring population demographics by entirely removing certain cohorts of birds.

What Does the Future Hold?

Multiple sources of information are showing convincingly that the Arctic climate and sea-ice are changing, and this is particularly apparent in Hudson Bay (McDonald et al. 1997; Gagnon and Gough 2005; Stirling and Parkinson 2006, this volume). What our work indicates is that the effects of these physical changes in the Hudson Bay marine environment will have quite different effects on the two key marine birds of this region.

Thick-billed murres are already experiencing dietary changes over the past 2 decades, and this has led to reduced chick growth and adult body mass, as well as altering their exposure to contaminants. However, many other factors influence murre populations, most notably human harvest and environmental conditions in wintering areas (Gaston and Hipfner 2000; Gaston 2003; Irons et al. 2008). Currently, there is no evidence that murres are being deleteriously affected at the population level by ice changes in Hudson Bay, although evidence from many murre colonies suggests a pattern of long-term population cycles in response to regime shifts in marine conditions (Irons et al. 2008). However, recent climate change may be disrupting this cycle, leading to long-term, massive changes in sea-ice patterns (Serreze et al. 2007; Comiso et al. 2008). As such, our expectation is that with continued changes to sea-ice and the prey base in the breeding and wintering areas, thick-billed murre colonies at the southern limit of their breeding range (i.e., northern Hudson Bay) will eventually decline (Gaston et al. 2005).

In contrast, Hudson Bay common eiders currently experience population-level bottlenecks or collapses in very cold years that result in persistent and extensive sea-ice cover. Not only is this a potential wildlife conservation concern, but it is an important issue for aboriginal residents of southern Hudson Bay, who rely on eiders as a key source of country food in the winter (Nakashima and Murray 1988; McDonald et al. 1997; Robertson and Gilchrist 1998; Priest and Usher 2004). Clearly this eider population has survived such natural, possibly cyclical challenges for thousands of years. Nonetheless, we expect that the non-migratory eiders of southern Hudson Bay will likely benefit by long-term reductions in sea-ice during the winter, as they should have more reliable access to open water for feeding. Then again, there are some substantial "unknowns" still to be resolved.

Despite decades of research on marine birds in Hudson Bay, there are some major gaps in our knowledge of the effects of other stressors on existing marine ecosystems that may affect future bird populations of this region, in concert with climate-induced changes. First, our knowledge of the potential sublethal effects of various contaminants on these species remains limited (e.g., Wayland et al. 2001), and requires further investigation. This may become particularly important if hydroelectric activities in Quebec alter inputs of contaminants to Hudson Bay, or alter currents that move this pollution around the Bay (McDonald et al. 1997). Second, reduced ice cover in the Bay will probably lead to increased shipping activity for community supply, industrial activity including offshore hydrocarbon exploration, and tourism (Arctic Council 2009). The effects of these activities and their potential environmental disturbance and damage (e.g., oil spills) has not been adequately assessed. Third, climate warming will allow other competitor species to move north, and will also permit the northward movement of novel parasites and diseases into these bird populations. For example, outbreaks of avian cholera have recently appeared in northern common eiders nesting in Hudson Strait and northern Hudson Bay where they were not found previously. These have caused dramatic declines in the size of breeding colonies (e.g., Descamps et al. 2009). Finally, ongoing changes in winter habitats may be altering where, and in what proportion, Hudson Bay birds spend the non-breeding season. This can have major implications for population-level changes through anthropogenic impacts of disturbance, harvest and pollution (e.g., Merkel 2004; Wiese et al. 2004; Mosbech et al. 2006).

References

Abraham, K. F., and G. H. Finney. 1986. Eiders of the eastern Canadian Arctic. In A. Reed (Ed.) Eider ducks in Canada, pp. 55–73. Canadian Wildlife Service Report Series No. 47.

Arctic Council. 2009. Arctic Marine Shipping Assessment 2009 Report. Protection of the Marine Environment (PAME) and the Arctic Council. Alaska, USA.

Arrigo, K. R., and G. L. van Dijken. 2004. Annual changes in the sea-ice, chlorophyll *a*, and primary production in the Ross Sea, Antarctica. Deep-Sea Research Part II: Topical Studies in Oceanography 51: 117–138.

Boyd, H. 1996. Arctic temperatures and the long-tailed ducks shot in North America. Wildlife Biology 2: 113–117.

Bradstreet, M. S. W., and W. E. Cross. 1982. Trophic relationships at high Arctic ice- edges. Arctic 35: 1–12.

Braune, B. 2009a. Spatial patterns of contaminants in arctic seabirds, pp. 264–268. In Synopsis of Research Conducted Under the 2008–2009 Northern Contaminants Program. Department of Indian Affairs and Northern Development, Ottawa.

Braune, B. 2009b. Effect of climate change on diet and contaminant exposure in seabirds breeding in northern Hudson Bay, pp. 269–272. In Synopsis of Research Conducted Under the 2008–2009 Northern Contaminants Program. Department of Indian Affairs and Northern Development, Ottawa.

Braune, B. M., G. M. Donaldson, and K. A. Hobson. 2002. Contaminant residues in seabird eggs from the Canadian Arctic: spatial trends and evidence from stable isotopes for intercolony differences. Environmental Pollution 117: 133–145.

Bustnes, J. O., and K. E. Erikstad. 1993. Site fideilty in breeding Common Eider *Somateria mollissima* females. Ornis Fennica 70: 11–16.

Cairns, D. K. 1987. Seabirds as indicators of marine food supplies. Biological Oceanography 5: 261–271.

Chardine, J. W., G. J. Robertson, and H. G. Gilchrist. 2008. Seabird harvest in Canada. In F. Merkel, and T. Barry (Eds.) Seabird harvest in the Arctic, pp. 20–29. Circumpolar Seabird Group (CBird), CAFF Technical Report No. 16. CAFF International Secretariat, Akureyri, Iceland.

Comiso, J. C., C. L. Parkinson, R. Gersten, and L. Stock. 2008. Accelerated decline in the Arctic sea ice cover. Geophysical Research Letters 35: L01703.

Descamps, S., H. G. Gilchrist, J. Bety, E. I. Buttler, and M. R. Forbes. 2009. Costs of reproduction in a long-lived bird: large clutch size is associated with low survival in the presence of a highly virulent disease. Biology Letters 5: 278–281.

Elliott, K. H., and A. J. Gaston. 2008. Mass-length relationships and energy content of fishes and invertebrates delivered to nestling thick-billed murres *Uria lomvia* in the Canadian Arctic, 1981–2007. Marine Ornithology 36: 25–34.

Elliott, K. H., G. K. Davoren, and A. J. Gaston. 2007. Influence of buoyancy and drag on the dive behaviour of an arctic seabird, the thick-billed murre. Canadian Journal of Zoology 85: 352–361.

Elliott, K. H., K. Woo, A. J. Gaston, S. Benvenuti, L. Dall'Antonia, and G. K. Davoren. 2008. Seabird foraging behaviour indicates prey type. Marine Ecology Progress Series 354: 289–303.

Furness, R.W., and J. D. Greenwood (Eds.). 1993. Birds as monitors of environmental change. Chapman & Hall, London.

Gagnon, A. S., and W. A. Gough. 2005. Trends in the dates of ice freeze-up and breakup over Hudson Bay, Canada. Arctic 58: 370–382.

Ganter, B., and H. Boyd. 2000. A tropical volcano, high predation pressure, and the breeding biology of Arctic waterbirds: a circumpolar review of breeding failure in the summer of 1992. Arctic 53: 289–305.

Gaston, A. J. 1985. The diet of thick-billed murre chicks in the eastern Canadian Arctic. Auk 102: 727–734.

Gaston, A. J. 2003. Synchronous fluctuations of thick-billed murre (*Uria lomvia*) colonies in the eastern Canadian Arctic suggest population regulation in winter. Auk 120: 362–370.

Gaston, A. J., and J. M. Hipfner. 1998. The effects of ice conditions in northern Hudson Bay on breeding by thick-billed murres. Canadian Journal of Zoology 76: 480–492.

Gaston, A. J., and J. M. Hipfner. 2000. Thick-billed murre *Uria lomvia*. In A. Poole and F. Gill (Eds.) The Birds of North America, No. 497. The Birds of North America, Inc., Philadelphia, PA.

Gaston, A. J., and K. Woo. 2008. Razorbills (*Alca torda*) follow subarctic prey into the Canadian Arctic: colonization results from climate change? Auk 125: 939–942.

Gaston, A. J., J. M. Hipfner, and D. Campbell. 2002. Heat and mosquitoes cause breeding failures and adult mortality in an Arctic-nesting seabird. Ibis 144: 185–191.

Gaston, A. J., H. G. Gilchrist, and J. M. Hipfner. 2005. Climate change, ice conditions and reproduction in an Arctic nesting marine bird: Brunnich's guillemot (*Uria lomvia* L.). Journal of Animal Ecology 74: 832–841.

Gaston, A. J., H. G. Gilchrist, M. L. Mallory, and P. A. Smith. 2009. Changes in seasonal events, peak food availability, and consequent breeding adjustment in a marine bird: a case of progressive mismatching. Condor 111: 111–119.

Gaston, A. J., K. Woo, and J. M. Hipfner. 2003. Trends in forage fish populations in northern Hudson Bay since 1981, as determined from the diet of nestling thick- billed murres *Uria lomvia*. Arctic 56: 227–233.

Gaston, A. J., D. F. Bertram, A. W. Boyne, J. W. Chardine, G. Davoren, A. W. Diamond, A. Hedd, J. M. Hipfner, M. J. F. Lemon, M. L. Mallory, W. A. Montevecchi, J. - F. Rail, and G. J. Robertson. 2009. Changes in Canadian seabird populations and ecology since 1970 in relation to changes in oceanography and food webs. Environmental Reviews 17: 267–286.

Gilchrist, H. G., and G. J. Robertson. 2000. Observations of marine birds and mammals wintering at polynyas and ice edges in the Belcher Islands, Nunavut, Canada. Arctic 53: 61–68.

Gilchrist, H. G., J. Heath, L. Arragutainaq, G. Robertson, K. Allard, S. Gilliland, and M. L. Mallory. 2006. Combining science and local knowledge to study common eider ducks wintering in Hudson Bay. In R. Riewe and J. Oakes (Eds.) Climate change: linking traditional and scientific knowledge, pp. 284–303. Aboriginal Issues Press, Winnipeg, MB.

Goudie, R. I., G. J. Robertson, and A. Reed. 2000. Common eider *Somateria mollissima*. In A. Poole and F. Gill (Eds.) The Birds of North America, No. 546. The Birds of North America, Inc., Philadelphia, PA.

Heath, J. P., H. G. Gilchrist, and R. C. Ydenberg. 2006. Regulation of stroke pattern and swim speed across a range of current velocities: diving by common eiders wintering in polynyas in the Canadian Arctic. Journal of Experimental Biology 209: 3974–3983.

Heath, J. P., H. G. Gilchrist, and R. C. Ydenberg. 2007. Can dive cycle models predict foraging behaviour? Diving by common eiders in an Arctic polynya. Animal Behaviour 73: 877–884.

Irons, D. B., T. Anker-Nilssen, A. J. Gaston, G. V. Byrd, K. Falk, H. G. Gilchrist, M. Hario, M. Hjernquist, Y. V. Krasnov, A. Mosbech, J. Reid, G. Robertson, B. Olsen, A. Petersen, H. Strom, and K. D. Wohl. 2008. Fluctuations in circumpolar seabird populations linked to climate oscillations. Global Change Biology 14: 1455–1463.

Johnston, D. W., A. S. Friedlaender, L. G. Torres, and D. M. Lavigne. 2005. Variation in sea ice cover on the east coast of Canada from 1969 to 2002: climate variability and implications for harp and hooded seals. Climate Research 29: 209–222.

Latour, P. B., J. Leger, J. E. Hines, M. L. Mallory, D. L. Mulders, H. G. Gilchrist, P. A. Smith, and D. L. Dickson. 2008. Key migratory bird terrestrial habitat sites in the Northwest Territories and Nunavut. Canadian Wildlife Service Occasional Paper No. 114.

Le Fouest, V., B. Zakardijian, F. J. Saucier, and M. Starr. 2005. Seasonal versus synoptic variability in planktonic production in a high-latitude marginal sea: the Gulf of St. Lawrence (Canada). Journal of Geophysical Research – Oceans 110: C09012.

Mallory, M. L., and A. J. Fontaine. 2004. Key marine habitat sites for migratory birds in Nunavut and the Northwest Territories. Canadian Wildlife Service Occasional Paper No. 109.

Mallory, M. L., B. M. Braune, M. Wayland, H. G. Gilchrist and D. L. Dickson. 2004. Contaminants in common eiders (*Somateria mollissima*) of the Canadian Arctic. Environmental Reviews 12: 197–218.

Mallory, M. L., J. Akearok, and H. G. Gilchrist. 2006. Local ecological knowledge of the Sleeper Islands and Split Island: comparisons among individuals and previous studies. In R. Riewe and J. Oakes (Eds.) Climate change: linking traditional and scientific knowledge, pp. 304–310. Aboriginal Issues Press, Winnipeg, MB.

McDonald, M., L. Arragutainaq, and Z. Novalinga. 1997. Voices from the bay, p. 98. Canadian Arctic Resources Committee, Ottawa.

Merkel, F. R. 2004. Impact of hunting and gillnet fishery on wintering eiders in Nuuk, southwest Greenland. Waterbirds 27: 469–479.

Merkel, F. R., and T. Christensen. 2008. Seabird harvest in Greenland. In F. Merkel and T. Barry (Eds.) Seabird harvest in the Arctic, pp. 41–49. Circumpolar Seabird Group (CBird), CAFF Technical Report No. 16, CAFF International Secretariat, Akureyri, Iceland.

Merkel, F. R., A. Mosbech, C. Sonne, A. Flagstad, K. Falk, and S. E. Jamieson. 2006. Local movements, home ranges and body condition of common eiders *Somateria mollissima* wintering in Southwest Greenland. Ardea 94: 639–650.

Montevecchi, W. A. 1993. Birds as indicators of change in marine prey stocks. In R. W. Furness and J. D. Greenwood (Eds.) Birds as monitors of environmental change. Chapman & Hall, London.

Mosbech, A., G. Gilchrist, F. Merkel, C. Sonne, A. Flagstad, and H. Nyegaard. 2006. Year- round movements of northern common eiders *Somateria mollissima borealis* breeding in Arctic Canada and west Greenland followed by satellite telemetry. Ardea 94: 651–665.

Nakashima, D. J., and D. J. Murray. 1988. The common eider (*Somateria mollissima sedentaria*) of eastern Hudson Bay: a survey of nest colonies and Inuit ecological knowledge. Ottawa, Environmental Studies Revolving Fund Report No. 102.

Parsons, M., I. Mitchell, A. Butler, N. Ratcliffe, M. Frederiksen, S. Foster, and J.B. Reid. 2008. Seabirds as indicators of the marine environment. ICES Journal of Marine Science 65: 1520–1526.

Priest, H., and P. J. Usher. 2004. The Nunavut wildlife harvest study. August 2004, p. 822. Nunavut Wildlife Management Board, Iqaluit, Nunavut.

Robertson, G. J. 1995a. Annual variation in common eider egg size: effects of temperature, clutch size, laying date, and laying sequence. Canadian Journal of Zoology 73: 1579–1587.

Robertson, G. J. 1995b. Factors affecting nest site selection and nesting success in the Eider *Somateria mollissima*. Ibis 137: 109–115.

Robertson, G. J., and H. G. Gilchrist. 1998. Evidence of populations declines among common eiders (*Somateria mollissima sedenteria*) breeding in the Belcher Islands, Northwest Territories. Arctic 51: 378–385.

Robertson, G. J., A. Reed, and H. G. Gilchrist. 2001. Clutch, egg and body size variation among common eiders breeding in Hudson Bay, Canada. Polar Research 20: 85–94.

Serreze, M. C., M. M. Holland, and J. Stroeve. 2007. Perspectives on the Arctic's shrinking sea-ice cover. Science 315: 1533–1536.

Stirling, I., and C. L. Parkinson. 2006. Possible effects of climate warming on selected populations of polar bears (*Ursus maritimus*) in the Canadian Arctic. Arctic 59: 261–275.

Wang, J., G. F. Cota, and J. C. Comiso. 2005. Phytoplankton in the Beaufort and Chukchi seas: distribution, dynamics, and environmental forcing. Deep-Sea Research Part II: Topical Studies in Oceanography 52: 3355–3368.

Wayland, M., H. G. Gilchrist, D. L. Dickson, T. Bollinger, C. James, R. A. Carreno, and J. Keating. 2001. Trace elements in king eiders and common eiders in the Canadian Arctic. Archives of Environmental Contamination and Toxicology 41: 491–500.

Welch, H. E., M. A. Bergmann, T. D. Siferd, K. A. Martin, M. F. Curtis, R. E. Crawford, R. J. Conover, and H. Hop. 1992. Energy flow through the marine ecosystem of the Lancaster Sound region, Arctic Canada. Arctic 45: 343–357.

Williams, T. D. 1994. Intraspecific variation in egg size and egg composition in birds: effects on offspring fitness. Biological Reviews of the Cambridge Philosophical Society 68: 35–39.

Wiese, F. K., G. J. Robertson, and A. J. Gaston. 2004. Impacts of chronic marine oil pollution and the murre hunt in Newfoundland on thick-billed murre *Uria lomvia* populations in the eastern Canadian Arctic. Biological Conservation 116: 205–216.

Woo, K. J., K. H. Elliott, M. Davidson, A. J. Gaston, and G. K. Davoren. 2008. Individual specialization in diet by a generalist marine predator reflects specialization in foraging behaviour. Journal of Animal Ecology 77: 1082–1091.

Temporal Trends in Beluga, Narwhal and Walrus Mercury Levels: Links to Climate Change

A. Gaden and G.A. Stern

Abstract The exposure of Arctic marine mammals to contaminants may change via ecological dynamics in response to climate change. For example, changes to the structure of the food web or shifts in regional foraging could affect dietary exposure. We examined the temporal variation of total mercury (THg) concentrations in Hudson Bay beluga (*Delphinapterus leucas*) and Foxe Basin walrus (*Odobenus rosmarus rosmarus*) and narwhal (*Monodon monoceros*) with $\delta^{15}N$ and $\delta^{13}C$ signatures (beluga only) and the North Atlantic Oscillation. We found THg concentrations in female Arviat beluga muscle tissue decreased significantly from the early 1980s to 2008. Similarly $\delta^{13}C$ signatures in beluga sampled from Arviat declined over the same time period. $\delta^{15}N$ and the NAO index did not appear to significantly change over time nor strongly influence THg concentrations. Results suggest beluga summering in Arviat may forage in more offshore areas upon less contaminated prey in response to the increasing ice-free season over the last couple of decades. As sea ice continues to recede, dietary mercury exposure may continually decrease in beluga and other marine mammals.

Keywords Marine mammals • Contaminants • $\delta^{15}N$ • $\delta^{13}C$ • NAO index • Diet

Introduction

Monitoring contaminant concentrations in marine mammals is necessary to regulate consumption guidelines for the safety, health, and well-being of northern communities and to protect marine mammal health. In a time of accelerating warming in the Arctic (Overpeck et al. 1997), these monitoring efforts are particularly important to document what types of changes, if any, marine mammals are experiencing.

G.A. Stern (✉)
Fisheries and Oceans Canada, 501 University Crescent, Winnipeg, MB R3T 2N6, Canada
e-mail: Gary.Stern@dfo-mpo.gc.ca

S.H. Ferguson et al. (eds.), *A Little Less Arctic: Top Predators in the World's Largest Northern Inland Sea, Hudson Bay*, DOI 10.1007/978-90-481-9121-5_10,
© Springer Science+Business Media B.V. 2010

197

Mercury (Hg) is both a natural element and an anthropogenic pollutant that infiltrates all biotic and abiotic components in aquatic ecosystems. Sources of Hg to Hudson Bay in order of decreasing magnitude are rivers, resuspension from glacial till on the seafloor, the atmosphere, the Arctic Ocean and coastal erosion (Hare et al. 2008). A mass balance model for Hudson Bay demonstrates these influxes are near equilibrium with exportation mechanisms (sedimentation and transport to the North Atlantic Ocean, Hare et al. 2008). A net of ~0.1 tonne Hg is added to the waterbody annually. Like the Arctic Ocean, Hudson Bay seawater is saturated with Hg (Outridge et al. 2008), so imports or exports to the overall Hg pool would not noticeably affect the 1 tonne (~1%) of Hg bound up in the biota (Hare et al. 2008).

Of the various species of Hg that exist in the marine environment, it is the methylated species (methyl mercury; MeHg) which is both bioavailable and toxic to organisms. Methylation, the process of converting inorganic Hg to an organic form, is mediated by sulfate-reducing bacteria in estuaries and coasts (Sunderland et al. 2004). Recently, high MeHg concentrations have been measured in the lower depths and in the sub-surface thermocline portion of the water column (Monperrus et al. 2007; Kirk et al. 2008; Cossa et al. 2009; Sunderland et al. 2009). MeHg provides no biological use to organisms (World Health Organization 2002), yet in marine ecosystems it is readily adsorbed from the water column by microorganisms and organic matter (Morel et al. 1998; Ravichandran 2004) and is then consumed by (and correlated to) higher trophic levels (Atwell et al. 1998; Stern and Macdonald 2005; Dehn et al. 2006; Pazerniuk 2007; Loseto et al. 2008a). Therefore, due to different diets (Stewart and Lockhart 2005), marine mammals at higher trophic levels such as beluga (*Delphinapterus leucas*) and narwhal (*Monodon monoceros*) tend to have higher contaminant concentrations than lower trophic level marine mammals like walrus (*Odobenus rosmarus rosmarus*; i.e. Wagemann et al. 1995).

The elimination rate of MeHg in animals is slower compared to its rate of intake and thus bioaccumulates in the body (Wagemann et al. 1996, 1998; Carrier et al. 2001; Nigro et al. 2002). Because of their high trophic level and long life spans, marine mammals accrue large quantities of MeHg. In mammals, MeHg accumulates primarily in the liver where it undergoes demethylation in the presence of selenium, forming a less toxic, inorganic compound (mercuric selenide (HgSe), Koeman et al. 1973; Ikemoto et al. 2004). This demethylation process leaves about only 15% MeHg in the liver (Wagemann et al. 1998). Nonetheless, marine mammals face ongoing exposure to MeHg from feeding on MeHg-contaminated prey (Loseto et al. 2008a, b), so MeHg is continually entering into their tissues. In muscle tissue, for example, MeHg is the exclusive species of Hg (Wagemann et al. 1998) and its concentration in muscle reflects relatively recent dietary exposure (Loseto et al. 2008b; Gaden et al. 2009). In this text we refer to total mercury (THg) which was analyzed in both liver and muscle tissues, but the composition of the term is different for each tissue as noted above: liver THg is a product of a majority of inorganic mercury and some MeHg; muscle THg is made up almost entirely of MeHg.

The dominant effect resulting from high MeHg exposure to mammals ranging from mink (*Mustela vison*) to polar bear (*Ursus maritimus*) is hindered functioning of the brain and central nervous system (Ronald et al. 1977; Basu et al. 2005, 2007, 2009). The immune systems of beluga whales (De Guise et al. 1996) and harbour porpoises (*Phocoena phocoena*; Bennett et al. 2001) have also been observed to function abnormally from MeHg-contaminated diets. Few studies have determined threshold concentrations for health effects in marine mammals, although Law (1996) reported 60 µg/g THg (wet weight) in the liver of marine mammals was damaging to hepatic processes. In humans, health effects from high MeHg exposure include subdued neurological functions in young children (via prenatal exposure) and high blood pressure (Grandjean et al. 1992, 1997, 2003; Sørensen et al. 1999).

Few studies have found clear temporal trends in THg among Canadian Arctic marine biota from the 1980s to 2000s. Lockhart et al. (2005) documented a strong trend in Mackenzie Delta beluga in which age-adjusted liver THg concentrations increased throughout the 1980s to mid-1990s. THg content in the eggs of northern fulmars (*Fulmarus glacialis*) and thick-billed murres (*Uria lomvia*) significantly increased from 1975 to 2003 (Braune 2007) as well. Using a global scale model with a time period of 100 years, Booth and Zeller (2005) noted an increase in MeHg levels in all marine organisms, including marine mammals. However, the majority of studies investigating THg in Arctic marine mammal populations have shown considerable variation in concentrations over time (Muir and Kwan 2003; Muir et al. 2001, 2006; Lockhart et al. 2005; Gaden et al. 2009).

Contaminant concentrations can also vary spatially among marine mammals. Spatial variation can result from differences in geography and ambient environmental concentrations. For example, in the late 1980s to early 1990s western Canadian Arctic marine mammals were observed to have higher concentrations of mercury compared to their counterparts in the eastern Canadian Arctic (Wagemann et al. 1996). At a smaller scale, Hare et al. (2008) found seawater THg concentrations were higher in Hudson Bay compared to Foxe Basin. Contaminant concentrations among spatially segregated populations may also emanate from differences in food-web structure or length (Hoekstra et al. 2003; Braune 2007; Pazerniuk 2007).

How contaminant exposure to marine mammals will change over time in association to climate change is uncertain although some progress had been made (Gaden et al. 2009; McKinney et al. 2009). Hudson Bay has experienced earlier break-up and later freeze-up over the last few decades (Gagnon and Gough 2005; Ford et al. 2009), and the length of the ice-free season in Hudson Bay and Foxe Basin has increased by 1.3 and 2.1 days/year, respectively, from 1979 to 2006 (Rodrigues 2009). Such large-scale environmental change could trigger ecological responses including shifts in the availability, abundance and types of prey species, influencing MeHg exposure to marine mammals (Tynan and DeMaster 1997; Macdonald et al. 2005; Learmonth et al. 2006; Simmonds and Isaac 2007; Burek et al. 2008).

We investigate the THg concentrations in Hudson Bay beluga and Foxe Basin walrus and narwhal harvested throughout the 1980s–2000s. THg concentrations are analyzed both over time, with length of the ice-free season, and the North Atlantic Oscillation (NAO). We include stable isotope ratios of nitrogen and carbon in our investigation to determine changes in trophic level ($\delta^{15}N$) and regional foraging ($\delta^{13}C$) (Hobson and Welch 1992; Cherel and Hobson 2007) which may influence contaminant exposure.

Methods

Field Sampling

Each year hunters from communities in the Canadian Arctic lend invaluable skills and expertise in harvest-based monitoring programs. Most marine mammal samples analyzed in this study were harvested in the open-water season. Liver and muscle tissues were collected from beluga at Arviat (1984–2008) and Sanikiluaq (1994–2008). From Sanikiluaq there were no female beluga harvested in 2002, nor were there any male muscle samples from 2004. No liver was available from 2005 female Arviat beluga. There were fewer beluga muscle samples available than liver, particularly for Arviat. Liver samples were harvested from walrus at Hall Beach (1988–2008) and Igloolik (1982–2008) and narwhal from Repulse Bay (1993–2001). Specific years and sample sizes are presented in Tables 1 and 2.

Sex and length were recorded from the samples, and the lower jaws (excluding narwhal) were collected for determining age estimates. Tissues were frozen soon

Table 1 Annual mean ages, lengths and THg (in liver, μg/g wet weight) ± standard errors in Hudson Bay beluga (separated by sex), walrus and narwhal (Liver THg values by Wagemann et al. 1983, 1995, 1996, 1998; Wagemann and Stewart 1994; Lockhart et al. 2005; Stern and Lockhart 2009)

Species, location	Year	N	Age	Length (cm)	THg in liver
Beluga,	1984	18	23.3 ± 3.0	337 ± 14	6.8 ± 1.5
Arviat (females)	1986	10	21.8 ± 5.7	306 ± 17	7.5 ± 3.0
	1997	5	33.2 ± 4.0	367 ± 19	20.6 ± 6.3
	1999	17	21.2 ± 2.4	321 ± 10	12.8 ± 2.1
	2003	13	19.7 ± 2.4	324 ± 7	17.7 ± 4.5
	2007	3	31.0 ± 11.6	230 ± 14	14.9 ± 5.0
	2008	6	23.7 ± 10.1	363 ± 16	11.6 ± 4.4
Beluga,	1984	3	24.0 ± 9.9	299 ± 40	10.2 ± 6.0
Arviat (males)	1986	5	25.4 ± 5.5	343 ± 29	7.6 ± 3.2
	1997	4	27.8 ± 5.1	371 ± 16	12.0 ± 6.9
	1999	15	22.8 ± 2.5	350 ± 13	12.4 ± 3.2
	2003	19	19.3 ± 2.6	343 ± 12	5.8 ± 1.4
	2005	9	21.3 ± 3.9		15.1 ± 5.9
	2007	9	7.9 ± 3.3	329 ± 20	4.7 ± 1.4
	2008	6	17.3 ± 6.5	363 ± 15	8.9 ± 2.9

(continued)

Table 1   (continued)

Species, location	Year	N	Age	Length (cm)	THg in liver
Beluga, Sanikiluaq	1994	12	26.4 ± 3.4	307 ± 12	8.1 ± 1.5
(females)	1995	7	24 ± 5.5	383 ± 23	11.3 ± 2.1
	1998	9	28.7 ± 3.9		21.1 ± 4.4
	2003	4	17.8 ± 8.0	273 ± 16	7.6 ± 3.6
	2004	8	40.9 ± 7.2	375 ± 15	13.3 ± 1.5
	2005	4	27.8 ± 11.6		18.2 ± 11.0
	2007	4	16.3 ± 5.1	305 ± 14	8.4 ± 4.1
	2008	5	18.5 ± 3.7	328 ± 2.6	10.2 ± 2.9
Beluga,	1994	15	29.9 ± 2.8	376 ± 14	17.2 ± 22.8
Sanikiluaq (males)	1995	7	38.1 ± 6.1	386 ± 49	42.9 ± 15.8
	1998	10	23.1 ± 3.3		24.1 ± 11.3
	2002	6	22.5 ± 3.9	372 ± 32	10.2 ± 4.3
	2003	7	17.9 ± 3.3	353 ± 33	5.8 ± 1.2
	2005	8	22.5 ± 2.8		15.9 ± 4.1
	2007	9	18.3 ± 5.2	366 ± 12	10.1 ± 4.0
	2008	7	16.0 ± 2.7	362 ± 10	8.2 ± 2.1
Walrus, Hall Beach and	1982	11	12.7 ± 1.1	287 ± 15	1.1 ± 0.26
Igloolik	1983	23	13.0 ± 0.9	284 ± 5	1.2 ± 0.18
	1987	15	9.4 ± 1.2	268 ± 12	1.2 ± 0.26
	1988	28	9.1 ± 0.9	256 ± 11	1.4 ± 0.21
	1992	20	11.1 ± 1.5	257 ± 14	1.2 ± 0.29
	1993	5	18.6 ± 1.4	320 ± 5	2.5 ± 0.43
	1996	29	15.7 ± 1.5	287 ± 11	1.9 ± 0.30
	2004	15	14.0 ± 0.9	302 ± 5	1.5 ± 0.2
	2007	7		294 ± 29	4.9 ± 1.5
	2008	7		300 ± 9	1.8 ± 1.1
Narwhal, Repulse Bay	1993	4		398 ± 7	8.9 ± 4.6
	1999	16		365 ± 16	12.0 ± 1.8
	2001	10		432 ± 11	9.8 ± 1.3

upon retrieval and shipped to the Freshwater Institute (Winnipeg, MB) for contaminant analysis. Individual growth layer groups (GLG) in the dentine portion of beluga and walrus teeth were counted as 1 year each (Garlich-Miller et al. 1993; Stewart et al. 2006).

Chemistry Techniques

Approximately 0.2 and 0.1 g of muscle and liver tissue, respectively, were used to analyze THg content. The samples were submerged in acids overnight and then placed on a heating block for 2 h the following morning. After samples were diluted with deionised water, they were analyzed for THg using Cold Vapour

Table 2 Annual mean ages, lengths and THg (in muscle, μg/g wet weight) ± standard errors in Hudson Bay beluga by location and sex

Location, sex	Year	N	Age	Length (cm)	THg in muscle
Arviat, female	1984	11	21.6 ± 3.8	326 ± 22	1.1 ± 0.1
	1986	10	21.8 ± 5.7	306 ± 17	1.2 ± 0.3
	1997	5	33.2 ± 4.0	367 ± 19	1.1 ± 0.2
	1999	10	19.8 ± 1.9	329 ± 10	0.9 ± 0.08
	2003	13	19.7 ± 2.4	324 ± 7	1.1 ± 0.1
	2005	2	17.0 ± 13.0		0.84 ± 0.47
	2007	3	31.0 ± 11.6	230 ± 14	0.86 ± 0.2
	2008	6	23.7 ± 10.1	363 ± 16	0.75 ± 0.08
Arviat, male	1984	3	24.0 ± 9.9	299 ± 40	1.1 ± 0.3
	1986	4	30.5 ± 2.7	372 ± 5	1.3 ± 0.069
	1997	4	27.8 ± 5.1	371 ± 16	0.88 ± 0.13
	1999	7	24.4 ± 3.6	362 ± 16	0.84 ± 0.047
	2003	16	18.3 ± 3.1	334 ± 13	0.68 ± 0.049
	2005	9	21.3 ± 3.9		1.1 ± 0.32
	2007	9	7.9 ± 3.3	329 ± 20	0.78 ± 0.15
	2008	6	17.3 ± 6.5	363 ± 15	0.98 ± 0.20
Sanikiluaq, female	1994	12	26.4 ± 3.4	307 ± 12	0.71 ± 0.037
	1995	7	24 ± 5.5	383 ± 23	1.0 ± 0.15
	1998	9	28.7 ± 3.9		1.2 ± 0.17
	2003	4	17.8 ± 8.0	273 ± 16	0.58 ± 0.12
	2004	8	40.9 ± 7.2	375 ± 15	1.2 ± 0.17
	2005	4	32 ± 11.7		1.2 ± 0.38
	2007	3	16.3 ± 5.1	304 ± 20	0.49 ± 0.057
	2008	5	18.5 ± 3.7	328 ± 2.6	0.63 ± 0.022
Sanikiluaq, male	1994	15	29.9 ± 2.8	376 ± 14	1.3 ± 0.17
	1995	6	38.7 ± 7.2	403 ± 55	1.8 ± 0.53
	1998	10	23.1 ± 3.3		1.5 ± 0.40
	2002	6	22.5 ± 3.9	372 ± 32	0.89 ± 0.28
	2003	7	17.9 ± 3.3	353 ± 33	0.70 ± 0.14
	2005	6	22.2 ± 1.6		0.92 ± 0.13
	2007	6	18.3 ± 5.2	355 ± 16	0.95 ± 0.16
	2008	7	16.0 ± 2.7	362 ± 10	0.73 ± 0.11

Atomic Absorption Spectroscopy (CVAAS; Armstrong and Uthe 1971). LUTS-1, TORT-2 and CRM 2976 were used as standard reference material (80–90% recovered). The limit of detection was 0.005 μg/g wet weight. All data are reported in wet weight.

Liver samples were analyzed for stable isotope (SI) ratios ($\delta^{15}N$ and $\delta^{13}C$). We chose to analyze $\delta^{15}N$ and $\delta^{13}C$ from liver tissue because SI ratios appear to have a shorter turnover time there, reflecting relatively recent prey consumption and incorporation into tissues (Loseto et al. 2008b). No liver samples from 1994 Sanikiluaq were available for this analysis. Prior to SI analysis, lipids were removed from the samples to reduce the bias in the $\delta^{13}C$ values (Kurle and Worthy 2002). A 2:1 chloroform/methanol solution was added to 0.2 g freeze-dried liver in test tubes followed by mixing, centrifuging, and removing and replacing the solvent (three times; based on Folch et al. 1957). Dried samples were packaged and sent to the University of Winnipeg Stable Isotope Laboratory for SI analysis by Continuous Flow Ion Ratio Mass Spectroscopy (CFIR-MS). PeeDee Belemnite and IAEA-N-1 were used as standards for ^{13}C and ^{15}N analysis, respectively. The calculation for SI ratios is given in Loseto et al. (2008a, b). Here units of the stable isotope ratios are given in per mil (‰) with δ notation. Average error reported in the SI analysis was 0.17‰ for ^{13}C and 0.30‰ for ^{15}N.

Climate Analysis

The length of each ice-free season in Hudson Bay and Foxe Basin over the last several decades was observed to significantly increase over the study period (data adapted from Rodrigues 2009). Because 'year' and 'length of the ice-free season' were significantly correlated, we used 'year' only in the statistical analysis and focused on temporal trends. The annual wintertime NAO index (December–March, Gough et al. 2004) in the winter prior to the summer harvest was calculated from monthly data available on the National Oceanic and Atmospheric Administration website (NOAA, http://www.cpc.noaa.gov/products/precip/CWlink/pna/nao.shtml). We note that the NAO index was not significantly correlated to year.

Statistical Analysis

After log-transforming all data, we tested for the effects of age, length, day of year samples were harvested, location, sex and the interaction of these variables separately for each marine mammal species. Raw $\delta^{15}N$ and $\delta^{13}C$ values were normally distributed so we did not log-transform these for the statistical analyses. A student's t-test examined differences in THg between the sexes for each year. For the time-series analysis we first used a general linear model (GLM) to assess which variables were significantly associated with THg, $\delta^{15}N$ and $\delta^{13}C$ (including the NAO index, year, $\delta^{15}N$, $\delta^{13}C$, and a categorical variable defining individuals by sex and location). The temporal variation of raw THg concentrations was illustrated with boxplots. Finally we tested for differences between mean annual (log-transformed) THg within species using an analysis of covariance (ANCOVA) and Bonferroni post-hoc tests. All statistical analyses were carried out with SYSTAT 11.

Results

Mercury Concentrations in Three Marine Mammals

Beluga

Due to observations in our time-series analysis, we report mean data of the sexes separately in Tables 1 and 2. The interval of mean annual liver THg concentrations in Arviat beluga (5–21 µg/g) was half that of Sanikiluaq beluga (5–43 µg/g). The large range of liver THg values is likely a result of the wide range of ages. Mean THg values presented here agree with values obtained from other Canadian Arctic beluga in the 1990s and early 2000s (Muir 2005, 6–44 µg/g ww). Mean muscle THg concentrations in beluga fell within 0.68–1.3 µg/g in Arviat and 0.49–1.8 µg/g in Sanikiluaq and are similar to values reported in Beaufort Sea beluga (Loseto et al. 2008a, 0.67–1.70 µg/g ww[1]).

Sex and location were not significant factors influencing beluga liver THg concentrations. In muscle, however, mean THg concentrations in female Arviat animals were higher than those of Sanikiluaq females (F = 2.88, p = 0.036, age as covariate). This may be attributed to the longer length of the food web (which results in higher contaminant exposure to top predators) in western Hudson Bay in comparison to eastern Hudson Bay (Pazerniuk 2007). When 'year' was accounted for, the significance dropped (p = 0.078).

Age was significantly related to liver and muscle THg in beluga from both Arviat (r = 0.60, p < 0.001; r = 0.58, p < 0.001) and Sanikiluaq (r = 0.60, p < 0.001; r = 0.69, p < 0.001). In addition, length was related to THg in both tissues for both locations (p < 0.05). Positive relationships between age, length and THg in liver and muscle have been previously documented (Hansen et al. 1990; Wagemann et al. 1995; Siebert et al. 1999; Woshner et al. 2001). In beluga, age tends to be strongly associated with THg in liver whereas length is more closely associated with muscle THg (Loseto et al. 2008b). In this study there was no length data for 2005 Arviat or 1998 and 2005 Sanikiluaq beluga. However, length and age were significantly related (Arviat r = 0.56, p < 0.001; Sanikiluaq r = 0.51, p < 0.001). Taken together, these findings lead us to use age as a covariate to control variation in our temporal analysis of both liver and muscle THg.

Walrus

No significant effects were observed with sex, day of year, or location, so we pooled sexes and samples from both communities at Igloolik and Hall Beach. Liver

---

[1]Conversion of THg concentration in dry weight to wet weight was calculated assuming 74% moisture content in beluga muscle (DFO archives).

THg concentrations fell within a small range of 1–5 µg/g (Table 1) which is similar to concentrations reported for walrus at Inukjuaq (2.64 µg/g ww in liver, Muir et al. 2000). Age was significantly related to THg in liver ($r = 0.56$, $p < 0.001$), as was length to THg ($r = 0.46$, $p < 0.001$). However, without age data for either community in 2007 and 2008, and with walrus age and length significantly related ($r = 0.87$, $p < 0.001$), we chose to use length as a covariate in controlling variation of data in the temporal analysis.

Narwhal

Narwhal liver samples were collected at Repulse Bay during 1993, 1999 and 2001. We pooled the sexes since there were fewer samples. Mean annual liver THg values fell between 8 and 12 µg/g (Table 1) and coincided with concentrations measured in narwhal from western Greenland (2–17 µg/g ww, Dietz et al. 2004) and in Pond Inlet (mean THg in liver: 12.9 µg/g ww, Wagemann et al. 1983). Narwhal length was positively related to concentrations of liver THg ($r = 0.37$, $p = 0.048$), so it was selected as a covariate for the time series analysis.

Time Series Analysis

Beluga

Female beluga from Arviat had significantly higher liver THg content than males in 2003 ($t = -2.173$, $p = 0.038$), and 1994 Sanikiluaq male beluga had significantly higher liver THg than females ($t = 2.377$, $p = 0.025$). Due to these observations we plotted male and female concentrations of THg in liver separately (Fig. 1a–d). Regression coefficients of age-adjusted liver THg by year are presented in Table 3. Only the female beluga from Arviat had liver THg concentrations that significantly changed (increased) over the years of collection. There were no significant differences in annual mean liver THg between sex or location. Apart from age, the GLM detected no other significant factors associated with liver THg concentrations.

Males from Sanikiluaq had significantly higher THg concentrations in muscle compared to females in 2007 ($t = 2.877$, $p = 0.024$) (Fig. 2c, d). In contrast to liver, age-adjusted muscle THg concentrations declined over time in all beluga and the trend was significant in Arviat female animals (Table 3). THg concentrations in Sanikiluaq female beluga were significantly higher in 1998 compared to 1994 (Bonferroni $p = 0.018$). The GLM for muscle THg revealed that age was the only significant factor (Table 4).

The importance of $\delta^{15}N$ and $\delta^{13}C$ driving beluga THg concentrations was assessed with GLMs. $\delta^{13}C$ was the only variable to significantly influence $\delta^{15}N$ values (Table 4). The GLM for $\delta^{13}C$ detected four significant variables: $\delta^{15}N$, year,

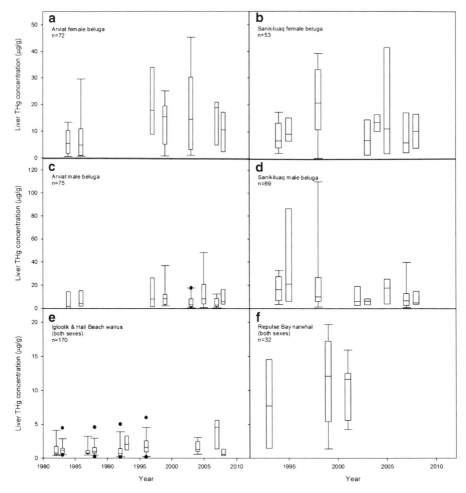

Fig. 1 Box and whisker plots of yearly THg concentrations (μg/g ww) in beluga liver sampled in Arviat (females (**a**), males (**c**)) and Sanikiluaq (females (**b**), males (**d**)) and walrus (**e**) and narwhal (**f**) sampled in Foxe Basin. *Lines in boxes* represent mean concentrations and *whiskers* represent 5th and 95th percentiles. Outliers are indicated by black dots

NAO index and sex/location. An ANCOVA (year as covariate) revealed no significant difference in $\delta^{13}C$ between sexes for Arviat and Sanikiluaq beluga. However, Arviat beluga were significantly enriched in $\delta^{13}C$ compared to Sanikiluaq beluga (F = 32.52, p < 0.001). Mean annual $\delta^{13}C$ was regressed as a function of both year and NAO by location (sex pooled) to illustrate these specific relationships (Fig. 3). $\delta^{13}C$ decreased over time in beluga sampled from Arviat (r = −0.70, p = 0.05). Only in Sanikiluaq animals did $\delta^{13}C$ appear to respond to the NAO index in a positive relationship (r = 0.77, p = 0.03). After removal of the

Table 3 Time-series analysis of THg in liver and muscle tissues of Hudson Bay marine mammals via linear regression. Prior to analysis THg concentrations in beluga were adjusted to age; THg concentrations in walrus and narwhal were adjusted to length

Species, location	Regression coefficient	
	Liver	Muscle
Beluga, Arviat (females)	0.87*	−0.80*
Beluga, Arviat (males)	0.46	−0.23
Beluga, Sanikiluaq (females)	−0.18	−0.34
Beluga, Sanikiluaq (males)	−0.42	−0.31
Walrus, Hall Beach and Igloolik (both sexes)	0.51	–
Narwhal, Repulse Bay (both sexes)	0.59	–

*$p < 0.05$

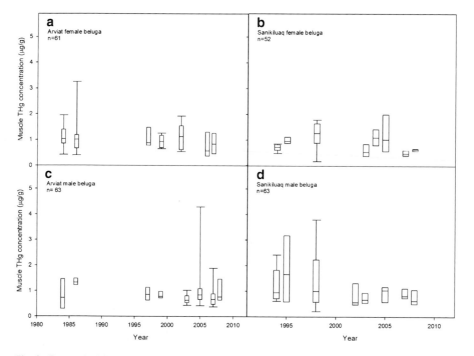

Fig. 2 Box and whisker plots of annual THg concentrations in beluga muscle (μg/g ww) sampled at Arviat (females (**a**), males (**c**)) and Sanikiluaq (females (**b**), males (**d**)). *Lines in boxes* represent mean concentrations and *whiskers* represent fifth and 95th percentiles

highest mean δ¹³C in Sanikiluaq beluga in 1995 (NAO 1.34), the significance of both year and the NAO index was lost from the GLM. We also tried removing 1995 Sanikiluaq samples from the GLM, but the significance of 'year' in the model was retained.

Table 4 Summary of three generalized linear models of THg in muscle, $\delta^{15}N$ and $\delta^{13}C$ for Hudson Bay beluga. Bold variables explain observed variance in each model at a significant level of $\alpha \leq 0.05$

Response variable	Explanatory variable	Sum of squares	Degrees of freedom	Mean-square	F-ratio	p
THg in muscle	Sex_Location	0.04	3	0.01	0.41	0.74
	Year	0.41	1	0.41	13.70	<0.001
	Age	3.57	1	3.57	118.20	<0.001
	$\delta^{15}N$	0.03	1	0.03	1.07	0.30
	$\delta^{13}C$	0.002	1	0.002	0.06	0.81
	NAO	0.05	1	0.05	1.65	0.20
	Error	5.38	178	0.03		
$\delta^{15}N$	Sex_Location	13.38	3	4.46	1.91	0.13
	Year	3.12	1	3.12	1.33	0.25
	Age	1.90	1	1.90	0.81	0.37
	$\delta^{13}C$	17.71	1	17.71	7.57	0.007
	NAO	8.35	1	8.35	3.57	0.06
	Error	451.61	193	2.34		
$\delta^{13}C$	**Sex_Location**	26.53	3	8.84	7.222	<0.001
	Year	6.74	1	6.74	5.50	0.02
	Age	1.27	1	1.27	1.04	0.31
	$\delta^{15}N$	9.27	1	9.27	7.57	0.007
	NAO	7.62	1	7.62	6.22	0.01
	Error	236.32	193	1.22		

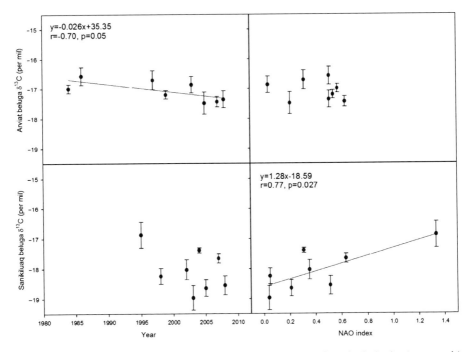

Fig. 3 Annual means and standard error bars of $\delta^{13}C$ in beluga from both Arviat (*top panels*) and Sanikiluaq (*lower panels*) regressed as a function of year and the NAO index

Walrus

Length-adjusted THg concentrations in Igloolik and Hall Beach walrus liver did not change significantly over the 26-year period (Fig. 1e, Table 3). Liver THg concentrations in samples from 2007, however, were significantly higher than those from 1982, 1983 and 1992 (length as covariate; Bonferroni p = 0.02, 0.024, 0.011, respectively). The NAO index was not significantly associated with liver THg concentrations as assessed with the GLM.

Narwhal

Length-adjusted liver THg concentrations in Repulse Bay narwhal also did not show any significant temporal trend (Fig. 1f, Table 3). There were no significant differences between THg concentrations analyzed from the 3 years samples were collected. Again the NAO index was not significantly associated with liver THg concentrations in narwhal.

Discussion

Trends in Liver Mercury Concentrations

An increasing trend in liver THg was observed only in female beluga from Arviat. The lack of a trend in male belugas may suggest habitat segregation and differences between prey selection between the sexes (Loseto et al. 2006, 2008b). Liver THg increased in Mackenzie Delta beluga from 1984 to 1996, although sexes were pooled in the analysis because there was no effect of gender (Lockhart et al. 2005). However, similar to this study, the remaining Canadian Arctic beluga populations examined showed no strong trends in liver THg from the early 1980s to 2000s. Walrus THg did not change significantly, corresponding to previous work in Northwest Greenland (Riget et al. 2007), nor did THg in narwhal. Because the half-life of Hg in liver is relatively long (10 years in humans; Friberg et al. 1979) and MeHg is ultimately converted to HgSe in marine mammal liver (Koeman et al. 1973; Ikemoto et al. 2004), liver THg may be better suited to monitoring longer term trends of Hg in the environment compared to muscle.

Beluga: Trends in Muscle Mercury and Stable Isotopes

Mercury concentrations in muscle may be more useful for dietary related trend analysis since these concentrations reflect relatively recent MeHg exposure (Loseto et al. 2008b; Gaden et al. 2009). The THg concentrations in muscle of female beluga from Arviat declined from 1984 to 2008, suggesting a decrease in dietary exposure. Although little is known about the Hudson Bay beluga diet, capelin (*Mallotus villosus*) was found to be an important prey species in the 1980s (Kelley et al. this volume), however Arctic cod (*Boreogadus saida*) is thought to dominate the diet of other circumpolar beluga populations (e.g. Loseto et al. 2009; Dahl et al. 2000). There are no temporal fish data sets in Hudson Bay to support trends in dietary exposure. A long term temporal data set for High Arctic land locked char (*Salvelinus alpinus*) showed a decrease in muscle THg in 2006 relative to 1992 that was related to diet (measured via $\delta^{15}N$ levels) (Gantner et al. 2009). Opposite trends in muscle THg were recorded in female adult hooded seals (*Cystophora cristata*) from the Greenland Sea where 1999 concentrations were significantly higher than 1985 values (Brunborg et al. 2006). Few data sets have found significant increasing or decreasing trends largely due to a few data points in which to extract significant time trends (Arctic Monitoring and Assessment Programme 2007). The inclusion of SI ratios can significantly alter results in THg trend analysis (Riget et al. 2007). The relatively unchanged $\delta^{15}N$ values in this study imply that beluga in Hudson Bay are not necessarily consuming different species or at least not switching to higher or lower trophic level prey.

$\delta^{13}C$, on the other hand, significantly decreased in Arviat animals, but $\delta^{13}C$ was not significantly related to THg in muscle. In beluga, concentrations of muscle THg are associated with SI ratios in liver, although $\delta^{13}C$ may have a longer turnover rate compared to THg (Loseto et al. 2008b), which may explain why THg in muscle in this study was not significantly related to $\delta^{13}C$ in liver even though both variables declined over time. The downward, increasingly negative trend in $\delta^{13}C$ signatures in Arviat beluga samples suggest that belugas that summer in Arviat may be spending more time foraging in further offshore regions or that they are more dependent on pelagic sources of prey where $\delta^{13}C$ are typically more depleted (Hobson and Welch 1992; Cherel and Hobson 2007).

Dietary changes have been observed elsewhere in Hudson Bay. Thick-billed murres (*Uria lomvia*) in northern Hudson Bay gradually began favouring capelin and sandlance in place of Arctic cod and sculpin from the early 1980s to 2002 (Gaston et al. 2003; Braune 2009; Mallory et al. this volume). Braune (2009) reported lower $\delta^{15}N$ in the eggs of thick-billed murres on Coats Island (northern Hudson Bay) from 1998 to 2007 in comparison to 1993, but there was no clear temporal trend in THg. The lower $\delta^{15}N$ in the seabirds in the more recent years was likely a result of the lower trophic position of sandlance and capelin in comparison to Arctic cod (Braune 2009). The shift in prey species may be the response of increasingly longer ice-free seasons (Gaston and Hipfner 1998; Gaston et al. 2009; Mallory et al. this volume). McKinney et al. (2009) also observed a climate-related dietary shift in the western Hudson Bay subpopulation of polar bear: with longer open water periods, the proportion of open-water seals (harbour and harp seals; *Phoca vitulina* and *Phoca groenlandica*, respectively) increased and the fraction of seals living in pack ice (bearded seals; *Erignathus barbatus*) decreased in the diet.

We believe the decreasing trends in THg and $\delta^{13}C$ in female beluga from Arviat are driven primarily by ecological shifts influenced by warming in Hudson Bay

(Gagnon and Gough 2005; Rodrigues 2009) – not by changing ambient THg concentrations. To illustrate this point, we refer to a study by Kirk and St. Louis (2009): river flows and Hg export to Hudson Bay were particularly high from the Nelson and Churchill Rivers in 2005 (Kirk and St. Louis 2009), yet we observed no parallel increases in THg in Arviat or Sanikiluaq beluga the same or following years.

Other ecological factors which may have contributed to the declining THg concentrations in female Arviat beluga are (1) elevated primary production levels associated with warming temperatures and (2) shifts in the migrations of the beluga. The first example involves high volumes of pelagic organic matter produced as a result of the lengthening ice-free season. The large quantity of organic matter may have adsorbed a higher quantity of MeHg from the water column, sunk to the benthos, and overall lowered the contaminant concentration for bioaccumulation (Outridge et al. 2005, 2007, 2008; Kuzyk et al. in press). Additionally, the warming climate may have affected the migrations of beluga, either in the timing or routes, due to altered prey distribution or travel accessibility (Tynan and DeMaster 1997; Learmonth et al. 2006; Stern and Macdonald 2005). Accounting for migration patterns in relation to diet shifts (and associated Hg exposure) in Hudson Bay beluga presents a challenge. The seasonal distribution and migration patterns appear variable even at the individual level (Richard and Orr 2003) paralleling genetic studies showing beluga stocks can mix amongst each other (De Marche and Postma 2003). Whereas we cannot submit evidence to suggest Hudson Bay beluga may have changed their migration patterns over the study period, we also cannot overlook the potential this possibility holds for explaining our observations.

The NAO index did not appear to influence THg in any marine mammals, nor $\delta^{15}N$ or $\delta^{13}C$ in beluga; its impacts to dietary exposure of MeHg and prey selection may be indirect and more complex. However, we did observe relatively high $\delta^{13}C$ signatures in Sanikiluaq beluga in 1995 which had the maximal NAO index (1.34) for the sampling years here (Fig. 3). High positive phases of the NAO tend to be associated with low temperatures and more ice in Hudson Bay (Qian et al. 2008). The Canadian Ice Service (http://ice-glaces.ec.gc.ca/) also reports above-average ice cover in 1994 and 1995 for eastern Hudson Bay. Thus the high $\delta^{13}C$ signatures in 1995 Sanikiluaq beluga may indicate that animals were foraging upon more inshore or benthic prey that year due to heavy ice conditions (Hobson and Welch 1992; Cherel and Hobson 2007).

Conclusion

Decreasing trends in muscle THg and $\delta^{13}C$ in Arviat beluga indicate that with increasing temperatures and longer duration of the ice-free period over time in Hudson Bay (Gagnon and Gough 2005), foraging activities of beluga summering in western Hudson Bay may be shifting spatially in favour of offshore, less MeHg-contaminated prey. Additionally, dietary MeHg exposure may have declined over time in response to a transfer of pelagic MeHg to the benthic environment (Outridge et al. 2005, 2007, 2008; Kuzyk et al. in press). Progressive warming in the Arctic and sub-arctic (Walsh 2008) may continue to ultimately lower THg concentrations in beluga muscle tissue. It is imperative that temporal and spatial studies on beluga and the food webs that support them continue to be able to identify changes within the marine ecosystem that are responding to climate change.

References

Arctic Monitoring and Assessment Programme (AMAP) (2007) AMAP workshop on statistical analysis of temporal trends of mercury in biota Arctic: Report of the workshop, Swedish Museum of Natural History, Stockholm, 30 October to 3 November 2006. www.amap.no

Armstrong JAJ, Uthe JF (1971) Semi-automated determination of mercury in animal tissue. At Abs News 10:101–103

Atwell L, Hobson KA, Welch HE (1998) Biomagnification and bioaccumulation of mercury in an arctic marine food web: insights from stable N isotope analysis. Can J Fish Aquat Sci 55:1114–1121

Basu N, Klenavic K, Gamberg M, O'Brien M, Evans RD, Scheuhammer AM, Chan HM (2005) Effects of mercury on neuro-chemical receptor binding characteristics in wild mink. Environ Toxicol Chem 24:1444–1450

Basu N, Scheuhammer AM, Rouvinen-Watt K, Grochowina N, Evans RD, O'Brien M, Chan HM (2007) Decreased N-methyl-d-aspartic acid (NMDA) receptor levels are associated with mercury exposure in wild and captive mink. Neurotoxicology 28:587–593

Basu N, Scheuhammer AM, Sonne C, Letcher RJ, Borne EW, Dietz R (2009) Is dietary mercury of neurotoxicological concern to wild polar bears (Ursus maritimus)? Environ Toxicol Chem 28:133–140

Bennett PM, Jepson PD, Law RJ, Jones BR, Kuiken T, Baker JR, Rogan E, Kirkwood JK (2001) Exposure to heavy metals and infectious disease mortality in harbour porpoises from England and Wales. Environ Pollut 112:33–40

Booth S, Zeller D (2005) Mercury, food webs, and marine mammals: implications of diet and climate change for human health. Environ Health Perspect 113:521–526

Braune BM (2007) Temporal trends of organochlorines and mercury in seabird eggs from the Canadian Arctic, 1975–2003. Environ Pollut 148:599–613

Braune BM (2009) Effect of climate change on diet and contaminant exposure in seabirds breeding in northern Hudson Bay. In: Synopsis of research conducted under the 2008–2009 Northern Contaminants Program; Smith S, Stow J, Edwards J (eds.) Indian and Northern Affairs Canada, Ottawa, pp. 269–272

Brunborg LA, Graff EI, Frøyland L, Julshamn K (2006) Levels of non-essential elements in muscle from harp seal (Phagophilus groenlandicus) and hooded seal (Cystophora cristata) caught in the Greenland Sea area. Sci Total Environ 366:784–798

Burek KA, Gulland FMD, O'Hara TM (2008) Effects of climate change on Arctic marine mammal health. Ecol Appl 18 (Suppl.):S126–S134

Canadian Ice Service. http://ice-glaces.ec.gc.ca/ (accessed November 10, 2009)

Carrier G, Brunet RC, Caza M, Bouchard M (2001) A toxicokinetic model for predicting the tissue distribution and elimination of organic and inorganic mercury following exposure to methylmercury in animals and humans. I. Development and validation of the model using experimental data in rats. Toxicol Appl Pharm 171:38–49

Cherel Y, Hobson KA (2007) Geographical variation in carbon stable isotope signatures of marine predators: a tool to investigate their foraging areas in the Southern Ocean. Mar Ecol Prog Ser 329: 281–287

Cossa D, Averty B, Pirrone N (2009) The origin of methylmercury in open Mediterranean waters. Limnol Oceanogr 54:837–844

Dahl TM, Lydersen C, Kovacs KM, Falk-Petersen S, Sargent J, Gjertz I, Gulliksen B (2000) Fatty acid composition of the blubber in white whales (*Delphinapterus leucas*). Polar Biol 23: 401–409

De Guise S, Bernier J, Martineau D, Béland P, Fournier M (1996) Effects of in vitro exposure of beluga whale splenocytes and thymocytes to heavy metals. Environ Toxicol Chem 15:1357–1364

De Marche BGE, Postma LD (2003) Molecular genetic stock discrimination of belugas (*Delphinapterus leucas*) hunted in Eastern Hudson Bay, Northern Quebec, Hudson Strait, and Sanikiluaq (Belcher Islands), Canada, and comparisons to adjacent populations. Arctic 56:111–124

Dehn L-A, Follmann EH, Thomas DL, Sheffield GG, Rosa C, Duffy LK, O'Hara TM (2006) Trophic relationships in an Arctic food web and implications for trace metal transfer. Sci Tot Environ 362:103–123

Dietz R, Riget F, Hobson KA, Heide-Jørgensen MP, Møller P, Cleemann M, de Boer J, Glasius M (2004) Regional and inter annual patterns of heavy metals, organochlorines and stable isotopes in narwhals (*Monodon monoceros*) from West Greenland. Sci Total Environ 331: 83–105

Folch J, Lees M, Stanley GHS (1957) A simple method for the isolation and purification of total lipids from animal tissues. J Biol Chem 226:497–509

Ford JD, Gough WA, Laidler GJ, MacDonald J, Irngaut C, Qrunnut K (2009) Sea ice, climate change, and community vulnerability in northern Foxe Basin, Canada. Clim Res 38:137–154

Friberg L, Nordberg GF, Vouk V (eds) (1979) *Handbook on the toxicology of metals*. Elsevier/North-Holland Biomedical Press, Amsterdam, the Netherlands

Gaden A, Ferguson SH, Harwood L, Melling H, Stern GA (2009) Mercury trends in ringed seals (*Phoca hispida*) from the Western Canadian Arctic since 1973: associations with length of the ice-free season. Environ Sci Technol 43:3646–3651

Gagnon AS, Gough WA (2005) Trends in the dates of ice free-up and breakup over Hudson Bay, Canada. Arctic 58:37–382

Gantner N, Power M, Babaluk JA, Reist JD, Köck G, Lockhart LW, Solomon KR, Muir DCG (2009) Temporal trends of mercury, cesium, potassium, selenium, and thallium in Arctic char (*Salvelinus alpines*) from Lake Hazen, Nunavut, Canada: effects of trophic potion, size and age. Environ Toxicol Chem 28:254–263

Garlich-Miller JL, Stewart REA, Stewart BE, Hiltz EA (1993) Comparison of mandibular with cemental growth-layer counts for ageing Atlantic walrus (*Odobenus rosmarus rosmarus*). Can J Zool 71:163 167

Gaston AJ, Hipfner M (1998) The effect of ice conditions in northern Hudson Bay on breeding by thick-billed murres (*Uria lomvia*). Can J Zool 76:480–492

Gaston AJ, Woo K, Hipfner JM (2003) Trends in forage fish populations in northern Hudson Bay since 1981, as determined from the diet of nestling thick-billed murres *Uria lomvia*. Arctic 56:227–233

Gaston AJ, Gilchrist HG, Mallory ML, Smith PA (2009) Changes in seasonal events, peak food availability, and consequent breeding adjustment in a marine bird: a case of progressive mismatching. Condor 111:111–119

Gough WA, Cornell AR, Tsuji LJS (2004) Trends in seasonal sea ice duration in southwestern Hudson Bay. Arctic 57: 299–305

Grandjean P, Weihe P, Videroe T, Joergensen PJ, Clarkson T, Cernichiari E (1992) Impact of maternal seafood diet on fetal exposure to mercury, selenium, and lead. Arch Environ Health 47:185–195

Grandjean P, Weihe P, White RF, Debes F, Araki S, Yokoyama K, Murata K, Sørensen N, Dahl R, Jørgensen PJ (1997) Cognitive deficit in 7-year-old children with prenatal exposure to methylmercury. Neurotoxicol Terratol 19:417–428

Grandjean P, White R, Weihe P, Jørgensen P (2003) Neurotoxic risk caused by stable and variable exposure to methylmercury from seafood. Ambul Pediatr 3:18–23

Hansen CT, Nielsen CO, Dietz R, Hansen MM (1990) Zinc, cadmium, mercury and selenium in minke whales, belugas and narwhales from West Greenland. Polar Biol 10:529–539

Hare A, Stern GA, Macdonald RW, Kuzyk ZZ, Wang F (2008) Contemporary and preindustrial mass budgets of mercury in the Hudson Bay Marine System: the role of sediment recycling. Sci Tot Environ 406:190–204

Hobson KA, Welch HE (1992) Determination of trophic relationships within a high Arctic marine food web using $\delta^{13}C$ and $\delta^{15}N$ analysis. Mar Ecol Prog Ser 84:9–18

Hoekstra PF, O'Hara, TM, Fisk AT, Borgå K, Solomon KR, Muir DCG (2003) Trophic transfer of persistent organochlorine contaminants (OCs) within an Arctic marine food web from the southern Beaufort-Chukchi Seas. Environ Pollut 124:509–522

Ikemoto T, Kunito T, Tanaka H, Baba N, Miyazaki N, Tanabe S (2004) Detoxification mechanism of heavy metals in marine mammals and seabirds: interaction of selenium with mercury, silver, copper, zinc, and cadmium in liver. Arch Environ Contam Toxicol 47:402–413

Kirk JL, St. Louis VL (2009) Multiyear total and methyl mercury exports from two major sub-arctic rivers draining into Hudson Bay, Canada. Environ Sci Technol 43:2254–2261

Kirk JL, St. Louis VL, Hintelmann H, Lehnherr I, Else B, Poissant L (2008). Methylated mercury species in marine waters of the Canadian high and sub arctic. Environ Sci Technol 42:8367–8373

Koeman JH, Peeters WHM, Koudstaal-Hol CHM, Tjioe PS, de Goeij JJM (1973) Mercury-selenium correlations in marine mammals. Nature 254:385–386

Kurle CM, Worthy GAJ (2002) Stable nitrogen and carbon isotope ratios in multiple tissues of the northern fur seal *Callorhinus ursinus*: implications for dietary and migratory reconstructions. Mar Ecol Prog Ser 236:289–300

Kuzyk ZZ, Macdonald RW, Johannessen SC, Stern GA (in press) Biogeochemical controls on PCB deposition in Hudson Bay. Env Sci Technol

Law RJ (1996) Metals in marine mammals. In: *Environmental contaminants in wildlife: interpreting tissue concentrations*; Beyer WN, Heinz GH, Redmon-Norwood AW (eds.) Lewis Publishers, Boca Raton, FL, pp. 357–376

Learmonth JA, MacLeod CD, Santos MB, Pierce GJ, Crick HQP, Robinson RA (2006) Potential effects of climate change on marine mammals. Oceanogr Mar Biol 44:431–464

Lockhart WL, Stern GA, Wagemann R, Hunt RV, Metner DA, DeLaronde J, Dunn B, Stewart REA, Hyatt CK, Harwood L, Mount K (2005) Concentrations of mercury in tissues of beluga whales (*Delphinapterus leucas*) from several communities in the Canadian Arctic from 1981 to 2002. Sci Total Environ 351–352:391–412

Loseto LL, Stern GA, Connelly TL, Deibel D, Gemmill B, Propokpowicz A, Fortier L, Ferguson SH (2009) Summer diet of beluga whales inferred by fatty acid analysis of the eastern Beaufort Sea food web. J Exp Mar Biol Ecol 374: 12–18

Loseto LL, Stern GA, Deibel D, Connelly TL, Prokopowicz A, Lean DRS, Fortier L, Ferguson SH (2008a) Linking mercury exposure to habitat and feeding behaviour in Beaufort Sea beluga whales. J Mar Syst 74:1012–1024

Loseto LL, Stern GA, Ferguson SH (2008b) Size and biomagnification: how habitat selection explains beluga mercury levels. Environ Sci Technol 42:3982–3988

Loseto LL, Richard P, Stern GA, Orr J, Ferguson SH (2006) Segregation of Beaufort Sea beluga whales during the open-water season. Can J Zool 84: 1743–1851

Macdonald RW, Harner T, Fyfe J (2005) Recent climate change in the Arctic and its impact on contaminant pathways and interpretation of temporal trend data. Sci Total Environ 342:5–86

McKinney MA, Peacock E, Letcher RJ (2009) Sea ice-associated diet change increases the levels of chlorinated and brominated contaminants in polar bears. Environ Sci Technol 43:4334–4339

Monperrus M, Tessier E, Amouroux D, Leynaert A, Huonnic P, Donard OFX (2007) Mercury methylation, demethylation and reduction rates in coastal and marine surface waters of the Mediterranean Sea. Mar Chem 107:49–63

Morel FMM, Kraepiel AML, Amyot M (1998) The chemical cycle and bioaccumulation of mercury. Annu Rev Ecol Syst 29:543–655

Muir D (2005) Metals in marine mammals In: *AMAP Assessment 2002: Heavy Metals in the Arctic*. Arctic Monitoring and Assessment Programme (AMAP), Oslo, Norway, p. 234

Muir D, Kwan M (2003) Temporal trends of persistent organic pollutants and metals in ringed seals from the Canadian Arctic. In: *Synopsis of research conducted under the 2001–2003 Northern Contaminants Program*; Kalhok S (ed.) Indian and Northern Affairs Canada, Ottawa, pp. 318–327

Muir D, Kwan M, Lampe J (2000) Spatial trends and pathways of persistent organic pollutants and metals in fish, shellfish and marine mammals of northern Labrador and Nunavik. In: *Synopsis of research conducted under the 1999–2000 Northern Contaminants Program*; Kalhok S (ed.) Indian and Northern Affairs Canada, Ottawa, pp. 191–201

Muir D, Fisk A, Kwan M (2001) Temporal trends of persistent organic pollutants and metals in ringed seals from the Canadian Arctic. In: *Synopsis of research conducted under the 2000–2001 Northern Contaminants Program*; Kalhok S (ed.) Indian and Northern Affairs Canada, Ottawa, pp. 208–214

Muir D, Kwan M, Evans M (2006) Temporal trends of persistent organic pollutants and metals in ringed seals from the Canadian Arctic. In: *Synopsis of research conducted under the 2005–2006 Northern Contaminants Program*; Smith S, Stow J. (eds.) Indian and Northern Affairs Canada, Ottawa, pp. 208–214

Nigro M, Campana A, Lanzillotta E, Ferrara R (2002) Mercury exposure and elimination rates in captive bottlenose dolphins. Mar Pollut Bull 44:1071–1075

NOAA. Climate Prediction Centre-Teleconnections: North Atlantic Oscillation. http://www.cpc.noaa.gov/products/precip/CWlink/pna/nao.shtml (accessed October 6, 2009)

Outridge PM, Stern GA, Hamilton PB, Percival JB, McNeely R, Lockhart WL (2005) Trace metal profiles in the varved sediments of an Arctic lake. Geochim Cosmochim Acta 69, 4881. doi:10.1016/J.GCA.2005.06.009

Outridge PM, Sanei H, Stern GA, Hamilton PB, Goodarzi F (2007) Evidence for control of mercury accumulation rates in Canadian High Arctic lake sediments by variations of aquatic primary productivity. Environ Sci Technol 41: 5259–5265

Outridge PM, Macdonald RW, Wang F, Stern GA, Dastoor AP (2008) A mass balance inventory of mercury in the Arctic Ocean. Environ Chem 5:89–111

Overpeck J, Hughen K, Hardy D, Bradley R, Case R, Douglas M, Finney B, Gajewski K, Jacoby G, Jennings A, Lamoureux S, Lasca A, MacDonald G, Moore J, Retelle M, Smith S, Wolfe A, Zielinski G (1997) Arctic environmental change of the last four centuries. Science 278: 1251–1256

Pazerniuk M (2007) Total and methyl mercury levels in Hudson Bay zooplankton; spatial trends and processes. Department of Environment and Geography. M.Sc. University of Manitoba, Winnipeg, MB

Qian M, Jones C, Laprise R, Caya D (2008) The influences of NAO and the Hudson Bay sea-ice on the climate of eastern Canada. Clim Dyn 31: 169–182

Ravichandran M (2004) Interactions between mercury and dissolved organic matter – a review. Chemosphere 55:319–331

Richard PR, Orr JR (2003) Study of the use of the Nelson Estuary and adjacent waters by beluga whales equipped with satellite-linked radio transmitters. Unpublished report prepared by Fisheries and Oceans Canada, Winnipeg, MB for Manitoba Hydro, Winnipeg, MB

Riget F, Dietz R, Born EW, Sonne C, Hobson KA (2007) Temporal trends of mercury in marine biota of west and northwest Greenland. Mar Pollut Bull 54: 72–80

Rodrigues J (2009) The increase in the length of the ice-free season in the Arctic. Cold Reg Sci Technol 59:78–101

Ronald K, Tessaro SV, Uthe JF, Freeman HC, Frank R (1977) Methylmercury poisoning in the harp seal (*Pagophilus groenlandicus*). Sci Total Environ 8:1–11

Siebert U, Joiris C, Holsbeek L., Benke H, Failing K, Frese K, Petzinger E (1999) Potential relation between mercury concentrations and necropsy findings in cetaceans from German waters of the North and Baltic Seas. Mar Pollut Bull 38:285–295

Simmonds MP, Isaac SJ (2007) The impacts of climate change on marine mammals: early signs of significant problems. Oryx 41:19–26

Sørensen N, Murata K, Budtz-Jørgensen E, Weihe P, Grandjean P (1999) Prenatal methylmercury exposure as a cardiovascular risk factor at seven years of age. Epidemiology 10:370–375

Stern GA, Macdonald RW (2005) Biogeographic provinces of total and methyl mercury in zooplankton and fish from the Beaufort and Chukchi Seas: results from the SHEBA drift. Environ Sci Technol 39:4707–4713

Stern GA, Lockhart L (2009) Mercury in beluga, narwhal and walrus from the Canadian Arctic: status in 2009. In: *Synopsis of research conducted under the 2008–2009 Northern Contaminants Program*; Smith S, Stow J, Edwards J (eds.) Indian and Northern Affairs Canada, Ottawa, pp. 108–130

Stewart DB, Lockhart WL (2005) An overview of the Hudson Bay marine ecosystem. Can Tech Rep Fish Aquat Sci 2586, pp. 17–22

Stewart REA, Campana SE, Jones CM, Stewart BE (2006) Bomb radiocarbon dating calibrates beluga (*Delphinapterus leucas*) age estimates. Can J Zool 84:1840–1952

Sunderland EM, Gobas FAPC, Heyes A, Branfireun BA, Bayer AK, Cranston RE, Parsons MB (2004) Speciation and bioavailability of mercury in well-mixed estuarine sediments. Mar Chem 90:91–105

Sunderland EM, Krabbenhoft DP, Moreau JW, Strode SA, Landing WM (2009) Mercury sources, distribution and bioavailability in the North Pacific Ocean: insights from data and models. Global Biogeochem Cycles 23: GB2010, doi:10.1029/2008GB003425

Tynan C, DeMaster D (1997) Observations and predictions of Arctic climate change: potential effects on marine mammals. Arctic 50:308–322

Wagemann R, Stewart REA (1994) Concentrations of heavy metals and selenium in tissues and some foods of walrus (*Odobenus rosmarus rosmarus*) from the eastern Canadian Arctic and sub-arctic, and associations between metals, age, and gender. Can J Fish Aquat Sci 51:426–436

Wagemann R, Snow NB, Lutz A, Scott DP (1983) Heavy metals in tissues and organs of the narwhal (*Monodon monoceros*). Can J Fish Aquat Sci 40(Suppl. 2):206–214

Wagemann R, Lockhart WL, Welch H, Innes S (1995) Arctic marine mammals as integrators and indicators of mercury in the Arctic. Water Air Soil Pollut 80:683–693

Wagemann R, Innes S, Richard PR (1996) Overview and regional and temporal differences of heavy metals in Arctic whales and ringed seals in the Canadian Arctic. Sci Total Environ 186:41–66

Wagemann R, Trebacz E, Boila G, Lockhart WL (1998) Methylmercury and total mercury in tissues of Arctic marine mammals. Sci Total Environ 218:19–31

Walsh JE (2008) Climate of the arctic marine environment. Ecol Appl 18 (Supplement):S3–S22

World Health Organization (2002) *Environmental Health Criteria 228: Principles and Methods for the Assessment of Risk from Essential Trace Elements*. World Health Organization, Geneva

Woshner VM, O'Hara TM, Bratton GR, Suydam RS, Beasley VR (2001) Concentrations and interactions of selected essential and non-essential elements in bowhead and beluga whales of Arctic Alaska. J Wildlife Dis 37:693–710

Hudson Bay Ecosystem: Past, Present, and Future

C. Hoover

Abstract In order to gain insight into the dynamics of the Hudson Bay ecosystem as well as past and future states, an ecosystem model was created using a static Ecopath model to represent the present day ecosystem in Hudson Bay. Simulations of past and future ecosystem states were used to gain insight to key trophic linkages within the system, with focus on marine mammal populations. The past ecosystem was simulated by increasing ice algae and removing killer whales (*Orcinus orca*) from the system, which led to an increased biomass of all other groups within the model, excluding pelagic producers. Future states of Hudson Bay are presented in three scenarios representing various degrees of reported and predicted ecosystem changes including climate change and increased hunting pressure. All future scenarios show an overall decrease in species biomass, although some species are positively impacted by the changes in the system. Model simulations suggest bottom up forcing of ice algae is an important factor driving marine mammal biomass.

Keywords Hudson Bay • Ecosystem modelling • Ecopath with Ecosim • Food web • Climate change

Introduction

Hudson Bay has been an important region to native cultures beginning with prehistoric Inuit roughly 4,000 years ago, and continuing up to the current Inuit and Cree communities which still inhabit the region today (Stewart and Lockhart 2005; Henri et al. this book). Aboriginal people have depended on the use of natural resources available to them, including birds, fish, plants, and marine mammals.

C. Hoover (✉)
Fisheries Centre, University of British Columbia, 2202 Main Mall,
Vancouver, BC V6T1Z4, Canada
e-mail: c.hoover@fisheries.ubc.ca

S.H. Ferguson et al. (eds.), *A Little Less Arctic: Top Predators in the World's Largest Northern Inland Sea, Hudson Bay*, DOI 10.1007/978-90-481-9121-5_11,
© Springer Science+Business Media B.V. 2010

Marine mammals generally found in the high Arctic latitudes are also found in relatively high abundances in Hudson Bay (Maxwell 1986), mostly due to the high Arctic climate being present at this lower latitude. Recently, marine mammals in the Arctic have been under increased stress caused by direct and indirect stressors such as; climate change, environmental contaminants, off-shore oil and gas activities, shipping, hunting, and commercial fisheries (Huntington 2009).

The average annual temperature in the Arctic has increased at a rate nearly double that of temperature increases in the rest of the world, and is expected to increase 4–7°C in the next 100 years (Arctic Climate Impact Assessment 2004). More specifically, climate models for Hudson Bay predict increases in annual precipitation, temperature from 3.9°C–4.5°C, and the length of the ice free period (Gagnon and Gough 2005). The combined effects of these changes to weather patterns impact ecosystems around the globe, with increased sensitivity in polar regions (Arctic Climate Impact Assessment 2004).

There are likely to be large scale changes within the ecosystem caused by increases in temperature. Perhaps most important will be changes to the extent and temporal dynamics of sea ice, which comprises the crux of the ecosystem. The extent of summer sea ice cover had decreased by 15–20% in the last 30 years in the Arctic, with the breakup of ice in Hudson Bay advancing at least 3 days per decade during 1971–2003 (Gough et al. 2004). As temperatures are predicted to continue increasing, ice cover in the Arctic is expected to nearly disappear later this century (Arctic Climate Impact Assessment 2004). This chapter addresses potential consequences these changes may have on the Hudson Bay ecosystem.

The Hudson Bay ecosystem, as it is referred to throughout this chapter, includes Hudson Bay and James Bay, and excludes Hudson Strait and Foxe Basin (Stewart and Barber this volume). Hudson and James Bays are rather shallow in contrast to the deeper and more dynamic Hudson Strait and Foxe Basin. Most of the marine species within Hudson Bay and James Bay complete their entire life cycle in this area, with the exception of some marine mammals.

The annual sea ice cycle in Hudson Bay begins with freeze up by mid December, with the ice being the thickest from April to May and beginning to break up in June (Markham 1986; Stewart and Lockhart 2005). When the sea ice forms the phytoplankton and zooplankton species found within the water are frozen within the ice. Many species of phytoplankton and zooplankton have adapted to survive the winter frozen within the ice, and be returned to the water column the following summer to complete their life cycle (Horner et al. 1992). In the winter, this ice algae, which is concentrated at the ice–water interface, sustains the upper pelagic food web, which in turn provides nutrition for fish, seals, and polar bears (*Ursus maritimus*). Ice algae is an important source of energy, contributing up to 25–57% of annual primary production in some areas of Hudson Bay (Gosselin et al. 1997; Legendre et al. 1996).

Gammaridean amphipods are the dominant macrofauna feeding under ice on detritus that include ice algae, bacteria and crustacean remains (Poltermann 2001). The gammaridean amphipod is an important link between the lower trophic levels of sea ice based food webs and higher trophic level predators that include Arctic cod (*Boreogadus saida*), ringed seals (*Pusa hispida*) and birds (Bradstreet and Cross 1982; Lonne and Gabrielsen 1992). In other polar ecosystems such as the Antarctic, ice algae is used throughout the winter by invertebrates such as Antarctic krill (*Euphausia superba*), which scrape the algae out of the ice as their main food source in winter (Marschall 1988, 1998; Nicol 2006). When the ice melts in the spring, the remaining algae are released into the water column, where it is available to copepods, krill, and other zooplankton. Portions of the Hudson Bay zooplankton community may also be able to consume the ice algae in late fall or early winter in much the same manner. Copepods are an abundant zooplankton in Hudson Bay, and an important food source to many species of fish, birds, and marine mammals (Estrada et al. 2008; Harvey et al. 2001). Copepod species located in Hudson Bay are able to utilize ice algae located at the ice-water interface before the annual phytoplankton bloom (Runge and Ingram 1987, 1991), which is believed to sustain the ecosystem in winter through to early spring (Freeman et al. 1982; Stewart and Lockhart 2005). In addition, copepods have adapted to aggregate under the melting sea ice in the spring to feed on the released algae (Conover et al. 1986; Runge and Ingram 1987, 1991), further demonstrating the importance of this trophic interaction.

The release of algae and some zooplankton species from the ice in the spring transport energy to the benthic food web which sustains molluscs, bearded seals (*Erignathus barbatus*), walrus (*Odobenus rosmarus*), and the Hudson Bay eider (*Somateria mollissima*) (Arctic Climate Impact Assessment 2004). Marine mammals in the Arctic have also adapted to survive with the sea ice cycle. Ringed seals build breeding dens on the sea ice under snow cover in the winter and early spring to protect their young from the elements (Chambellant this volume). Polar bears take advantage of this food source, by seeking out breeding dens and hunting young pups (Peacock et al. this volume). This winter/spring feeding period represents the majority of annual caloric intake for polar bears, building up their fat reserves for the summer months (Stirling and Derocher 1993). Sea ice provides walruses an alternative platform to land for pupping from April to June (NAMMCO 2005), and allows young calves a place to rest, decreasing travel distances and stress when they are learning to swim (Cooper et al. 2006). Bowhead whales (*Balaena mysticetus*) live most of their lives along the edge of the sea ice following it along their migration routes (Dyke et al. 1996), into Hudson Bay around June and back out in September (Stewart and Lockhart 2005). While the whales in the ecosystem generally migrate in and out every year, pinnipeds are year round inhabitants. Figure 1 shows a comparison of marine mammals and sea ice throughout the year to give an overview of which species might be most vulnerable to changes in sea ice.

Human residents of Hudson Bay have learned to utilize the numerous resources in order to survive in the region (Henri et al. this volume). Marine mammals have provided a food source to northern people since prehistoric times. In addition to food, Inuit continue to use the available natural resources to provide themselves with clothing and tools. Hunting also has a significant cultural importance to each community, as in the case of belugas, which brings the community together for a common goal and allows successful hunters to share their harvest with other families (Tyrrell 2007).

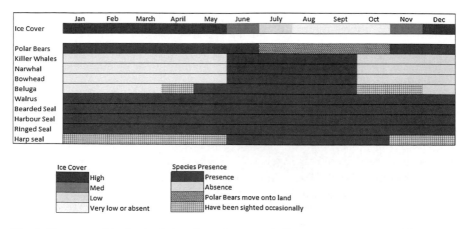

Fig. 1 Freeze-up of ice begins in late November or early December and break-up of ice occurs from late June to late July (Gagnon and Gough 2005). Presence and sightings of marine mammals provided in Stewart and Lockhart (2005)

Currently narwhal, beluga, walrus, polar bears, seals, and bowhead whales are hunted along with select species of fish, invertebrates, and marine plants. Beluga, narwhal and bowhead are hunted for muktak (muktuk, maqtaq), the layer of skin and blubber, which is dried and eaten as a favorite food among many Inuit. Male narwhal possess an ivory tusk which is often made into carvings. Polar bears are hunted throughout Hudson Bay, mainly for their fur which is used to make clothing (Stewart and Lockhart 2005). Walrus are hunted for meat and their ivory tusks which are also used for carvings or sold whole (NAMMCO 2005). Historically walrus skins were used to make tents and ropes and their tusks were used to make harpoons (Stewart and Lockhart 2005). Seals provide a large amount of the protein consumed by native peoples, with their furs also being used for clothing.

Many Marine fish such as Arctic char (*Salvelinus alpinus*) and Greenland cod (*Gadus ogac*) are harvested for subsistence, but this harvest is currently unregulated, and no comprehensive stock assessments have been completed to estimate the amount harvested (Stewart and Lockhart 2005). Commercial fishery operations have been attempted; however, they were neither profitable nor productive. In some communities invertebrates such as green sea urchins (*Strongylocentrotus droebachiensis*) and blue mussels (*Mytilus edulis*) and marine plants are harvested for local consumption (Stewart and Lockhart 2005). The resources provided by this ecosystem continue to be a critical source for meeting the socio-economic needs to pursue the cultural integrity to the communities dependent on them.

An ecosystem modelling approach is used to gain insight into the dynamics of the Hudson Bay ecosystem at present as well as past and into the future. Here, a general representation of a sophisticated modelling exercise (detailed in Hoover et al. 2010) is presented. First a static model of the present day ecosystem in Hudson Bay was created. Then, based on knowledge of the past Hudson Bay conditions, an ecosystem model was simulated that incorporated top down and bottom up forcing. Lastly, three simulations of the future HB ecosystem were constructed to gain insight into key trophic linkages within the system, with focus on marine mammal populations.

Methods

Model simulations representing the Hudson Bay ecosystem were created using the Ecopath with Ecosim suite of software (Buszowski et al. 2009; Christensen et al. 2007). This ecosystem modelling framework allows all species or species groups within the ecosystem to be connected through trophic linkages as defined by the user. Models are constructed in Ecopath, under a mass balance assumption to give a snapshot of the ecosystem in a particular time frame, and then projected through time using Ecosim, the dynamic portion of the modelling software. Full model structure, parameters, data sources, and trophic links for the Hudson Bay model are reported in Hoover et al. (2010). The three main analyses in this chapter are evaluations of the present, past and future states of the ecosystem.

Present

The model constructed for the Hudson Bay ecosystem includes 40 functional groups representing all species or species groups found within the ecosystem. In this model-ling exercise the current ecosystem state is represented as a static Ecopath model, and serves as a baseline to compare past and future ecosystem states. The model was constructed using published literature values, with all parameters available in Hoover et al. (2010). Because there are no comprehensive assessments of fish abundance or biomass for Hudson Bay the model was used to estimate the biomass of these groups. Biomass was estimated through a pedigree analysis, whereby parameters are ranked according to the credibility of their source, and then subjected to a Monte Carlo simu-lation to provide ranges of values. Due to the capabilities of Ecopath, each species group is required to have three of four required parameters (biomass, production, consumption, and ecotrophic efficiency), however, through the use of linear equa-tions and trophic interactions the model can estimate one missing parameter per spe-cies group based on the inputs for other species groups. In essence there must be enough prey species to support a given biomass of predator with known growth and consumption rates. For full details on this please see Christensen et al. (2007). Unknown parameters are then estimated by solving for values which will fit into all sets of equations for the model, and repeated 1,000 times.

Hunting mortality was also incorporated into the model through the use of har-vest statistics from 1989 to 1995 (DFO 1990, 1991, 1992, 1993, 1994, 1995, 1996, 1997, 1998, 1999; Stewart and Lockhart 2005). In addition, Harvest trends were used for future scenario C (increased hunting) in order to establish harvest mortalities.

Past

Human populations in communities on the Nunavut portion of Hudson Bay (Arviat, Baker Lake, Chesterfield Inlet, Coral Harbor, Rankin Inlet, Repulse Bay, Sanikiluaq, Whale Cove) rose from nearly 4,700 in 1980 to over 9,000 in 2006 (Nunavut Bureau of Statistics 2008).[1] While this does not include population changes from the Nunavik or Ontario portions of Hudson Bay, the general trend has been high growth rates at most communities. While hunting methods may have improved through use of technology in recent years, targeted species and their use have not changed drastically from traditional use. Inuit still harvest animals for the same purposes as their ancestors; primarily as food and clothing, but also for trade and income.

Attempting to recreate the ecosystem around 1900 we note a few key differences. The bowhead whale population declined significantly from the late 1800s to the early 1900s due to commercial whaling and increased again towards the end of the twentieth century (Higdon and Ferguson this volume). In addition, ice cover throughout this time would have been higher then present levels, providing an increased source of ice algae. Finally, killer whales (*Orcinus orca*) would almost certainly be absent from the region, as their occurrence in Hudson Bay has only been documented since the 1950s according to a review of published literature and local knowledge (Higdon and Ferguson 2009).

In order to represent the past ecosystem (roughly 1900) the biomass of bowhead whales was increased (100%) along with ice associated algae (50%), while killer whales were removed from this simulation. Dynamic simulations were used to manipulate the killer whale, bowhead whale, and ice algae species groups. Forced biomass changes to these groups within the model simulation resulted in alterations to the rest of the ecosystem through trophic links, by altering prey available or predation on other species groups. Final values for simulations were taken as an average biomass for the last 5 years of the simulation, and were then compared to the baseline Ecopath model (or present day ecosystem) in Hoover et al. (2010) to give a relative increase or decrease.

Future

In the past 20 years the extent of sea ice in the northern hemisphere has declined at a rate of about 3% per year (Parkinson et al. 1999). In Hudson Bay, analysis of ice trends from 1971 to 2003 show sea ice forming later in the fall and breaking up earlier in the spring (Gagnon and Gough 2005; Gough et al. 2004). Because the sea ice reflects solar radiation back into the atmosphere, its reduction caused by warming temperatures may increase solar radiation to the ocean, delaying the freeze up in successive years,

---

[1] Prior to 1981 statistics for Nunavut were combine with Northwest Territories.

creating a positive feedback in the ecosystem (Gagnon and Gough 2005). This delay in sea ice formation would likely lead to a reduction in ice formed and an overall reduction of ice algae available to the food web. However, an increase in temperature is likely to increase primary production in the open water pelagic ecosystem in the form of spring blooms (Melnikov 2000). The switch in dominant phytoplankton from ice algae to pelagic phytoplankton blooms will most likely cause a restructuring of the food web, causing shifts in the abundance of zooplankton, benthos, and ultimately fish, birds, and marine mammals.

Direct effects to higher trophic level organisms are expected to occur. For example, bowhead whale migration routes follow the edge of the sea ice (Dyke et al. 1996) and changes to ice patterns may cause bowhead whales to shift their migration routes into this ecosystem (Ferguson et al. in review). This may potentially alter feeding opportunities at decreased ice edges, or allow for greater exposure to predation by killer whales. Earlier spring break-up of sea ice together with a change in snow trends can cause ringed seal dens to collapse through melting, exposing the young pups to harsh climates thereby reducing pup survival in western Hudson Bay (Ferguson et al. 2005). The breakdown of dens also exposes pups to polar bears, making them easier to find, thus further increasing the population mortality (Stirling and Parkinson 2006). The feeding season of the polar bear has been altered in the last 35 years (Stirling and Parkinson 2006), as a decreasing ice season impairs their ability to hunt and build energy stores needed to survive the ice free period (Peacock et al. this volume). This may also lead to declining body condition, lower reproductive rates, and decreased survival of polar bear cubs (Stirling and Derocher 1993). Whereas bears used to remain on the ice for much of the spring and summer they now travel closer to settlements in order to seek out food, thus increasing their interactions with humans (Stirling and Parkinson 2006).

Perhaps the most immediate effect of sea ice loss to the food web will be a change in primary production. By removing or altering the physical structure of sea ice, the flow of energy and carbon from ice algae to higher trophic level organisms will be reduced spatially and temporally thus limiting energy and nutrient transfer to higher trophic levels.

Considering the increases in human population over the last 30 years, it is likely that there will be continued growth. This will put additional pressure on resources that will include marine mammals if future generations continue to hunt and follow traditional lifestyles. Although estimates of current harvest levels are not always accurate due to difficulty in obtaining harvest levels for all species, it is assumed that increasing populations will consume more resources than the present day.

Based on predicted changes described above the following three scenarios were used for the future simulations of Hudson Bay ecosystem:

(A) A 50% reduction in sea ice algae biomass
(B) A 50% reduction in sea ice algae biomass, a 50% reduction in copepod biomass, a 50% reduction in polar bear biomass, a 25% reduction in ringed seal biomass, a 25% reduction in bird biomass, and a 25% increase in killer whale biomass
(C) Scenario B plus a 100% increase in hunting-based mortality

The increased hunting scenario incorporates the same ecological changes as future scenario B plus a doubling of harvest rates on all marine mammals to account for a future doubling in human population size. Again, the biomass of specified species groups was altered to identify the consequent changes to the ecosystem through trophic linkages. For scenario C catches were forced to double their current rate. Average biomass for the last 5 years of the simulation was used as the final value, and compared to present day values.

Results and Discussion

Within each section the results for each scenario are presented and discussed. Figures 2, 4, and 5 illustrate the Hudson Bay food web for present, past, and future scenarios respectively while Fig. 6 presents mean changes in past and future biomass relative to present biomass. Key species were selected for figures leaving birds, benthos, and some individual species groups missing from the figures. These groups were included in the full model, but excluded for graphical purposes due to the large number of species groups in the model.

Present Ecosystem

Figure 2 shows the simplified Hudson Bay ecosystem, the trophic links as they are believed to exist today, and serves as a reference point for past and future ecosystem states. In this food web both sea ice algae and pelagic phytoplankton represent the autotrophic primary producers, whereby seasonality and ice cover largely determines prevalence of each one. Copepods being the dominant zooplankton, provide an essential link between producers and consumers, fuelling fish and other zooplankton species. Capelin are an important forage fish to marine mammals and birds which annually migrate through the regional. A full list of species found within the ecosystem is available from Stewart and Lockhart (2005), along with the modeled functional groups of species Hoover et al. (2010).

While there are no surveys of fish to estimate abundance, as little is known (Stewart and Lockhart 2005), pedigree ranking for Hudson Bay as provided in Hoover (2008), and Monte Carlo simulations were used to estimate fish biomass. By constraining the ecosystem through the abundance of predators and the amount of production available at the base of the food web, a biomass estimate of all fish groups is provided in Fig. 3. In comparison to other ecosystems, the biomass of fish in Hudson Bay is relatively low, as one might expect considering unsuccessful attempts of commercial operations within the ecosystem. For example on average Hudson Bay has 2.4 tonnes per km² for all fish species, which as expected is substantially less than reported values for other ecosystems: Ionian Sea 6.43 tonnes per km² (Piroddi 2008), or the Antarctic peninsula 4.32 tonnes per km²

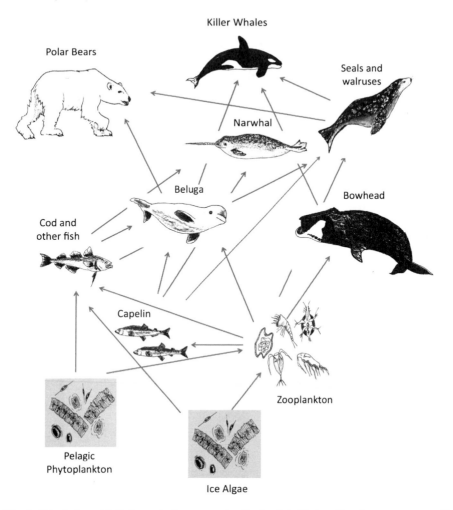

Fig. 2 Principal trophic linkages in the present day food web in Hudson Bay. Images courtesy of Megan Bailey

(Hoover and Pitcher 2009). Thus, these preliminary estimates appear to be within a reasonable range given comparisons for other ecosystems. Because these values are contingent upon the ability of the ecosystem to produce enough prey, the needs of predators, and the food web links, they are sensitive to the input parameters of other modeled groups. For example an underestimate of marine mammal biomass will require less prey (fish biomass) within the model, and vice versa. Overestimates of primary production can cause overestimates in higher trophic level organisms. However, input parameters for primary producers and marine mammals were obtained from published literature and subjected to Monte Carlo routines to estimate errors.

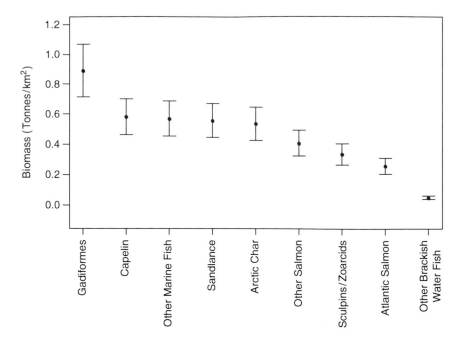

Fig. 3 Biomass estimates for fish species in the Hudson Bay ecosystem. Monte Carlo simulations provide 95% confidence Limits based on pedigree ranking in Hoover 2008. Refer to Appendix for a full list of species within each fish grouping

Past

For this simulation, ice associated algae were increased by 50% of their present day biomass, in order to account for the increased sea ice in the past. In addition, bowhead whale abundance was doubled to represent the highest abundance in the late 1800s (Higdon and Ferguson this volume). These changes, combined with the removal of killer whales as top predators resulted in an increase in biomass in every species group within the model, with the exception of pelagic primary production (Fig. 4). The decrease in pelagic primary production is due to the limitation of nutrients through the detritus functional group, which become increasingly scarce as they are utilized by the ice algae groups which is forced to increase in this simulation. Zooplankton groups increased the most, up to 50% of their present day biomass, most likely attributed to the increased ice algae, an important food source. This was propagated up the food web, where fish biomasses increased from 20% to 30% depending on the species group. Seal biomass increased 30% on average with harbor seals (*Phoca vitulina*) increasing the most at 38%, and harp seals (*Phoca groenlandica*) increasing the least at 28%. In addition, polar bears and beluga (*Delphinapterus leucas*) groups both increased 38%, while narwhal (*Monodon monoceros*) biomass increased by 59%. It should be noted that such

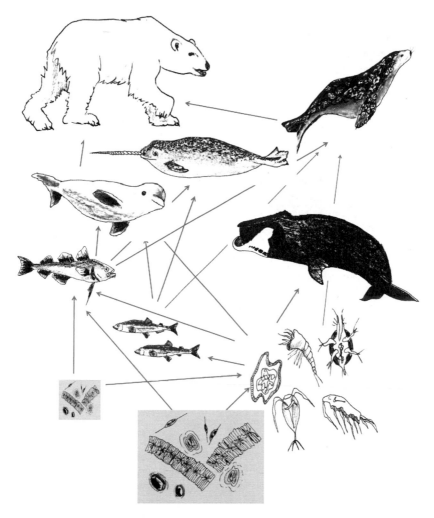

Fig. 4 Representation of Hudson Bay Ecosystem in the past accounting for increased ice-algae, increased bowhead whales, and the absence of killer whales. Images courtesy of Megan Bailey

large increases are not necessarily believed to be representative of the past ecosystem, but rather to identify what sort of shifts may have occurred when comparing the past and present ecosystems.

Due to multiple perturbations to the system, it is difficult to identify which factors are having significant effects on each species group. The removal of killer whales from the system should allow certain marine mammal species (narwhal, bowhead, beluga, walrus, and seals) to increase their biomass through reduced predation (note this does occur to all species except bowhead which were forced to increase in this simulation). However, the increased ice algae impacted the ecosystem through bottom-up trophic interactions which is likely responsible for the increases in zooplankton and fish biomass. It is difficult to tease out which interactions are affected most by each change within the ecosystem unless multiple combinations of perturbations are run. Nevertheless, insight into the future ecosystem scenarios may shed some light on these processes.

Future

Killer whale abundance in Hudson Bay has increased exponentially as they migrate into Hudson Bay to take advantage of the populations of other marine mammals as food (Ferguson et al. this volume). Their appearance in the last 50 years has been linked to decreasing sea ice, which has allowed them easier access to the food resources in Hudson Bay (Higdon and Ferguson 2009). Heavy sea ice cover prevents the narwhal from overwintering in Hudson Bay. However, a decrease in ice might allow them to remain in Hudson Bay longer each year, and possibly overwinter, although other factors such as available prey and exposure to predators will also influence movements (Laidre and Heide-Jørgensen 2005).

Three future ecosystem simulations were performed (Fig. 5). For Scenario A, the 50% reduction in ice-algae biomass was based on a 30% reduction already observed and the possibility of sea ice disappearing in the next century (Arctic Climate Impact Assessment 2004). This scenario was created to represent an esti- mate of food web changes, assuming a loss of sea ice would result in a loss of ice algae and lead to cascading ecosystem changes. The resulting impacts to the eco- system show averages of 30% decrease in biomass for all marine mammals, 25% decrease in biomass for all fish, 40% decrease for zooplankton biomass, 20% decrease in benthic biomass, and a 15% increase in pelagic primary production. Increases in biomass to species groups in the past ecosystem scenario were larger than the decreases under future scenario A, indicating that ice algae is important, but not the sole reason for biomass increases in the past simulation.

For scenario B, the same changes to the ecosystem were observed with the excep- tion of a further decrease to narwhal, caused by the increased killer whales (Fig. 6). Seal biomass did increase slightly, about 8%, from scenario A, likely due to the further decrease in their predator, the polar bear. Some of the largest changes between scenarios A and B were due to species groups being forced. For example polar bears only declined by 29% of present values under scenario A, but were forced to a 50% reduction in scenario B indicating that although sea ice may be contributing to their decline through alterations of the food web, other interactions are also resulting in substantial declines.

As expected, scenario C shows the greatest declines in biomass for many species groups, in particular the hunted marine mammal species. Results show there are further reductions of marine mammal biomasses, with bowhead, harbor seals, and ringed seals having minor further reductions to population; decreases less than 10% of future scenario B (Fig. 6). The likely reason for this is current harvest rates for these species (bowhead, harbor seals, and ringed seals) are low relative to their population

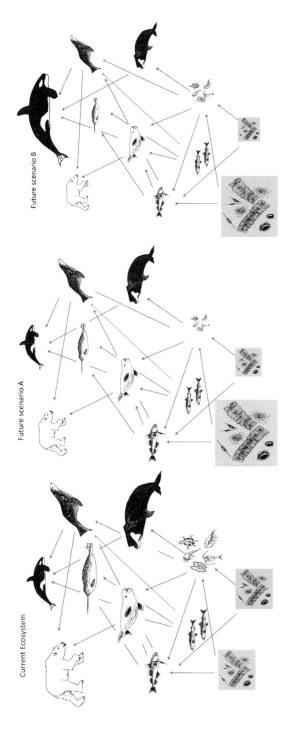

Fig. 5 Present ecosystem compared to future scenarios A and B. Images courtesy of Megan Bailey

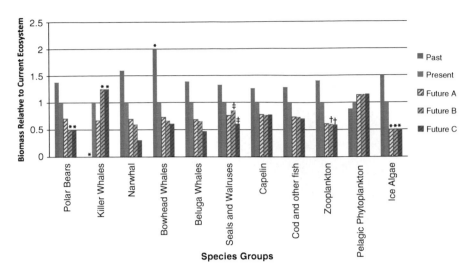

Fig. 6 Comparison of select species group biomass for the following scenarios; past, future A, future B, and future C presented as a percentage increase or decrease from the present day biomass. For species groups representing multiple species within the full model are averaged within the figure. Species groups where biomass was forced in specific scenario are indicated (*). Groups with multiple species, where only one component was forced are marked as ‡ (for ringed seals within "Seals and Walruses") and † (for copepods within "Zooplankton")

sizes, thus even a doubling of harvest does not significantly alter the population biomass in the model. Narwhal were the most severely impacted marine mammal species with an overall decline of 75% from present day biomass, a further decline of 45% from scenario B demonstrating this species' sensitivity to increased mortality rates.

Scenarios of past, present, and future ecosystems of Hudson Bay reveal the importance of ice-algae to the food web dynamics and as an important driver of ecosystem productivity. The increase in ice algae for the past ecosystem shows an increase across nearly every species group within the model, while reductions of ice algae for future scenario A showed declines in nearly all species groups. Future scenarios incorporating increased mortality through increased predation (i.e. killer whales) or hunting activities, generally did not have as strong of an impact on the higher trophic level species as did the reductions in ice algae. These results indicate the ecosystem is sensitive to bottom-up forcing.

While models can provide an overview of changes to a system, and assess perturbations, it is unlikely that any of these scenarios will truly reflect the future state of the Hudson Bay ecosystem. By altering species or species groups within the model, the impact to the food web can be studied, thereby identifying new research directions and contributing to the understanding of the food web dynamics. The author does not assume to have an all-inclusive knowledge of the system, but rather intends to provide some insight to ecosystem links that are important to user groups, researchers, and management, by using the data and research available at present.

The approaches taken here did not account for physiological effects of warming, loss of habitat, or feedback loops resulting from sea ice reduction and increased temperature. In addition, alterations to the food web and replacement of trophic levels by invasive species are likely to occur in Hudson Bay and throughout the Arctic as temperatures increase. These considerations are important components when assessing ecosystem responses to stressors such as climate change and hunting. However, these were not accounted for within the current modelling framework, and should be considered in additional studies as they would provide value to the overall understanding of the ecosystem.

There have been few ecosystem studies in Hudson Bay, so understanding the changes in other systems in relation to climate change will help establish plausible futures for Hudson Bay. The scenarios presented show a linear progression of the ecosystem under predetermined varying conditions. However, longer term studies of populations over hundreds of years have shown cyclical patterns of abundance through ecosystem changes. For example a 2000 year time series of anchovy and sardine abundance in the Pacific Ocean reveals changes in the abundance of these species, portraying a waxing and waning pattern (Hayword 1997). These regime shifts, or oscillations in biomass, have allowed either the anchovy or the sardine to increase its abundance in the ecosystem, only to be replaced by the other through the next regime shift. The consequences are observed in higher trophic level species, such as seabirds which primarily consume anchovies (Chavez et al. 2003), as their abundance also fluctuates in accordance with the regime shifts. In the Bering Sea, regime shifts are thought to be caused by atmospheric oscillations which drive changes in the physical environment such as wind, ocean upwelling, and extent of freshwater plumes; these were hypothesized to result in the population crash of Stellar seal lions (*Eumetopias jubatus*) (Benson and Trites 2002). These changes in the physical system have affected species at various trophic levels beginning with the spatio-temporal changes in primary production, to a shift in forage fish species and to ones that may not have had the nutritional content to sustain top predators such as Stellar sea lion (Trites and Donnelly 2003; Trites et al. 2007).

In light of information on regime shifts, and given the ability of ecosystems to switch between dominant species, multiple different futures for Hudson Bay are possible. The ecosystems studied with known shifts are quite "data rich" when compared to systems like Hudson Bay. Sediment cores from southern Hudson Bay roughly 8000 BP indicate ostracods were the dominant zooplankton (Bilodeau et al. 1990). Perhaps, the reduction of sea ice and increases in seasonal pelagic phytoplankton have allowed for reorganization of dominant zooplankton, and in the future will allow different zooplankton species and forage fish to dominate the ecosystem. Warming of Hudson Strait and Foxe Basin may allow new species to cross these currently colder waters, and enter Hudson Bay. Invasive species may well lead to the local extirpation of some species, and will likely result in restructuring of the food web. Adaptations in diet are not exclusive to marine organisms, as humans will likely be forced to adjust their diets to the resources available to them.

As shown through bottom-up forcing of the Hudson Bay ecosystem model, and supported by bottom-up induced regime shifts, the effects of climate change on the physical environment easily cascade to have far reaching consequences throughout the entire ecosystem.

Summary

Assessment of the current ecosystem structure in conjunction with past and future scenarios indicate the Hudson Bay ecosystem is sensitive to bottom-up forcing, specifically from ice algae. Future simulations show a decrease in ice algae affecting nearly every species group within the Hudson Bay ecosystem model. While controlling hunting mortality may be an important factor to managing marine mammal stocks, decreases in ice algae appear to have a more significant effect on most species of marine mammals. Future work on specific effects of sea ice loss in relation to individual species is imperative to improve modelling techniques as well as an overall understanding of the Hudson Bay ecosystem and climate change.

Appendix

Fish groupings for the Hudson Bay ecosystem modeled after Stewart and Lockhart (2005). Under the classification, only species found within Hudson and James Bays which spend at least part of their life in areas where they would be available to marine mammals were included. In Stewart and Lockhart (2005) these are classified as; Marine, Brackish, Estuarine, and some Diadromous.

Atlantic Salmon	
Atlantic salmon	*Salmo salar*
Arctic Char	
Arctic Char	*Salvelinus alpines*
Capelin	
Capeli	*Mallotus villosus*
Gadiformes	
Arctic cod	*Boreogadus saida*
Greenland cod	*Gadus ogac*
Polar cod	*Arctogadus glacialis*
Other Brackish Water Fish	
Arctic shanny	*Stichaeus punctatus*
Canadian plaice	*Hippoglossoides platessoides*
Slender eelblenny	*Lumpenus fabricii*
Other Marine Fish	
Alligator poacher	*Leptagonus decagonus*

(continued)

(continued)

Atlantic alligatorfish	*Ulcina olriki*
Atlantic Herring	*Clupea harengus*
Atlantic spiny lumpsucker	*Eumicrotremus spinosus*
Banded gunnel	*Pholis fasciata*
Daubed shanny	*Leptoclinus maculatus*
Dusky snailfish	*Liparis gibbus*
Fourline snakeblenny	*Eumesogrammus praecisus*
Gelatinous snailfish	*Liparis fabricii*
Kelp snailfish	*Liparis tunicatus*
Leatherfin lumpsucker	*Eumicrotremus derjugini*
Lumpfish	*Cyclopterus lumpus*
Sea tadpole	*Careproctus reinhardti*
Stout eelblenny	*Anisarchus medius*

Other Salmon

Arctic Char	*Salvelinus alpinus*
Brook trout	*Salvelinus fontinalis*
Lake cisco	*Coregonus artedi*
Lake whitefish	*Coregonus clupeaformis*
Round whitefish	*Prosopium cylindraceum*

Sandlance

Northern sand lance	*Ammodytes dubius*
Stout sand lance	*Ammodytes hexapterus*

Sculpins/Zoarcids

Arctic eelpout	*Lycodes reticulatus*
Arctic sculpin	*Myoxocephalus scorpiodes*
Arctic staghorn	*Gymnocanthus tricuspis*
Fish doctor	*Gymnelus viridis*
Fourhorn sculpin	*Myoxocephalus quadricornis*
Moustache sculpin	*Triglops murrayi*
Pale eelpout	*Lycodes pallidus*
Ribbed sculpin	*Triglops pingelli*
Shorthorn sculpin	*Myoxocephalus scorpius*
Spatulate sculpin	*Icelus spatula*
Twohorn sculpin	*Icelus bicornis*

References

Arctic Climate Impact Assessment. 2004. Impacts of a warming Arctic, Cambridge University Press New York, p. 139.

Benson, A.J., and Trites, A.W. 2002. Ecological effects of regime shifts in the Bering Sea and eastern Pacific Ocean. Fish and Fisheries **3**: 95–113.

Bilodeau, G., De Vernal, A., and Hillaire-Marcel, C. 1990. Postglacial paleoceanography of Hudson Bay: stratigraphic, microfaunal, and palynological evidence. Canadian Journal of Earth Sciences **27**: 946–963.

Bradstreet, M.S.W, and Cross, W.E. 1982. Trophic relationships at high Arctic ice edges. Arctic **35**: 1–12.

Buszowski, J., Christensen, V., Gao, F., Hui, J., Lai, S., Steenbeek, J., Walters, C., and Walters, W. 2009. Ecopath with Ecosim 6. University of British Columbia Fisheries Centre, Vancouver, BC, Canada.

Chavez, F.P., Ryan, J., Lluch-Cota, S.E., and Niquen, M.C. 2003. From anchovies to sardines and back: Multidecadal change in the Pacific Ocean. Science **299**: 217–221.

Christensen, V., Walters, C., Pauley, D., and Forrest, R. 2007. Ecopath with Ecosim version 6.0. User manual/ help files guide. Lenfest Oceans Futures Project 2007. University of British Columbia, Vancouver, BC, Canada. version 2.0.

Conover, R.J., Herman, A.W., Prinsenberg, S.J., and Harris, L.R. 1986. Distribution of and feeding by the copepod *Pseudocalanus* under fast ice during Arctic spring. Science **232**(4755): 1245–1247.

Cooper, L.W., Ashjian, C.J., Smith, S.L., Codispoti, L.A., Grebmeier, J.M., Campbell, R.G., and Sherr, E.B. 2006. Rapid seasonal sea-ice retreat in the Arctic could be affecting Pacific Walrus (*Odobenus rosmarus divergens*) recruitment. Aquatic Mammals **31**(1): 98–102.

DFO. 1990. Annual summary of fish and marine mammal harvest data for the Northwest Territories 1988–1989, Volume 1, Freshwater Institute, Central and Arctic Region, Winnipeg, MB, p. 59.

DFO. 1991. Annual summary of fish and marine mammal harvest data for the Northwest Territories 1989–1990, Volume 2, Freshwater Institute, Central and Arctic Region, Winnipeg, MB, p. 61.

DFO. 1992. Annual summary of fish and marine mammal harvest data for the Northwest Territories 1990–1991, Volume 3, Freshwater Institute, Central and Arctic Region, Winnipeg, MB, p. 67.

DFO. 1993. Annual summary of fish and marine mammal harvest data for the Northwest Territories 1991–1992, Volume 4, Freshwater Institute, Central and Arctic Region, Winnipeg, MB, p. 69.

DFO. 1994. Annual summary of fish and marine mammal harvest data for the Northwest Territories 1992–1993, Volume 5, Freshwater Institute, Central and Arctic Region, Winnipeg, MB, p. 104.

DFO. 1995. Annual summary of fish and marine mammal harvest data for the Northwest Territories 1993–1994, Volume 6, Freshwater Institute, Central and Arctic Region, Winnipeg, MB, p. 86.

DFO. 1996. Annual summary of fish and marine mammal harvest data for the Northwest Territories 1994–1995, Volume 7, Freshwater Institute, Central and Arctic Region, Winnipeg, MB, p. 85.

DFO. 1997. Annual summary of fish and marine mammal harvest data for the Northwest Territories 1995–1996, Volume 8, Freshwater Institute, Central and Arctic Region, Winnipeg, MB, p. 80.

DFO. 1998. Hudson Bay narwhal, *Stock status report E5–44*. Department of Fisheries and Oceans Canada. Central and Arctic Region p. 4.

DFO. 1999. Hudson Bay-Foxe Basin bowhead whales, *Stock status report E5–52*. Department of Fisheries and Oceans Canada. Central and Arctic Region p. 8.

Dyke, A.S., Hooper, J., and Savelle, J.A. 1996. A history of Sea Ice in the Canadian Arctic archipelago based on postglacial remains of the bowhead whale (*Balaena mysticeus*). Arctic **49**(3): 235–255.

Estrada, R., Harvey, M., Gosselin, M., and Starr, M. 2008. Late summer zooplankton community structure, abundance, and distribution in the Hudson Bay system and their relations with environmental conditions, 2003-2006. *In* Arctic Change Conference 2008. ArcticNet, Quebec City, Quebec.

Ferguson, S.H., Stirling, I., and Mcloughlin, P. 2005. Climate change and ringed seal (*Phoca hispida*) recruitment in western Hudson Bay. Marine Mammals Science **21**(1): 121–135.

Ferguson, S.H., Dueck L., and Loseto L.L., in review. Bowhead whale (*Balaena mysticetus*) seasonal selection of sea ice. Marine Ecology Progress Series.

Freeman, N.G., Flemming, B.M., and Anning, J.L. 1982. La Grande/Great Whale plume measurements data report, winter 1979. *In* Unpublished Manuscript Data Report Series No. 82-3. Department of Fisheries and Oceans, Bayfield Laboratory, Burlington, ON. p. ix+259.

Gagnon, A.S., and Gough, W.A. 2005. Trends in the dates of ice freeze-up and breakup over Hudson Bay, Canada. Arctic **58**(4): 370–382.

Gosselin, M., Levasseur, M., Wheeler, P.A., Horner, R.A., and Booth, B.C. 1997. New measurements of phytoplankton and ice algal production in the Arctic Ocean. Deep Sea Research II **44**(8): 1623–1644.

Gough, W.A., Cornwell, A.R., and Tsuji, L.J.S. 2004. Trends in seasonal sea ice duration in southwestern Hudson Bay. Arctic **57**(3): 299–305.

Harvey, M., Therriault, J.-C., and Simard, N. 2001. Hydrodynamic control of late summer species composition and abundance of zooplankton in Hudson Bay and Hudson Strait (Canada). Journal of Plankton Research **23**(5): 481–496.

Hayword, T.L. 1997. Pacific Ocean climate change: atmospheric forcing, ocean circulation and ecosystem response. TREE **12** (4): 150–153.

Higdon, J.W., and Ferguson, S.H. 2009. Loss of Arctic sea ice causing punctuated change in sightings of killer whales (*Orcinus Orca*) over the past century. Ecological Applications **19**(5): 1365–1375.

Hoover, C. 2008. Preliminary results from a Hudson Bay ecosystem model: Estimating mid trophic levels. Available from http://www.arcticnet.ulaval.ca/pdf/talks2008/hooverCarie.pdf [accessed 09/23/2009].

Hoover, C., and Pitcher, T.J. 2009. Unpublished manuscript. Ecosystem model of the Antarctic Peninsula, University of British Columbia, Vancouver, BC.

Hoover, C., Pitcher, T.J., and Christensen, V. (2010). An Ecosystem model of Hudson Bay, Canada. Fisheries Centre Research Reports. eds. Hoover, C. and Wabnitz, C. University of British Columbia Fisheries Centre, Vancouver, BC, in press.

Horner, R., Ackley, S.F., Dieckmann, G.S., Gulliksen, B., Hoshiai, T., Legendre, L., Melnikov, I.A., Reeburgh, W.S., Spindler, M., and Sullivan, C.W. 1992. Ecology of sea ice biota; Habitat, terminology, and methodology. Polar Biology **12**: 417–427.

Huntington, H.P. 2009. A preliminary assessment of the threats to arctic marine mammals and their conservation in the coming decades. Marine Policy **33**: 77–82.

Laidre, K.L., and Heide-Jørgensen, M.P. 2005. Arctic sea ice trends and narwhal vulnerability. Biological Conservation **121**: 509–517.

Legendre, L., Robineau, B., Gosselin, M., Michel, C., Ingram, R.G., Fortier, L., Therriault, J.C., Demers, S., and Monti, D. 1996. Impact of freshwater on a subarctic coastal ecosystem under seasonal sea ice (southeastern Hudson Bay) Canada II. Production and export of microalgae. Journal of Marine Systems **7**: 233–250.

Lonne, O.J., and Gabrielsen, G.W. 1992. Summer diet of sea birds feeding in sea-ice covered waters near Svalbard. Polar Biology **12**: 685–692.

Markham, W.E. 1986. The ice cover. *In* Canadian Inland Seas. *Edited by* I.P. Martini. Elsevier, New York, pp. 101–116.

Marschall, H.-P. 1988. The overwintering strategy of Antarctic krill under the pack-ice of the Weddell Sea. Polar Biology **9**: 129–135.

Maxwell, J.B. 1986. A climate overview of the Canadian inland seas. *In* Canadian Inland Seas. *Edited by* I.P. Martini. Elsevier, New York, pp. 79–100.

Melnikov, I.A. 2000. The Arctic sea ice ecosystem and global warming. *In* Impacts of changes in sea ice and other environmental parameters in the Arctic. *Edited by* R.H. Mattlin, J.E. Reynolds, H.P. Huntington and C. Pungowiyi. Marine Mammal Commission, Girdwood, Alaska, 15–17 February 2000.

NAMMCO. 2005. The Atlantic walrus. *Status of Marine Mammals in the North Atlantic*. North Atlantic Marine Mammal Commission, Polar Environmental Centre, Tromso, pp. 1–7.

Nicol, S. 2006. Krill, currents, and sea ice: *Euphausia superba* and its changing environment. BioScience **56**(2): 111–120.

Nunavut Bureau of Statistics. 2008. Nunavut population counts by community, 1981 Census to 2006 Census. Available from http://www.gov.nu.ca/eia/stats/census.html [accessed Sept 15, 2009].

Parkinson, C.L., Cavalieri, D.J., Gloerson, P., Zwally, J.H., and Cosimo, J.C. 1999. Arctic sea ice extents, areas, and trends 1978–1996. Journal of Geophysical Research Oceans **104**(C9 20): 837–856.

Piroddi, C. 2008. An ecosystem based approach to study two dolphin populations around the islands of Kalamos, Ionian Sea, Greece, Zoology, University of British Columbia, MSc. thesis. Vancouver, BC, Canada, 132.

Poltermann, M. 2001. Arctic sea ice as feeding ground for amphipods – food sources and strategies. Polar Biology **24**: 89–96.

Runge, J.A., and Ingram, R.G. 1987. Underice grazing by planktonic, calanoid copepods in relation to a bloom of ice microalgae in southeastern Hudson Bay. Limnology and Oceanography **33**(2): 280–286.

Runge, J.A., and Ingram, R.G. 1991. Under-ice feeding and diel migration by the planktonic copepods *Calanus glacialis* and *Pseudocalanus minutus* in relation to the ice algal production cycle in southeastern Hudson Bay, Canada. Marine Biology **108**: 217–225.

Stewart, D.B., and Lockhart, W.L. 2005. An overview of the Hudson Bay marine ecosystem. Can Tech. Rep. Fish. Aquat. Sci. 2586: vi+487 p.

Stirling, I., and Derocher, A.E. 1993. Possible impacts of climatic warming on polar bears. Arctic **46**(3): 240–245.

Stirling, I., and Parkinson, C.L. 2006. Possible effects of climate warming on selected populations of polar bears (*Ursus maritimus*) in the Canadian Arctic. Arctic **59**(3): 261–275.

Trites, A.W., Miller, A.J., Maschner, H.D.G., Alexander, M.A., Bograd, S.J., Calder, J.A., Capotondi, A., Coyle, K.O., Di Lorenzo, E., Finney, B.P., Gregr, E.J., Grosch, C.E., Hare, S.J., Hunt, S.J.J., Jahncke, J., Kachel, N.B., Kim, H.-J., Ladd, C., Mantua, N.J., Marzban, C., Maslowski, W., Mendelssohn, R., Neilson, D.J., Okkonen, S.R., Overland, J.E., Reedy-Maschner, K.L., Royer, T.C., Schwing, F.B., Wang, J.X.L., and Winship, A.W. 2007. Bottom-up forcing and the decline of Stellar sea lions (*Eumetopias jubatas*) in Alaska; assessing the ocean climate hypothesis. Fisheries Oceanography **16**(1): 46–67.

Trites, A.W., and Donnelly, C.P. 2003. The decline of the Stellar sea lions (*Eumetopias jubatus*) in Alaska: a review of the nutritional stress hypothesis. Mammal Review, **33**: 3–28.

Tyrrell, M. 2007. Sentient beings and wildlife resources: Inuit, beluga whales and management regimes in the Canadian Arctic. Human Ecology **35**: 575–586.

Population Genetics of Hudson Bay Marine Mammals: Current Knowledge and Future Risks

S.D. Petersen, M. Hainstock, and P.J. Wilson

Abstract Hudson Bay has experienced, and is predicted to further undergo, significant environmental changes that may affect the distribution of marine mammals. These changes will affect gene flow among regions within the greater Hudson Bay ecosystem, as well as between Hudson Bay and the rest of the Arctic. Currently, there are few genetic studies that include marine mammals from Hudson Bay, even though this area is critical to understanding how Arctic species will adjust to climate changes. Within this region, some marine mammals may become extirpated or isolated (e.g., southern Hudson Bay polar bears), while other species may expand their ranges (e.g., killer whales, harbour seals) as a result of warmer temperatures. Researchers and the public should view the greater Hudson Bay ecosystem as an early warning system for the larger Arctic ecosystem.

For population geneticists, marine mammals pose a unique challenge because they show little differentiation over large spatial scales due to: large historical population sizes; high mobility; seasonal migration; and breeding patterns that promote gene flow. Genetic monitoring programs need to take these factors into account in order to be effective. If designed carefully, these programs can be used to track changes in marine mammal populations that result from climate change.

We survey current genetic data collected from marine mammals in Hudson Bay and suggest possible trajectories that may result from temporal shifts in ice thaw and decreasing overall ice cover. We also comment on sampling strategies that will allow for the effective monitoring of genetic changes.

S.D. Petersen (✉)
Fisheries and Oceans Canada, 501 University Crescent, Winnipeg, Manitoba,
MB R3T 2N6, Canada
e-mail: Stephen.Petersen@dfo-mpo.gc.ca

M. Hainstock
Ducks Unlimited Canada, Oak Hammock Marsh Conservation Centre, Stonewall, Manitoba,
R0C 2Z0, Canada

P.J. Wilson
Natural Resources DNA Profiling and Forensic Centre, Biology Department & Forensic Science
Program, Trent University, Peterborough, Ontario, K9J 7B8, Canada

S.H. Ferguson et al. (eds.), *A Little Less Arctic: Top Predators in the World's
Largest Northern Inland Sea, Hudson Bay*, DOI 10.1007/978-90-481-9121-5_12,
© Springer Science+Business Media B.V. 2010

Keywords Gene flow • Marine mammals • Monitoring genetic changes • Population genetic structure

Introduction

Arctic mammals have evolved to survive long, harsh winters and to take advantage of brief, yet highly productive, summers. The close relationship between these species and environmental conditions suggests that climate change will lead to changes in the ecology, spatial and temporal distribution, and genetic structure of Arctic populations (Laidre et al. 2008; Post et al. 2009). These changes may result from adaptation to new conditions, or changing population sizes and increased isolation. Although there are a variety of ways that Arctic mammals may be affected by climate change, we are primarily concerned with population genetic structure (defined below) in this chapter. More specifically, this discussion is presented from the perspective of conservation and management, where the genetic delineation of populations, stocks, or management units[1] is of paramount importance. Therefore, this chapter summarizes the current level of knowledge regarding relevant patterns that have been observed, as well as proposed explanations for them. It will discuss the challenge of quantifying genetic variation in Arctic marine mammals and will suggest how studies should be designed to understand temporal and spatial genetic patterns. Finally, this chapter will consider how the distribution of neutral genetic variation (defined below) may change with a shifting environmental regime, in addition to highlighting areas where future research could be focused.

A Primer in Conservation Genetics and Population Genetics

Conservation is aimed at maintaining diversity at three levels: ecosystem, species, and genetic (McNeely et al. 1990). Within those levels, genetic diversity is fundamental to the conservation of biodiversity because the loss of this source of variability has been

---

[1] In this chapter, we use the terms management unit, population, and stock in two ways: when referring to one particular species and when speaking more generally about Arctic mammals. The first use of these terms reflects the fact that several stakeholder groups are involved in the conservation and management of species in the greater Hudson Bay ecosystem. Since each of these groups uses different objectives and analyses to define biologically meaningful units, the terms used to describe these units will vary from species to species. For example, polar bears are defined according to management units that may include more than one genetic population because they correspond primarily to demographic boundaries, while beluga whales are managed as stocks that roughly coincide with both demographic and genetic population boundaries and bowhead whales are managed as a single population because numerous data support this as a meaningful biological unit. The second use of these terms will be a more general classification, where management units are considered jurisdictional, populations are genetic populations, and stocks are groupings that are affected by humans (e.g., harvested) and exist below the population level (i.e., meta-populations, groups with a learned behaviour like summering ground locations). An interesting discussion regarding the definition of stock can be found in Stewart (2008).

linked to lower fitness, especially when animals are confronted with environmental stressors (Coltman et al. 1999; Kristensen and Sørensen 2005; Woodworth et al. 2002). Thus, loss of genetic diversity may reduce the ability of individuals and populations to survive and adapt to changing conditions. Therefore, it is important to try to maintain natural levels of genetic diversity in all components of an ecosystem.

Throughout the following paragraphs we will discuss a number of conservation genetics and population genetics concepts that may, or may not, be familiar to the reader. It is important to note that conservation genetics is concerned with maintaining gene flow in some cases, while preventing gene flow in others. Why can't one objective fit all conservation dilemmas? To answer this, one must understand that conservation genetics is concerned with maintaining natural levels of gene flow among regions, based on the assumption that this rate of flow is adaptive in some way. Population genetic structure (or genetic structure) is a quantification of the genetic differences among groups (differentiation), which is a reflection of the amount of gene flow among groups, where low gene flow will usually lead to higher genetic structure. For example, when physical barriers fragment landscapes, fewer animals move among groups, which reduces gene flow. When these isolated groups are small, this will lead to a rapid increase in genetic structure. In some cases, barriers to gene flow allow populations to adapt to local conditions and these adaptations would be lost if new genes, or new gene combinations, were constantly being introduced in the population. In other cases, anthropogenic activities will create barriers to gene flow that isolate a portion of a population, leaving too few individuals or too little genetic diversity to adapt to new conditions. Therefore, recognizing that one strategy (e.g., increase gene flow) will not be effective in all situations, conservation genetics aims to maintain natural levels of diversity and structure in hopes that these will give species the best chances of survival in the long term.

Genetic diversity is difficult to conserve because: it is created through mutation; mutations are lost or maintained through selective processes; and important diversity can be lost by genetic drift. Genetic drift is the random loss of mutations, which tends to be more significant in smaller populations (Frankham et al. 1999). When a population experiences a dramatic reduction in population size and a corresponding reduction in genetic diversity, this is known as a bottleneck event. Therefore, population size is correlated with genetic diversity (Frankham 1996) and managers can maintain genetic diversity and gene flow by maintaining population sizes and landscape connectivity, which facilitates natural levels of movement of individuals among groups. In the context of the greater Hudson Bay ecosystem (GHBE), the loss of ice can result in two simultaneous threats for many species. First, habitat loss may lead to an overall reduction in population size and a parallel reduction in genetic diversity. Second, the landscape may become fragmented, which may create even smaller isolated populations where genetic drift will further erode genetic diversity.

Determining the natural levels of genetic diversity and existing genetic structure is critical to assessing potential impacts of climate changes on wild populations. To accomplish this, a number of genetic markers are used to detect and quantify genetic variation, each of which has different characteristics and strengths. In this chapter, we mention two broad classes of markers: those that are functional and those that are neutral. Functional markers are found in genes that are under selection, such as genes involved in disease resistance. Although the development of functional markers has increased in recent years and will become more important in the future, we primarily discuss neutral markers in this chapter. Neutral markers are not (theoretically) linked to genes under selection and thus should provide a better estimate of the relationships among individuals, groups, and populations. For an example of why this is the case, imagine two groups of animals that are only distantly related but are both living in a similar environment. Convergent evolution may cause the functional markers to be similar, but neutral markers would reveal that they are only distantly related.

Neutral markers can be further categorized as nuclear DNA (nDNA) or mitochondrial DNA (mtDNA). As the names imply, nDNA is present in the nucleus of cells, whereas mtDNA is found in the mitochondria. Both parents pass nDNA to their offspring. This allows researchers to identify close relationships (e.g., parent–offspring), as well as more distant relationships by comparing the patterns in nDNA within a group to the population as a whole. Only mothers pass mtDNA to their offspring (maternal inheritance), so this marker is valuable for tracing female contributions to gene flow. Similarly, but to a much lesser degree, Y-chromosome markers have been used to trace male contributions to gene flow. The discussions that follow in this chapter will be primarily based on the use of neutral genetic markers and the analyses that these markers lend themselves to: that of identifying temporal and spatial genetic changes.

An important first step in the process of monitoring, conserving, or managing is that of delineating biologically meaningful units and obtaining baseline data against which future changes can be detected (Amos and Balmford 2001; Haig 1998; Moritz 1994). In recent years, the number of population genetic studies of Arctic species has increased. Given the importance of this information, this chapter reports the current population genetic data for mammalian species that regularly occur in the GHBE. This includes 11 mammals that commonly occur in and on the waters of the GHBE: two terrestrial carnivores (Arctic fox (*Vulpes* [=*Alopex*] *lagopus*) and polar bear (*Ursus maritimus*)); five pinnipeds (walrus (*Odobenus rosmarus*), ringed seal (*Pusa* [=*Phoca*] *hispida*), harbour seal (*Phoca vitulina*), harp seal (*Pagophilus groenlandicus*), and bearded seal (*Erignathus barbatus*)); and four cetaceans (narwhal (*Monodon monoceros*), beluga (*Delphinapterus leucas*), killer whale (*Orcinus orca*) and bowhead whale (*Balaena mysticetus*)). Currently, harp seals and killer whales are being observed with greater frequency in the GHBE (Higdon and Ferguson this volume). Although their changing distribution and demography may provide compelling examples of changes already occurring in the GHBE, we will omit these species from our summary because little is currently known about their genetics in the region.

General Characteristics and Patterns

In studies of population genetic structure in Arctic marine mammals, investigators commonly observe a surprising lack of genetic differentiation over large distances (Brown Gladden et al. 1999; Davis et al. 2008; Paetkau et al. 1999).

This pattern, known as panmixia, is characterized by a high degree of genetic similarity (homogeneity) among locations, and is derived from a combination of processes that function over different time scales. Long-distance dispersal movements among regions and/or short-distance gene flow within continuous distributions contribute to the maintenance of genetic homogeneity over recent time scales. Over longer time scales, post-glacial range expansion and genetic storage (the retention of genetic diversity in long-lived life stages) can contribute to the maintenance of genetic homogeneity.

Numerous telemetry and mark-recapture studies, in addition to more recent genetic studies (Crompton et al. 2008; Petersen 2008), suggest a higher degree of breeding site fidelity than previously assumed in Arctic mammals. For example, although adult polar bears may travel many hundreds of kilometers in the interval between initial capture and recapture, they are often recaptured within 100 km of their original capture location (Taylor et al. 2001). Likewise, individual narwhal move between summer and winter locations with great fidelity (Heide-Jørgensen et al. 2003). Although site fidelity can lead to genetic structure at various levels, this form of differentiation is generally low in species found within the GHBE. The following paragraphs discuss the ways in which this apparent contradictory result can occur.

Evolutionary Legacy

Current patterns of genetic variation in a particular species are dictated by their evolutionary history, which has been shaped by successive cool and warm periods. The last interglacial period, which ended approximately 117,000 BP (years before present, Kukla et al. 2002), was warmer than current conditions and may have reduced the ranges of most ice-obligate species. The genetic signature of a population expansion in Arctic foxes dates roughly to the end of this interglacial period (Dalén et al. 2005). More recently, during the last glacial period (ca. 24,000–14,000 BP), the entire GHBE was covered by the Laurentide ice sheet (Dyke et al. 2002), which likely excluded all marine mammals. Subsequent range expansions into this area, coincident with population size increases, have likely contributed to the observed genetic homogeneity in the species that occur in the GHBE. The reduction in gene flow between populations within the GHBE and populations outside of the GHBE has occurred so recently in evolutionary time that divergence may be difficult to detect. This is consistent with other studies that show patterns of reduced genetic structure in taxa that have experienced a recent range expansion (Ibrahim et al. 1996; Ray et al. 2003).

More recently (ca. 8,200 BP), a short-duration cold anomaly, which is hypothesised to be linked to an influx of freshwater following the last outburst flood of glacial Lake Agassiz (Alley and Ágústsdóttir 2005), may have reduced populations of some cetaceans. It has been suggested that cetaceans were excluded from parts of the Canadian Arctic during this time, based on a lack of bowhead whale bones at archaeological sites from the same period (Dyke and England 2003; Dyke et al. 1996). This 8,000 BP event, and the Little Ice Age of recent times (1300s–1800s), may have reduced population sizes and caused genetic bottlenecks in some populations. Large-scale ice entrapments could also have occurred during times of rapid cooling and this may explain the lack of mitochondrial diversity in narwhal (Palsbøll et al. 1997).

Following these cold events, it is likely that populations expanded in size and re-colonized the GHBE. However, each cycle probably caused a bottleneck because population sizes were reduced and few lineages were able to re-colonize the GHBE. Based on sighting records, Higdon and Ferguson (2009) concluded that killer whales have been expanding their range into Hudson Bay since the 1950s. We would predict that this likely represents an invasion by relatively few lineages; and an ice-free Hudson Bay would initially contain a population of killer whales with low genetic diversity.

Anthropogenic Bottlenecks

Some Arctic mammals have been subject to intense hunting pressure, which may have had an impact on genetic diversity. From the 1600s to 1900s, bowhead whales were targeted extensively by European and Yankee whalers (Reeves and Cosens 2003; Reeves et al. 1983). These harvests reduced their numbers from hundreds of thousands of animals to only hundreds (Reeves et al. 1983). Although this drop in population size should coincide with a parallel loss of genetic diversity, this has not been detected in bowhead whales (Borge et al. 2007). This could be due to their longevity and mobility, which enabled existing genetic variation to be maintained despite bottleneck events.

In the face of harvest pressure, marine mammals in the GHBE have generally been able to maintain higher population numbers compared to species that occur in temperate and southern oceans. This is due, in part, to the difficulties of accessing and harvesting GHBE marine mammals over their entire range (e.g., Nares Strait walrus COSEWIC 2006). Local exceptions, like the commercial over-harvest of Eastern Hudson Bay beluga and Ungava Bay beluga, do occur where animals are accessible. However, these represent only portions of the species total range and the corresponding loss of diversity due to local extirpations may be minimal in the context of the entire species.

Life History

The life history characteristics of species also play an important role in the observed level of genetic structure. For example, Stirling and Thomas (2003) suggest several life history characteristics, including natal breeding site philopatry, population specific vocalizations, and mating strategy, are important determinates of population structure for seals. We highlight three characteristics that play an important role in determining the genetic patterns observed, or hypothesized, in the marine mammals that occur in the GHBE.

Dispersal and Mobility

Many theoretical models for the dispersal of genes across a landscape assume, or look for, a pattern of isolation-by-distance. In these models, breeding among individuals is limited by how far an individual can move and this creates a pattern where genetic differences are positively correlated with geographic distances. In other words, the closer individuals are in space, the more closely related to each other they tend to be (Epperson 1995; Wright 1943). As we will point out in the species summaries below, this pattern may be challenging to detect because all of the marine mammals within the GHBE have recorded annual movements that would allow them to disperse between any points in the GHBE over a relatively short time span. For example, although beluga summer concentration areas are discrete, most are closer to each other than the distance these whales travel between summering and wintering areas (COSEWIC 2004a). In beluga whales, and all other marine mammals discussed here, individuals can travel significant distances over the course of a season; and they could disperse their genes to any other part of their range over their reproductive life span. Therefore, the observation of genetic structure in any of these species suggests that movement ability is not influencing genetic structure and that gene dispersal among locations is a function of some other parameter, such as natal fidelity or social structure.

Annual variability in both the timing and patterns of ice formation has contributed to species having large areas of occurrence and/or long migration routes. However, these characteristics may also buffer populations from poor conditions that are short in temporal duration or local in spatial extent. In the event of a local extinction, animals with high mobility have a higher probably of re-colonizing, which lowers their overall vulnerability (Roberts and Hawkins 1999). On the negative side, if this mobility is associated with migration events, it may also expose a greater segment of the population to risks that occur along the migration route (Robinson et al. 2009). Whatever the benefits or costs associated with high mobility, researchers must account for it during their experimental designs, to ensure that they obtain samples from the times and places that are most likely to reveal genetic structure (e.g., breeding season and breeding grounds).

Mating Strategy

Post-glacial expansion into the GHBE has been limited to those animals adapted to seasonal ice cover. This has placed reproductive constraints on marine mammals in the GHBE and has shaped the evolution of their mating strategies. In ringed seals, the annual variability in location of suitable breeding habitat may prevent strict natal fidelity, thus promoting mating systems that would result in a predictably panmictic genetic structure (Coltman et al. 2007; Stirling 1983). In cetaceans, the use of the GHBE involves migration into and out of the region each year. Migration along traditional routes may provide an opportunity to maximize the number of potential mates and thus would lead to genetic homogeneity if mating occurred during migration. For the marine mammals that inhabit the GHBE, data suggest that mating occurs in the late winter or early spring. However, our ability to study mating systems is extremely limited during this period because the animals are under ice and usually far off shore. There are also few, if any, daylight hours to work with.

In some species, the seasonal concentration of animals at predictable wintering areas has likely influenced the evolution of mating strategies. For example, in these circumstances, males can compete for and defend harems of females. Walrus, harbour seals, and bearded seals are associated with areas that are ice free throughout the winter and are likely candidates for exhibiting strong population structure. Genetic structure based on mating strategy has been predicted for bearded seals based on vocalization patterns associated with breeding (Stirling and Thomas 2003), and genetic results seem to support this in areas outside of the GHBE (Davis et al. 2008). In contrast to this pattern, ringed seals are able to maintain breathing holes in ice and thus have a more dispersed distribution during the breeding season. Although the ringed seal mating system has not been described in detail, males likely defend territories that contain several females (Reeves 1998). This arrangement will facilitate gene flow to a much greater degree and would be predicted to result in a pattern of isolation-by-distance (Davis et al. 2008; Petersen 2008).

Longevity

Long life spans and overlapping generations increase effective population size by allowing multiple cohorts to simultaneously be part of the gene pool. Long lives provide a greater opportunity to increase lifetime dispersal distance and, potentially, the distance over which gene flow occurs. In long-lived species, large-scale movements can occur in the sub-adult life stage before an individual starts breeding, or in response to changes in environmental conditions. These types of movements may act to increase homogeneity of the population over time in the absence of breeding site fidelity. Again, this should be factored into the sampling design of temporal studies of Arctic marine mammals, because long life spans may create a time lag between the disruption in gene flow and the ability to detect it.

Long life spans and overlapping generations may maintain genetic variability in a temporally fluctuating environment, through the storage of genotypes by certain life stages (Ellner and Hairston 1994; Gaggiotti et al. 1997). If selection pressure varies over time and is focused on a single life stage (e.g., juveniles), then long-lived species will retain genetic diversity in the stage that is not selected against (e.g., adults). This "storage of genotypes" (Gaggiotti et al. 1997) or "storage effect"

(Chesson 1983) may account for the maintenance of genetic diversity in bowhead whales following whaling because all of the genetic information that survived the bottleneck event has been maintained. In marine mammals, pup/calf mortality rates are usually higher than all other age classes, while adult mortality is generally lower (e.g., ringed seals, Ferguson et al. 2005). If the environmental conditions in one time period do not favour the genotype of the adult, they may still be alive in a subsequent period when their genotype may be selected for.

An additional benefit to having a long life span is that fewer generations experience a bottleneck event and thus genetic diversity is buffered over an extended period of time (Kuo and Janzen 2004; Lippé et al. 2006). Extended longevity may make it more challenging to distinguish genetic groups for management purposes, but from an evolutionary perspective it is beneficial. Longer-lived species have likely experienced climate change towards a warmer Arctic in their recent evolutionary history and may have the variability to adapt to the changes.

Possible Future Genetic Changes

Although speculative, a number of factors may impact the population structure of marine mammals in the GHBE. In general, decreases in population size and range will act to increase population genetic structure, whereas increases in population size will have the opposite effect. The following paragraphs will summarize the expected genetic impacts due to changes expected in the Arctic. Although the rate of change has been the topic of much discussion, it is generally expected that sea ice extent and duration will continue to decrease in the GHBE region (Gough and Wolfe 2001; Parry et al. 2007).

To date, research that has examined sea ice trends has shown increased open water over the last several decades in Hudson Bay and Foxe Basin (Gough et al. 2004; Heide-Jørgensen and Laidre 2004). This reduction in sea ice will have obvious consequences for ice-adapted species, which are expected to lose habitat and presumably population size (Harwood 2001; Learmonth et al. 2006; Simmonds and Isaac 2007; Tynan and DeMaster 1997). For example, ringed seals are obligate ice breeders (Furgal et al. 1996; Reeves 1998; Smith and Stirling 1975), and reduced ice and snow have been correlated with lowered recruitment in this species (Ferguson et al. 2005). Likewise, spring rain is known to increase pup mortality in ringed seals (Stirling and Smith 2004). Over several years in the Baltic Sea, the absence of sea ice for most of the winter resulted in high pup mortality during the late 1980s (Härkönen et al. 1998). Therefore, a rapid trend towards less ice could result in a series of years with low recruitment and would be expected to translate into decreases in ringed seal populations. It is unlikely that this species would go extinct, but they may become extirpated from the southern portions of their range and isolated in other areas. For all species, reductions in population sizes due to reduced resources or available habitat will present challenges to maintaining gene flow and levels of diversity.

Conversely, loss of ice is expected to have a positive effect on some species. Increases in population size and a concordant reduction in genetic structure may also be observed in species that can respond to an increase in productivity resulting from a longer ice-free season. The current low density of harbour seals in Hudson Bay may be expected to increase because they are limited by the availability of ice-free areas during winter (COSEWIC 2007). In harbour seals, nothing is currently known regarding the rate of gene flow among locations; however, it is likely that the loss of sea ice would provide the opportunity for increased dispersal and thus gene flow among locations. Locally adapted populations may be lost through genetic swamping, which would lead to an overall reduction in functional adaptations (García-Ramos and Kirkpatrick 1997; Kawecki and Holt 2002).

Decreases in the duration and extent of ice cover may have major impacts on the species compositions of the GHBE. There are indications that some predators are increasing (e.g., killer whales, Higdon and Ferguson 2009, Higdon and Ferguson this volume), which could reduce other species or increase the costs of dispersal, leading to isolation of some areas. New competitors may also invade (e.g., harp seals, minke whales), and the resulting equilibrium that is established may not favour the existing species. These changes will affect each species and each web of species in a unique way. Changes in ice and food distribution may lead to changes in predator distribution, and could isolate or connect populations. The seasonal structure in southern and western Hudson Bay polar bears indicates the potential for changes in sea ice thaw to increase genetic differentiation over time (Crompton et al. 2008). In contrast, lineages that have been isolated in the eastern and western Arctic will be reconnected. For example, bowhead and beluga whale populations will be the first to reconnect if an ice-free route is established across the Canadian Arctic.

Changes in the ecosystem and climate will not only modify patterns of gene flow, they will also change selection pressures with unknown outcomes. Researchers should attempt to document these changes, while developing predictive models to help mitigate negative impacts. Berteaux et al. (2006) have pointed out that care must be exercised in developing these models, to maximize their utility. The following section highlights promising genetic analyses that can help develop models directly applicable to conservation and management.

Current and Future Directions for Research

Analytical Directions

Recent advances in molecular ecology have emphasized landscape genetics (Holderegger and Wagner 2006, 2008; Manel et al. 2003; Storfer et al. 2007) and spatial genetic approaches (Guillot et al. 2005, 2009; Jombart et al. 2008) as ways to use genetic variation to understand the complex interaction between individuals and their environment. Landscape genetics combines the fields of landscape ecology with population genetics and has made significant progress in recent years in examining and explaining patterns of gene flow over large scales. Several analyses have emerged that incorporate geographic information systems (GIS) technology to model gene flow in more complex ways than simple isolation-by-distance. These include using least-cost paths where habitat types are assigned values based on relative permeability to gene flow (Adriaensen et al. 2003; Coulon et al. 2004; Wang et al. 2008). For example, water may restrict gene flow in terrestrial species and would be given a high cost value. The correct cost structure is inferred when the correlation between genetic distance and cost is highest, which allows for important areas (i.e., corridors) to be identified. Similarly, resistance models incorporate both the cost of the length of the path and the width of the path (McRae and Beier 2007). While these developments are facilitating genetic studies in terrestrial populations, the dramatic seasonality experienced in the GHBE (i.e., from sea ice to open water) may prevent these approaches from being applied to some marine mammals, such as polar bears and cetaceans. At the very least, the applicability of these approaches will depend on sample designs that account for temporal changes in distribution. In this respect, the difference between seasonal and year round residents of the GHBE is a valuable distinction to make. We would suggest that landscape and other spatial genetic methods will be promising approaches to examine gene flow in resident species, where isolation-by-distance would likely be occurring. However, these approaches will be most successful if sampling designs take factors such as seasonal mixing into account. Two further items should be kept in mind when designing studies and analysing data; both relate to the apparent panmixia observed in many taxa.

First, the overall panmixia of GHBE populations likely reflects high mobility over an annual cycle, while observations of site fidelity indicate that some genetic discontinuities may exist during the mating season. This breeding structure may be masked if the assumption of geographic versus seasonal structuring is incorrectly applied. This requires careful consideration of when samples are collected and of which specific population genetic analyses are required. A recent review of spatial genetics stresses the need for geographic-based Bayesian structure analyses (Guillot et al. 2009). Bayesian methods are statistical analyses that cluster data based on theoretical assumptions (priors) rather than predefined grouping. In genetic analyses, the priors that are used to cluster individuals into groups are based on the genetic assumptions of populations (i.e., Hardy–Weinberg equilibrium and linkage disequilibrium), while newer analyses include continuous spatial data as a prior to inform clustering algorithms. This strategy is less applicable for studies of seasonal residents to the GHBE because pooling samples collected throughout the year for each location may mask the genetic structure present at the critical time when this structure would be defined: the breeding season. When samples are known to be sampled during the breeding season, and can be partitioned in the analyses as such, then traditional genetic structure estimates (e.g., F_{ST}) can be applied. If this is not the case, applying Bayesian genetic structure models such as STRUCTURE (Pritchard et al. 2000; Falush et al. 2005), that do not incorporate spatial data, are more appropriate than programs like Geneland (Guillot et al. 2005) or TESS (Chen et al. 2007), which will inaccurately weigh the sampling site into the model.

The second aspect to keep in mind corresponds to overall low genetic differentiation in GHBE marine mammals, even when structure is detected. Although panmixia is a useful null model and is a common genetic pattern among species in the GHBE, a panmictic population is not necessarily entirely homogeneous, nor would it be predicted to be homogeneous. Various aggregations can lead to significant and important genetic structure (Latch and Rhodes 2006). For example, aggregations of related individuals such as whale pods, or learned behaviour relating to the locations of feeding sites or nursery areas, can create genetic heterogeneity within a population. This structure is important for management and conservation, and it needs to be understood. Therefore, analyses that characterize heterogeneity below the population level and examine those patterns are promising directions to move in. By focusing on the connections among individuals or locations, researchers will be able to identify important areas that work to maintain overall connectivity in the system. One approach to identify and model connectivity is to represent individuals in the context of a network and apply graph theory to explore that network.

Graph theory is a mathematical framework used to represent and model interconnected observations; in this case, genetic data. Graph theory has been used extensively in computing and social sciences. It has gained ground in biological sciences in recent years (May 2006; Proulx et al. 2005; Whitehead 2008) and has been more recently applied to genetic data (Dyer and Nason 2004; Garroway et al. 2008; Petersen 2008). Essentially, a network is constructed wherein the nodes are individuals or locations, and the links that connect nodes can be weighted to represent the strength of the relationships among them. When connections between nodes are strong or numerous, we can infer a high degree of relatedness if the nodes represent individuals, or we can infer high connectivity if the nodes represent locations. The network that can be modeled: is not dependent on the existence of population level genetic structure; does not require that Hardy–Weinberg equilibrium assumptions be met; and can characterize relationships even within a panmictic population. When this genetic network is compared to geographic data, sites of high connectivity can be identified. In addition, statistical tests can be performed to assess the relative importance of these sites, by removing their associated nodes or links to determine if panmixia will be maintained. If the removal of one element from the network results in a significant change in the network structure, we can infer that the removed element is critical for the maintenance of gene flow in the system. Furthermore, we can infer that negative impacts (e.g., habitat modification) in that area would result in increased differentiation and subsequent isolation. A monitoring program, with comprehensive baseline data, would allow for these critical areas to be identified in a network context and for proactive management decisions to be made if those connections were being degraded. Adding data regarding the location of sensitive areas, such as shipping lanes or greatest changes in ice concentration, will further enhance our ability to proactively mitigate the damage to the network as a whole, as opposed to being forced into remediation efforts after detecting the problem.

Technical Directions

To be most effective and proactive, our first priority should be to remove the gaps in our collective knowledge, as identified in the species summaries below. Following this, a genetic monitoring strategy (that includes a comprehensive baseline dataset) needs to be devised so researchers can track genetic change over time and in relation to climatic variables. Finally, there is a need for a long-term commitment to sampling strategies.

By committing to sampling strategies, researchers will be able to improve, develop, and validate models such as those generated using graph theory. The specifics of these strategies will vary among species, and one broad strategy will not be appropriate, given the variation in species distributions and life history characteristics (i.e., a sampling plan for seals will not be appropriate for studies of killer whales). Furthermore, as with all work in the Arctic, sample designs will need to be logistically feasible. As a broad guide, researchers need to factor in seasonality, and samples should be collected as close to the season that ultimately defines the genetic structure of interest as possible. For example, for resident species where population level genetic structure is sought, the breeding season should be targeted. For seasonal visitors, genetic structure likely occurs below the population level and sampling should avoid migrating animals. However, in the absence of an overarching sample design, simple measures like collecting accurate data on time of year (exact date so that ice charts can be cross-referenced) and geographic location (precise GPS data) will significantly improve researchers' ability to examine temporal and spatial genetic patterns.

In addition, fine-scale approaches that examine patterns of relatedness, kinship, or sex-biased movements will only be possible with sampling design improvements that include increases in sample size. This will require researchers to incorporate non-invasive or minimally invasive (e.g., biopsy sampling) techniques to increase samples without negatively affecting the numbers or behaviours of these species.

The logistical challenges and financial costs of conducting research in the Arctic will require people to develop strong partnerships with northern communities. Sample-sharing collaborations with other researchers will also maximize the value, in terms of scientific insights, that each sample provides. Development of sampling designs and strategies that can be maintained over time, such as community-based monitoring programs like the International Polar Year – Global Warming and Marine Mammals project, will be key to monitoring changes in the GHBE.

Conclusions

It is important to understand how ecosystems will respond in the face of changing conditions. These changes will be simultaneously environmental and anthropogenic, and will likely be most pronounced in the GHBE. Researchers should view this ecosystem as the "canary-in-the-coalmine" with respect to early responses to climatic changes, ecosystem shifts, and anthropogenic disturbances, and should attempt to characterize and monitor it. This is critical, not just for the populations inhabiting the GHBE, but also to establish a framework to apply to other Arctic ecosystems.

For marine mammals, one way to monitor changes at a variety of scales is through the use of genetic data. These data can be applied to fine-scale issues of individual dispersal patterns or modified social structure, as well as to large-scale issues of connectivity or evolution. Because GHBE mammals have large population sizes, high individual mobility, and an evolutionary history of recent expansion, researchers will need to use genetic data and novel analytic approaches that are effective below the population level to monitor the coming changes. One critical consideration in monitoring in the GHBE is that baseline data will be required immediately so connections within a naturally panmictic population can be measured over time. This will put researchers in a position to build predictive models and suggest proactive measures for minimizing impacts related to the changing conditions in the GHBE.

Species Summaries

In the following species summaries, we have highlighted a number of population and life history characteristics that have likely resulted in the genetic patterns discussed in this chapter. These include: distribution and recognized units (i.e., management units, populations, or stocks) in the GHBE; life history parameters, such as life span and age of maturity; and details of the species' annual cycle and mating strategy. These summaries also include the current available information regarding genetic population structure. Because data are lacking for many species within the region, we have assumed some similarities to conspecifics in other regions where appropriate.

Arctic Fox (Vulpes [=Alopex] lagopus)

Arctic foxes are year round residents of the GHBE and extend their distribution onto the ice as they scavenge polar bear kills and prey on ringed seal pups (Audet et al. 2002). These winter food sources can comprise a major component of their diet, especially in years of low lemming (*Dicrostonyx* and *Lemmus* spp.) abundance (Roth 2002, 2003). Breeding occurs in March and April, coincident with the establishment of summer territories (Audet et al. 2002). These territories break down in the fall, when Arctic foxes redistribute themselves to take advantage of seasonally available resources such as carrion (Audet et al. 2002). Arctic foxes have an average longevity of 3–4 years and their maximum life span is 12–14 years (Audet et al. 2002).

Dalén et al. (2005) observed that a significant amount of genetic variation in Arctic foxes was portioned between inland and coastal ecotypes in Scandinavia. Subsequent research, using mitochondrial and nuclear markers, found that the greatest single predictor of genetic differentiation in this species was the presence of sea ice (Geffen et al. 2007). The frequency of sea ice, which provides a transport or corridor among islands, explained 40–60% of the genetic variation in Arctic foxes (Geffen et al. 2007) among locations. Both Dalén et al. (2005) and Geffen et al. (2007) were only able to include one GHBE location (Churchill, Manitoba) in their studies. Carmichael et al. (2007) obtained samples from the entire region and observed little differentiation within the GHBE and across their circumpolar distribution. This is not surprising, given the large population size observed, as well as their long-distance dispersal ability (>1,000 km Wrigley and Hatch 1976) and the presence of a land-based dispersal route that allows circumnavigation of the GHBE.

Arctic foxes face some threats that may impact their genetic structure in the future (Carmichael et al. 2007). Studies in Scandinavia have shown that competition with red foxes (*V. vulpes*) can reduce the range of Arctic foxes (Hersteinsson and Macdonald 1992). This competition throughout GHBE, and possible range reduction in the south, has the potential to isolate some regions. For example, the loss of range in the James Bay area could reduce gene flow between Quebec and Manitoba. Ice transport from northern areas could maintain gene flow in the near future, although a complete loss of ice could isolate Arctic foxes residing in northern Quebec and Labrador, given that permanently open water reduces gene flow (Geffen et al. 2007). Competition, reduction in dispersal opportunities and increased mortality due to disease or human interactions may collectively act to reduce gene flow in the future. Furthermore, although not yet reported, the potential for novel hybridization between Arctic foxes and red foxes through range expansion is possible and has been recently reported in other species (Garroway et al. 2010).

Polar Bear (Ursus maritimus)

Polar bears are year round residents of the GHBE, with southern Hudson Bay representing the farthest south that this species regularly occurs (DeMaster and Stirling 1981). Established based on data from capture-mark-recapture and satellite telemetry, as well political boundaries, four management units occur in this region: Foxe Basin, Davis Strait, western Hudson Bay, and southern Hudson Bay (IUCN/SSC Polar Bear Specialist Group 2005). These management units are estimated to include approximately 5,000 individuals in total (IUCN/SSC Polar Bear Specialist Group 2005, Peacock et al. this volume).

Breeding occurs in the spring, while polar bears are on the ice. Although male–male competition has been inferred from sexual dimorphism, little is known about mating strategies. Individual bears, particularly females, show site fidelity

(Mauritzen et al. 2001). There is no evidence of territoriality (Ramsay and Stirling 1986). The maximum life span is up to 40 years in captivity, but is likely 25–30 years in the wild (DeMaster and Stirling 1981). Polar bears mature a year earlier in the eastern Arctic than in the western Arctic; at 4 years in females (Stirling and Kiliaan 1980) and 6 years in males (DeMaster and Stirling 1981).

Polar bears show little overall genetic differentiation among regions (mean FST 0.04 (Paetkau et al. 1999): from Table 4), although significant differences have been detected at some scales (Crompton et al. 2008; Paetkau et al. 1999). Within the GHBE, there is evidence supporting three genetic units that roughly correspond to an eastern, western, and southern region (Crompton et al. 2008). Based on analysis of nDNA markers, Crompton et al. (2008) suggested the possibility that polar bears from James Bay form a genetically distinct group, albeit with gene flow with other management units.

Because polar bears are adapted to the marine environment, particularly to life on sea ice, they are sensitive to seasonal changes in ice extent and duration (Amstrup 2003; Stirling et al. 1999). In most of the GHBE, sea ice melts each summer (Etkin 1991; Wang et al. 1994), which forces bears ashore. In southern Hudson Bay, polar bears spend 4 or 5 months on shore waiting for ice to reform (Stirling et al. 2004), at which time they live primarily on stored fat reserves (DeMaster and Stirling 1981; Stirling et al. 2004). In recent years, there has been a significant trend towards earlier spring break-up in both western Hudson Bay (Stirling et al. 1999, 2004) and eastern and southern Hudson Bay (Gagnon and Gough 2005; Gough et al. 2004). Less time spent on the ice has been correlated with lower fat reserves, which can put bears at risk of starvation (Obbard et al. 2006; Stirling and Parkinson 2006, Peacock et al. this volume). Crompton et al. (2008) suggested that a continued trend towards earlier ice break-up could lead to decreased gene flow among Hudson Bay units and potential isolation of southern polar bears.

Walrus (Odobenus rosmarus)

Walrus are year round residents of GHBE, but their distribution is not uniform and three stocks have been recognised in the past: Foxe Basin; southern and eastern Hudson Bay; and northern Hudson Bay and Hudson Straight (Born et al. 1995). More recently, it has been suggested that these can be further subdivided (Stewart 2002, 2008). A population estimate for the GHBE is not available but there are likely less than 10,000 individuals there, most of which are in Foxe Basin as well as northern Hudson Bay and Hudson Straight (COSEWIC 2006).

Walrus are polygynous and breed during the winter (Fay 1985). Females exhibit delayed implantation and a long gestation period, such that pups are born the following spring between April and June (Fay 1985). Pups are weaned at 2 years of age and females first breed when they are 5–10 years old, while males are probably upwards of 15 years old before they breed (Fay 1985). Maximum longevity has been estimated to be 40 years (Nowak 1999).

To date, genetic studies have used nDNA and mtDNA markers to demonstrate significant genetic differences between Baffin Bay and Foxe Basin stocks of walrus (Buchanan et al. 1998; de March et al. 2002). This research was primarily aimed at testing if the Foxe Basin stock should be further subdivided, and it did not include samples from most of the GHBE. De March et al. (2002) did not find genetic evidence of differentiation between samples harvested from two communities in Foxe Basin (Igloolik and Hall Beach). This contrasted research using lead isotope signatures, which were significantly different in walrus harvested in the same two areas, suggesting the presence of two stocks (Outridge et al. 2003).

A decrease in ice cover has the potential to increase gene flow among areas. However, because walrus mating occurs near maximum ice cover (winter), almost complete ice loss would be needed, at which point changes in gene flow may not be the most significant concern. Female walrus are associated with ice to a greater degree than males (COSEWIC 2006), and thus may be impacted sooner by changes in ice cover and duration.

Ringed Seal (Pusa [=Phoca] hispida)

Ringed seals are common in the Canadian Arctic and have a circumpolar distribution (Reeves 1998). Within the GHBE, ringed seals are treated as a single stock. As ice forms, ringed seals start establishing territories that they will maintain during ice-covered months (Reeves 1998; Smith and Hammill 1981). Breathing holes are maintained during the winter, and lairs for resting and giving birth are dug out in snowdrifts that accumulate on the ice (Calvert and Stirling 1985; Furgal et al. 1996; Smith and Stirling 1975). Breeding occurs in late spring, after pups from the previous breeding season are weened (Reeves 1998; Smith and Hammill 1981). Ringed seals exhibit delayed implantation of approximately 3 months (Reeves 1998). Males and females mature at 5–7 years of age and can live for approximately 40 years (McLaren 1958).

No stocks have been identified in ringed seals, and genetic research has found little differentiation across the entire species range (Davis et al. 2008; Palo et al. 2001). Petersen (2008) observed geographic-related genetic differentiation between ringed seals harvested in Chesterfield Inlet, Nunavut and other locations, suggesting that gene flow can be reduced in some areas. The cause of this differentiation was not known, but may have to do with the physical characteristic of the Inlet. Over a larger scale, a pattern of isolation-by-distance has been detected in ringed seals from locations across the GHBE and eastern Canadian Arctic (Davis et al. 2008; Petersen 2008). In ringed seals, genetic structure exists but is most pronounced during the ice-covered season. When Petersen et al. (data on file) divided their data set of Hudson Bay ringed seals by sex and by season, they found less genetic structure during the open-water season, implying that seals from different areas are more mixed at this time. They also observed that both male and female seals are more structured during the ice-covered season, suggesting natal site fidelity in both sexes. Their results highlight the importance of sampling during the appropriate season when attempting to determine genetic patterns in Arctic marine mammals.

Ringed seals have been assessed as having a lower sensitivity to climate-related changes in habitat, because of large population sizes and extensive circumpolar distribution (Laidre et al. 2008). However, southern breeding populations may become extirpated due to reduced sea ice cover during the pupping season. In the Baltic Sea, Meirer et al. (2004) suggested that most suitable ringed seal breeding habitat would be lost with air temperature increases of 2–3°C. Similarly, ringed seals in southern Hudson Bay may be affected to a greater degree by climate changes than conspecifics in the high Arctic. These impacts will likely be detected first in Hudson Bay, because climate models predict that the rate of warming will be greatest in this region (Gough and Wolfe 2001) and observations from the last 30 years have supported this (Gough et al. 2004). Decreases in ice cover are likely to increase movement and thus gene flow. However, loss of breeding habitat and increased predation pressure from killer whales may act to fragment the range and increase genetic drift in isolated areas.

Harbour Seal (*Phoca vitulina*)

Harbour seals are year round residents in Hudson Bay and are primarily associated with the mouths of rivers where ice-free areas exist throughout the winter (COSEWIC 2007). This is because they lack claws that would enable them to maintain breathing holes in the ice, as ringed seals do. The full distribution of harbour seals in the GHBE is known only from anecdotal reports and includes the near shore environment within Hudson Bay and Hudson Strait, but not Foxe Basin (COSEWIC 2007). Harbour seals occurring in the GHBE are part of the stock that includes all eastern Canadian marine harbour seals and there are no estimates of numbers of animals within the Hudson Bay and Arctic regions (COSEWIC 2007).

Relatively little research has been conducted on harbour seals in the GHBE. However, in other parts of their range, harbour seals mature between 4 and 5 years of age and have a maximum life span of 36 years (Härkönen and Heide-Jørgensen 1990). Pupping season is thought to start in June in Hudson Bay (COSEWIC 2007). In the western Atlantic (New England to Baffin Island), there is a cline in peak pupping date that is negatively correlated with latitude (Bigg 1969). Harbour seals tend to show fidelity to haul-out sites, although tracking data indicate that they can move over hundreds of kilometres (COSEWIC 2007).

Harbour seals have received relatively little attention in Hudson Bay due to their rare occurrence, as well as their similarity with the ubiquitous ringed seal. Stanley et al. (1996) examined range-wide mtDNA variation and included samples from Churchill, Manitoba. They determined that these seals grouped with eastern Atlantic harbour seals, and hypothesized a post-glacial expansion that moved north from eastern North America and then east to Europe. Subsequent genetic research has been focused on Sable Island, Nova Scotia (Coltman et al. 1998) and freshwater forms in Quebec (COSEWIC 2007). These results, however, do not address the genetic structure of populations in the GHBE.

There is archaeological evidence that the numbers and range of harbour seals can increase in warm periods (Woollett et al. 2000), and it is suggested that current climate trends are increasing the number of harbour seals in the GHBE (Derocher et al. 2004; Stirling and Derocher 1993). Although this could support the prediction that gene flow may increase in the future, there are no data with which to evaluate or monitor this process.

Bearded Seal (Erignathus barbatus)

Bearded seals are year round residents in the GHBE and are harvested at low rates by Inuit in most communities (Cleator 1996; Priest and Usher 2004). There are currently no data regarding their numbers in the GHBE, although 190,000 individuals are estimated to inhabit all Canadian waters (Cleator 1996). No stock boundaries have been identified for management and harvests have declined over time (Cleator 1996). The greatest concentrations of bearded seals occur near polynyas and areas where regular leads form (Stirling 1997).

Bearded seals mature between 5 and 7 years of age and can live up to 25 years (Cleator 1996; Kovacs 2002). Breeding takes place after the pups are weaned, and males display using vocalizations (Cleator 1996). Juveniles disperse widely, but there are significant differences in vocalizations among locations, suggesting some degree of site fidelity (Cleator et al. 1989; Kovacs 2002; Stirling and Thomas 2003). Pups are born on the ice in the spring and quickly take to the water with their mother (Cleator 1996; Kovacs 2002).

Little genetic information is available regarding bearded seals in the GHBE. Davis et al. (2008) surveyed their global range and included samples from Arviat, Nunavut. They found significant population structure among locations, using nDNA markers. This conformed to expectations based on vocalization patterns (Stirling and Thomas 2003) and suggests that more research is needed to fully understand population structure and gene flow in this species.

Bearded seal numbers may increase in response to climate warming if it allows for greater winter access to benthic feeding areas. Because bearded seal pups can enter the water soon after birth, the loss of sea ice may not be a significant source of mortality during the pupping period. However, threats (e.g., from predators) that would accompany more open water are unknown. Increased population size could lead to increases in connectivity and gene flow among regions.

Narwhal (Monodon monoceros)

Narwhal are seasonal residents in the GHBE and are concentrated in northern regions. Animals migrate in through the Hudson Strait in the spring and concentrate in Repulse Bay and Lyon Inlet, Nunavut (Westdal and Richard this volume).

This northern Hudson Bay population (also stock) migrates out in the fall, to winter at the intersection of the Hudson and Davis Straits. This population is thought to contain less than 5,000 individuals, while the Baffin Bay population (made of multiple stocks) likely contains more than 50,000 animals (COSEWIC 2004b; DFO 2008). Narwhal are also observed in the Fury and Hecla Strait area of northern Foxe Basin.

Challenges in ageing narwhal have lead to uncertainty regarding its life span and maturity ages. However, estimates indicate that their average life span likely extends beyond 30 years (COSEWIC 2004b) and could be close to 100 years (Garde et al. 2007). Little is known regarding dispersal in this species. However, annual migration distances, recorded using satellite telemetry, show that animals could potentially migrate between any points in the species range with relative ease. Breeding occurs in the spring, although there may be a great degree of variability based on when newborn calves are observed (COSEWIC 2004b). Little is known about mating strategies or behaviours in this species.

Based on mtDNA marker frequencies, differentiation has been found among regions within eastern and western Greenland, but not between Canadian high Arctic and northern Greenland locations (Palsbøll et al. 1997). Using a combination of mtDNA and nDNA markers, de March et al. (2003) observed significant differentiation between samples from northern Hudson Bay (Repulse Bay) and locations in Baffin Bay. Narwhal show little genetic variability in the control region (a commonly used mtDNA marker), which suggests that they have experienced a significant bottleneck event and/or recent population expansion, possibly because of increased availability of habitat (expansion) following the last glacial maxima (bottleneck) (Palsbøll et al. 1997). The low mtDNA diversity in narwhal (17 haplotypes from 360 animals) is in striking contrast to beluga whales [over 75 haplotypes from 500 animals collected in GHBE (DFO 1993–2002)], the sister group to narwhal, which has many of the same life history traits.

Nuclear diversity in narwhal is similar to that observed in beluga whales. Buchanan et al. (1996) found a mean observed heterozygosity[2] of 0.65 over 13 nDNA markers (loci), while mean observed heterozygosity in narwhal was 0.79 over nine loci (DFO 1993–2002). Although not directly comparable, due to differences in loci used and sample sizes, it is likely that nuclear diversity is similar in the two species. Differentiation among locations has been observed to be low in narwhal (F_{ST} less than 0.024 de March et al. 2003), suggesting recent isolation, ongoing gene flow, or low marker resolution.

Changes in ice cover will likely have significant impacts on narwhal populations. Increases in ice cover have been observed in Baffin Bay and could lead to an increase in the frequency of entrapments, where groups of whales are unable to breath due to complete ice formation at the surface (Laidre and Heide-Jorgensen 2005). Conversely, decreases in ice could expose them to increased killer whale

---

[2]Heterozygosity is a measure of genetic variability.

predation. Given the low level of differentiation among putative units, it is unlikely that climate changes will significantly modify genetic structure in this species although it may change the patterns of gene flow.

Beluga (*Delphinapterus leucas*)

Beluga whales are seasonal residents throughout the GHBE. Most individuals migrate into the region in the spring, from wintering grounds in the Hudson Strait and coast of Labrador, while a small number may be resident in James Bay (COSEWIC 2004a). Currently, the beluga whales in GHBE are divided into eastern Hudson Bay, western Hudson Bay, and Ungava Bay units (COSEWIC 2004a); with the possibility that the eastern Hudson Bay unit could be further divided to add a James Bay unit.[3] The most numerous stock in the GHBE is the western Hudson Bay stock, which may contain upwards of 57,000 animals concentrated in the Nelson, Churchill, and Seal River estuaries (Richard 2005).

Female beluga whales mature between 4 and 7 years, while males mature between 7 and 9 years (Stewart and Stewart 1989). Their average life span has been estimated to be 20–30 years (Stewart and Stewart 1989), although it should be noted that changes in the interpretation of toothaging assumptions may modify age estimates upwards (Stewart et al. 2006). Breeding occurs in the late winter and early spring, although this is variable among regions (COSEWIC 2004a; Stewart and Stewart 1989). Similar to other whales that occur in the GHBE, the distance covered during the annual beluga migration puts most portions of their range within dispersal distance.

Genetic data has suggested significant differentiation among some portions of the beluga whale range. Mitochondrial data suggest that they show matrilineal structure among summering areas, whereas nuclear data indicate that male-biased gene flow is having a homogenizing affect on genetic structure (Brown Gladden et al. 1999; de March and Postma 2003; Mancuso 1995). A large amount of genetic data has been produced for harvested beluga whales in the GHBE, although little has been published outside of federal government reports regarding harvests. The most recent examination of these data confirms the differentiation of eastern Hudson Bay and western Hudson Bay stocks (Turgeon et al. 2009). Ungava Bay and James Bay may also be distinct genetic units, but samples are currently not available to conduct these analyses.

Beluga whales are less adapted to ice than narwhal, and reductions in ice cover may not influence their genetic structure in the same way. One outcome may be the reconnection of Pacific and Atlantic sides of the Arctic Ocean, which could facilitate gene flow between these regions. Increases in the frequency of ice entrapments may have significant effects on population sizes in smaller stocks.

---

[3]The Committee on the Status of Endangered Wildlife in Canada (COSEWIC) considers these units to be populations while the Department of Fisheries and Oceans, Nunavut, and Quebec (Nunavik) considers them to be stocks.

Bowhead whale *(Balaena mysticetus)*

Bowhead whales are seasonal migrants into the GHBE. They are most regularly observed in Repulse Bay and further north in the Foxe Basin – Fury and Hecla Strait area. The latter area has been suggested to be a nursery ground, where females with calves are able to avoid killer whale predation on young (Cosens and Blouw 2003). Bowhead whales are considered to be part of a single stock, based on current genetic data (Postma et al. 2006) and telemetry evidence (COSEWIC 2009). There has been much debate and analysis of relatively few surveys, leading to the current population estimate of 6,000 whales or more (COSEWIC 2009).

Bowhead whales are extremely long-lived animals (over 100 years George et al. 1999), and they mature late in life (25 years). They breed in late winter and spring (COSEWIC 2009) and the calving interval is likely 3–5 years, based on western bowhead whale data (Rugh et al. 1992). Like all GHBE cetaceans, recorded annual movements put all portions of their eastern Arctic range within dispersal distance over the course of a relatively short period of time. Genetic information initially suggested the presence of two populations in the eastern Arctic: Hudson Bay and Baffin Bay (Maiers et al. 1999). However, with increased sampling and reanalysis, the data indicated a single population (Postma et al. 2006). Although bowhead whales were harvested intensively in the eighteenth to early twentieth centuries, there is no evidence that current genetic variation is lower than historical levels (Borge et al. 2007).

Given the longevity and mobility of this species, coupled with an increasing population size trend, it is unlikely that a genetic change in response to climate trends will be detectable within the GHBE. On a larger scale, it is highly likely that loss of ice cover will reconnect western and eastern Arctic populations of bowhead whales with unknown consequences.

Summary

The GHBE is predicted to experience environmental changes that will significantly affect the population size and distribution of marine mammals that occur there. Some marine mammals may become extirpated or isolated (e.g., southern Hudson Bay polar bears), while other species may expand their ranges (e.g., killer whales, harbour seals) as a result of warmer temperatures and reduced ice cover or duration. At the population level, these changes will modify patterns of gene flow and genetic structure. How will these changes affect the evolution of the species in the GHBE and, in turn, the ecosystem as a whole? Although this is difficult to predict, we can implement monitoring strategies that will allow for changes to be detected. In order to be effective, genetic monitoring programs need to take factors relating to mobility, seasonal migration, and breeding patterns into account. If designed carefully, these programs can also be used to mitigate negative impacts of climate changes on marine mammals. Given the southern latitudinal extent of the GHBE, major ecosystem changes will be observed there earlier that in other regions. This makes GHBE critical to understanding how Arctic species will adjust to climate changes, and researchers and the public should view this ecosystem as a bellwether for the larger Arctic ecosystem.

References

Adriaensen, F., J. P. Chardon, G. De Blust, E. Swinnen, S. Villalba, H. Gulinck, and E. Matthysen. 2003. The application of 'least cost' modelling as a functional landscape model. Landscape and Urban Planning 64:233–247.

Alley, R. B., and A. M. Ágústsdóttir. 2005. The 8k event: cause and consequences of a major Holocene abrupt climate change. Quaternary Science Reviews 24:1123–1149.

Amos, W., and A. Balmford. 2001. When does conservation genetics matter? Heredity 87:257–265.

Amstrup, S. C. 2003. Polar Bear, *Ursus maritimus*, pp. 587–610 in Wild Mammals of North America: biology, management, and conservation (G. A. Feldhamer, B. C. Thompson and J. A. Chapman, eds.). Hopkins University Press, Baltimore, MD.

Audet, A. M., C. B. Robbins, and S. Larivière. 2002. *Alopex lagopus*. Mammalian Species 713:1–10.

Berteaux, D., M. M. Humphries, C. J. Krebs, M. Lima, A. G. Mcadam, N. Pettorelli, D. Réale, T. Saitoh, E. Tkadlec, R. B. Weladji, and N. C. Stenseth. 2006. Constraints to projecting the effects of climate change on mammals. Climate Research 32:151–158.

Bigg, M. A. 1969. Clines in the pupping season of the harbour seal *Phoca vitulina*. Journal of the Fisheries Research Board of Canada 26:449–455.

Borge, T., L. Bachmann, G. Bjørnstad, and Ø. Wiig. 2007. Genetic variation in Holocene bowhead whales from Svalbard. Molecular Ecology 16:2223–2235.

Born, E. W., I. Gjertz, and R. R. Reeves. 1995. Population Assessment of Atlantic Walrus. Meddelelser 138:1–100.

Brown Gladden, J. G., M. M. Ferguson, M. K. Friesen, and J. W. Clayton. 1999. Population structure of North American beluga whales (*Delphinapterus leucas*) based on nuclear DNA microsatellite variation and contrasted with the population structure revealed by mitochondrial DNA variation. Molecular Ecology 8:347–363.

Buchanan, F. C., M. K. Friesen, R. P. Littlejohn, and J. W. Clayton. 1996. Microsatellites from the beluga whale *Delphinapterus leucas*. Molecular Ecology 5:571–575.

Buchanan, F. C., L. D. Maiers, T. D. Thue, B. G. E. De March, and R. E. A. Stewart. 1998. Microsatellites from the Atlantic walrus *Odobenus rosmarus rosmarus*. Molecular Ecology 7:1083–1090.

Calvert, W., and I. Stirling. 1985. Winter distribution of ringed seals (*Phoca hispida*) in the Barrow Strait area, Northwest Territories, determined by underwater vocalizations. Canadian Journal of Fisheries and Aquatic Science 42:1238–1243.

Carmichael, L. E., J. Krizan, J. A. Nagy, E. Fuglei, M. Dumond, D. Johnson, A. Veitch, D. Berteaux, and C. Strobeck. 2007. Historical and ecological determinants of genetic structure in arctic canids. Molecular Ecology 16:3466–3483.

Chen, C., E. Durand, F. Forbes, and O. François. 2007. Bayesian clustering algorithms ascertaining spatial population structure: a new computer program and comparison study. Molecular Ecology Notes 7:747–756.

Chesson, P. L. 1983. Coexistence of competitors in a stochastic environment: the storage effect, pp. 188–198 in Population Biology (Lecture notes in biomathmatics 52) (H. I. Freeman and C. Strobeck, eds.). Springer-Verlag, New York.

Cleator, H. J. 1996. The status of the bearded seal, *Erignathus barbatus*, in Canada. Canadian Field-Naturalist 110:501–510.

Cleator, H. J., I. Stirling, and T. G. Smith. 1989. Underwater vocalizations of the bearded seal (*Erignathus barbatus*). Canadian Journal of Zoology 67:1900–1910.

Coltman, D. W., W. D. Bowen, and J. M. Wright. 1998. Birth weight and neonatal survival of harbour seal pups are positively correlated with genetic variation measured by microsatellites. Proceedings of the Royal Society of London, B 265:803–809.

Coltman, D. W., J. G. Pilkington, J. A. Smith, and J. M. Pemberton. 1999. Parasite-mediated selection against inbreed Soay sheep in a free-living, island population. Evolution 53:1259–1267.

Coltman, D. W., G. Stenson, M. O. Hammill, T. Haug, C. S. Davis, and T. L. Fulton. 2007. Panmictic population structure in the hooded seal (*Cystophora cristata*). Molecular Ecology 16:1639–1648.

Cosens, S. E., and A. Blouw. 2003. Size-and-age class segregation of bowhead whales summering in northern Foxe Basin: a photogrammetric analysis. Marine Mammal Science 19:284–296.

Cosewic. 2004a. COSEWIC assessment and update status report on the beluga whale *Delphinapterus leucas* in Canada. Committee on the Status of Endangered Wildlife in Canada: 79.

Cosewic. 2004b. COSEWIC assessment and update status report on the narwhal *Monodon monoceros* in Canada. Committee on the Status of Endangered Wildlife in Canada: 50.

Cosewic. 2006. COSEWIC assessment and update status report on the Atlantic walrus *Odobenus rosmarus rosmarus* (Northwest Atlantic Population and Eastern Arctic Population) in Canada. Committee on the Status of Endangered Wildlife in Canada: 23.

Cosewic. 2007. COSEWIC assessment and update status report on the harbour seal Atlantic and Eastern Arctic subspecies *Phoca vitulina concolor* and Lacs des Loups Marins subspecies *Phoca vitulina mellonae* in Canada. Committee on the Status of Endangered Wildlife in Canada: 40.

Cosewic. 2009. COSEWIC assessment and update status report on the bowhead whale *Balaena mysticetus*, Bering-Chukchi-Beaufort population and Eastern Canada-West Greenland population, in Canada. Committee on the Status of Endangered Wildlife in Canada: 56.

Coulon, A., J. F. Cosson, J. M. Angibault, B. Cargnelutti, M. Galan, N. Morellet, E. Petit, S. Aulagnier, and A. J. M. Hewison. 2004. Landscape connectivity influences gene flow in a roe deer population inhabiting a fragmented landscape: an individual-based approach. Molecular Ecology 13:2841–2850.

Crompton, A., M. E. Obbard, S. D. Petersen, and P. J. Wilson. 2008. Population genetic structure in polar bears (*Ursus maritimus*) from Hudson Bay, Canada: implications for future climate change. Biological Conservation 141:2528–2539.

Dalén, L., E. Fuglei, P. Hersteinsson, C. M. O. Kapel, J. D. Roth, G. Samelius, M. Tannerfeldt, and A. Angerbjörn. 2005. Population history and genetic structure of a circumpolar species: the Arctic fox. Biological Journal of the Linnean Society 84:79–89.

Davis, C. S., I. Stirling, C. Strobeck, and D. W. Coltman. 2008. Population structure of ice-breeding seals. Molecular Ecology 17:3078–3094.

De March, B. G. E., and L. D. Postma. 2003. Molecular genetic stock discrimination of belugas (*Delphinapterus leucas*) hunted in eastern Hudson Bay, northern Quebec, Hudson Strait, and Sanikiluaq (Belcher Islands), Canada, and comparisons to adjacent populations. Arctic 56:111–124.

De March, B. G. E., L. D. Maiers, and R. E. A. Stewart. 2002. Genetic relationships among Atlantic walrus (*Odobenus rosmarus rosmarus*) in the Foxe Basin and the Resolute Bay-Bathurst Island area. Canadian Science Advisory Secretariat, 2002/092: 20.

De March, B. G. E., D. A. Tenkula, and L. D. Postma. 2003. Molecular genetics of narwhal (*Monodon monoceros*) from Canada and West Greenland (1982–2001). Canadian Science Advisory Secretariat, 2003/080: 23.

Demaster, D. P., and I. Stirling. 1981. *Ursus maritimus*. Mammalian Species 145:1–7.

Derocher, A. E., N. J. Lunn, and I. Stirling. 2004. Polar bears in a warming climate. Integrative and Comparative Biology 44:163–176.

DFO. 2008. Total allowable harvest recommendations for Nunavut narwhal and beluga populations. DFO Science Advisory Secretariat, Science Advisory Report 2008/035.: 7.

Dyer, R. J., and J. D. Nason. 2004. Population graphs: the graph theoretic shape of genetic structure. Molecular Ecology 13:1713–1727.

Dyke, A. S., and J. England. 2003. Canada's most northerly postglacial bowhead whales (*Balaena mysticetus*): holocene sea-ice conditions and polynya development. Arctic 56:14–20.

Dyke, A. S., J. Hooper, and J. M. Savelle. 1996. A history of sea ice in the Canadian Arctic Archipelago based on postglacial remains of the bowhead whale (*Balaena mysticetus*). Arctic 49:235–255.

Dyke, A. S., J. T. Andrews, P. U. Clark, J. H. England, G. H. Miller, J. Shaw, and J. J. Veillette. 2002. The Laurentide and Innuitian ice sheets during the Last Glacial Maximum. Quaternary Science Reviews 21:9–31.

Ellner, S., and N. G. Hairston Jr. 1994. Role of overlapping generations in maintaining genetic variation in a fluctuating environment. American Naturalist 143:403–417.

Epperson, B. K. 1995. Spatial distributions of geneotypes under isolation by distance. Genetics 140:1431–1440.

Etkin, D. A. 1991. Break-up in Hudson Bay: its sensitivity to air temperatures and implications for climate warming. Climatological Bulletin 25:21–34.

Falush, D., M. Stephens, and J. K. Pritchard. 2003. Inference of population structure using multilocus genotype data: linked loci and correlated allele frequencies. Genetics 164:1567–1587.

Fay, F. H. 1985. *Odobenus rosmarus*. Mammalian Species 238:1–7.

Ferguson, S. H., I. Stirling, and P. D. Mcloughlin. 2005. Climate change and ringed seal (*Phoca hispida*) recruitment in western Hudson Bay. Marine Mammal Science 21:121–135.

Frankham, R. 1996. Relationship of genetic variation to population size in wildlife. Conservation Biology 10:1500–1508.

Frankham, R., K. Lees, M. E. Montgomery, P. R. England, E. H. Lowe, and D. A. Briscoe. 1999. Do population size bottlenecks reduce evolutionary potential? Animal Conservation 2:255–260.

Furgal, C. M., S. Innes, and K. M. Kovacs. 1996. Characteristics of ringed seal, *Phoca hispida*, subnivean structures and breeding habitat and their effects on predation. Canadian Journal of Zoology 74:858–874.

Gaggiotti, O. E., C. E. Lee, and G. M. Wardle. 1997. The effect of overlapping generations and population structure on gene-frequency clines, pp. 355–756 in Structured-population models in marine, terrestrial, and freshwater systems (S. Tuljapurkar and H. Caswell, eds.). Chapman & Hall, Toronto.

Gagnon, A. S., and W. A. Gough. 2005. Trends in the dates of freeze-up and breakup over Hudson Bay, Canada. Arctic 58:370–382.

García-Ramos, G., and M. Kirkpatrick. 1997. Genetic models of adaptation and gene flow in peripheral populations. Evolution 51:21–28.

Garde, E., M. P. Heide-Jørgensen, S. H. Hansen, G. Nachman, and M. C. Forchhammer. 2007. Age specific growth and remarkable longevity in narwhals (*Monodon monoceros*) from West Greenland as estimated by aspartic acid racemization. Journal of Mammalogy 87:49–58.

Garroway, C. J., J. Bowman, D. Carr, and P. J. Wilson. 2008. Applications of graph theory to landscape genetics. Evolutionary Applications 1:620–630.

Garroway, C. J., J. Bowman, T. J. Cascaden, G. L. Holloway, C. G. Mahan, J. R. Malcom, M. A. Steele, G. Turner, and P. J. Wilson. 2010. Climate change induced hybridization in flying squirrels. Global Change Biology 16:113–121.

Geffen, E., S. Waidyaratne, L. Dalén, A. Angerbjörn, C. Vila, P. Hersteinsson, E. Fuglei, P. A. White, M. Goltsman, C. M. O. Kapel, and R. K. Wayne. 2007. Sea ice occurrence predicts genetic isolation in the Arctic fox. Molecular Ecology 16:4241–4255.

George, J. C., J. Bada, J. Zeh, L. Scott, S. E. Brown, T. O'hara, and R. Suydam. 1999. Age and growth estimates of bowhead whales (*Balaena mysticetus*) via aspartic acid racemization. Canadian Journal of Zoology 77:571–580.

Gough, W. A., and E. Wolfe. 2001. Climate change scenarios for Hudson Bay, Canada, from general circulation models. Arctic 54:142–148.

Gough, W. A., A. R. Cornwell, and L. J. S. Tsuji. 2004. Trends in seasonal sea ice duration in southwestern Hudson Bay. Arctic 57:299–305.

Guillot, G., F. Mortier, and A. Estoup. 2005. GENELAND: a computer package for landscape genetics. Molecular Ecology Notes 5:712–715.

Guillot, G., R. Leblois, A. Coulon, and A. C. Frantz. 2009. Statistical methods in spatial genetics. Molecular Ecology 18:4734–4756.

Haig, S. M. 1998. Molecular contributions to conservation. Ecology 79:413–425.

Härkönen, T., and M. P. Heide-Jørgensen. 1990. Comparative life histories of east Atlantic and other harbour seal populations. Ophelia 32:211–235.

Härkönen, T., O. Stenman, M. Jussi, I. Jussi, R. Sagitov, and M. Verevkin. 1998. Population size and distribution of the Baltic ringed seal (*Phoca hispida botnica*), pp. 167–180 in Ringed seals (*Phoca hispida*) in the North Atlantic (M. P. Heide-Jørgensen and C. Lydersen, eds.). North Atlantic Marine Mammal Commission (NAMMCO).

Harwood, J. 2001. Marine mammals and their environment in the twenty-first century. Journal of Mammalogy 83:630–640.

Heide-Jørgensen, M. P., and K. L. Laidre. 2004. Declining extent of open-water refugia for top predators in Baffin Bay and adjacent waters. Ambio 33:487–494.

Heide-Jørgensen, M. P., R. Dietz, K. L. Laidre, P. R. Richard, J. Orr, and H. C. Schmidt. 2003. The migratory behaviour of narwhals (*Monodon monoceros*). Canadian Journal of Zoology 81:1298–1305.

Hersteinsson, P., and D. W. Macdonald. 1992. Interspecific competition and the geographical distribution of red and Arctic Foxes *Vulpes vulpes* and *Alopex lagopus*. Oikos 64:505–515.

Higdon, J. W., and S. H. Ferguson. 2009. Loss of Arctic sea ice causing punctuated change in sightings of killer whales (*Orcinus orca*) over the past century. Ecological Applications 19:1365–1375.

Holderegger, R., and H. H. Wagner. 2006. A brief guide to landscape genetics. Landscape Ecology 21:793–796.

Holderegger, R., and H. H. Wagner. 2008. Landscape genetics. BioScience 58:199–207.

Ibrahim, K. M., R. A. Nickols, and G. M. Hewitt. 1996. Spatial patterns of genetic variation generated by different forms of dispersal during range expansion. Heredity 77:282–291.

IUCN/SSC Polar Bear Specialist Group. 2005. Polar bears, P. 188 pp in Polar Bears: Proceedings of the 14th working meeting of the IUCN/SSC polar bear specialist group (J. Aars, N. J. Lunn and A. E. Derocher, eds.). IUCN, World Conservation Union, Seattle, Washington.

Jombart, T., S. Devillard, A.-B. Dufour, and D. Pontier. 2008. Revealing cryptic spatial patterns in genetic variability by a new multivariate method. Heredity 101:92–103.

Kawecki, T. J., and R. D. Holt. 2002. Evolutionary consequences of asymmetrical dispersal rates. American Naturalist 160:333–347.

Kovacs, K. M. 2002. Bearded seal, pp. 84–87 in Encyclopedia of Marine Mammals (W. F. Perrin, B. Würsig and J. G. M. Thewissen, eds.). Academic, New York.

Kristensen, T. N., and A. C. Sørensen. 2005. Inbreeding – lessons from animal breeding, evolutionary biology and conservation genetics. Animal Science 80:121–133.

Kukla, G. J., M. L. Bender, J.-L. De Beaulieu, G. Bond, W. S. Broecker, P. Cleveringa, J. E. Gavin, Herbert,. T. D., J. Imbrie, J. Jouzel, L. D. Keigwin, K.-L. Knudsen, J. F. Mcmanus, J. Merkt, D. R. Muhs, H. Müller, R. Z. Poore, S. C. Porter, G. Seret, N. J. Shackleton, C. Turner, P. C. Tzedakis, and I. J. Winograd. 2002. Last interglacial climates. Quaternary Research 58:2–13.

Kuo, C.-H., and F. J. Janzen. 2004. Genetic effects of a persistent bottleneck on a natural population of ornate box turtles (*Terrapene ornata*). Conservation Genetics 5:425–437.

Laidre, K. L., and M. P. Heide-Jorgensen. 2005. Arctic sea ice trends and narwhal vulnerability. Biological Conservation 121:509–517.

Laidre, K. L., I. Stirling, L. F. Lowry, Ø. Wiig, M. P. Heide-Jørgensen, and S. H. Ferguson. 2008. Quantifying the sensitivity of Arctic marine mammals to climate-induced habitat change. Ecological Applications 18:S97–S125.

Latch, E. K., and O. E. Rhodes, Jr. 2006. Evidence for bias in estimates of local genetic structure due to sampling scheme. Animal Conservation 9:308–315.

Learmonth, J. A., C. D. Macleod, M. B. Santos, G. J. Pierce, H. Q. P. Crick, and R. A. Robinson. 2006. Potential effects of climate change on marine mammals. Oceanography and Marine Biology: an Annual Review 44:431–464.

Lippé, C., P. Dumont, and L. Bernatchez. 2006. High genetic diversity and no inbreeding in the endangered copper redhorse, *Moxostoma hubbsi* (Catostomidae, Pisces): the positive sides of a long generation time. Molecular Ecology 15:1769–1780.

Maiers, L. D., B. G. E. De March, J. W. Clayton, L. P. Dueck, and S. E. Cosens. 1999. Genetic variation among populations of bowhead whales summering in Canadian waters. Canadian Science Advisory Secretariat, 1999/134: 24.

Mancuso, S. J. 1995. Population genetics of Hudson Bay beluga whales (*Delphinapterus leucas*): an analysis of population structure and gene flow using mitochondrial DNA sequences and multilocus DNA fingerprinting, McMaster University, Hamilton.

Manel, S., M. K. Schwartz, G. Luikart, and P. Taberlet. 2003. Landscape genetics: combining landscape ecology and population genetics. Trends in Ecology and Evolution 18:189–197.

Mauritzen, M., A. E. Derocher, and Ø. Wiig. 2001. Space-use strategies of female polar bears in a dynamic sea ice habitat. Canadian Journal of Zoology 79:1704–1713.

May, R. M. 2006. Network structure and the biology of populations. Trends in Ecology and Evolution 21:394–399.

Mclaren, I. A. 1958. The biology of the ringed seal (*Phoca hispida* Schreber) in the eastern Canadian Arctic. Fisheries Research Board of Canada, Bulletin No 118: 97.

Mcneely, J. A., K. R. Miller, W. V. Reid, R. A. Mettermeier, and T. B. Werner. 1990. Conserving the world's biological diversity. IUCN Publication services Gland, Switzerland.

Mcrae, B. H., and P. Beier. 2007. Circuit theory predicts gene flow in plant and animal populations. Proceedings of the National Academy of Sciences of the United States of America 104:19885–19890.

Meirer, H. E. M., R. Döscher, and A. Halkka. 2004. Simulated distributions of Baltic sea-ice in warming climate and consequences for the winter habitat of the Baltic ringed seal. Ambio 33:249–256.

Moritz, C. 1994. Defining 'evolutionary significant units' for conservation. Trends in Ecology and Evolution 9:373–375.

Nowak, R. M. 1999. Walker's mammals of the World. Vol. 2, 6th ed. Johns Hopkins University Press, Baltimore, MD.

Obbard, M. E., M. R. L. Cattet, T. Moody, L. R. Walton, D. Potter, J. Inglis, and C. Chenier. 2006. Temporal trends in the body condition of southern Hudson Bay polar bears. Ontario Ministry of Natural Resources, No. 3: 8 pp.

Outridge, P. M., W. J. Davis, R. E. A. Stewart, and E. W. Born. 2003. Investigation of the stock structure of Atlantic walrus (*Odobenus rosmarus rosmarus*) in Canada and Greenland using dental Pb isotopes derived from local geochemical environments. Arctic 56:82–90.

Paetkau, D., S. C. Amstrup, E. W. Born, W. Calvert, A. E. Derocher, G. W. Garner, F. Messier, I. Stirling, M. K. Taylor, Ø. Wiig, and C. Strobeck. 1999. Genetic structure of the world's polar bear populations. Molecular Ecology 8:1571–1584.

Palo, J., H. S. Mäkinen, E. Helle, Stenman, O., and R. Väinölä. 2001. Microsatellite variation in ringed seals (*Phoca hispida*): genetic structure and history of the Baltic Sea population. Heredity 86:609–617.

Palsbøll, P. J., M. P. Heide-Jørgensen, and R. Dietz. 1997. Population structure and seasonal movements of narwhals, *Monodon monoceros*, determined from mtDNA analysis. Heredity 78:284–292.

Parry, M. L., O. F. Canziani, J. P. Palutikof et al. 2007. Technical Summary. Climate Change 2007: Impacts, Adaptation and Vulnerability, pp. 23–78 in Contribution of Working Group II to the Fourth Assessment Report of the Intergovernmental Panel on Climate Change (M. L. Parry, O. F. Canziani, J. P. Palutikof, P. J. van der Linden and C. E. Hanson, eds.). Cambridge University Press, Cambridge.

Petersen, S. D. 2008.Spatial genetic patterns of Arctic mammals: Peary caribou (*Rangifer tarandus pearyi*), polar bear (*Ursus maritimus*), and ringed seal (*Pusa [=Phoca] hispida*). Ph.D., Trent University, Peterborough, Ontario.

Post, E., M. C. Forchhammer, M. S. Bret-Harte, T. V. Callaghan, T. R. Christensen, B. Elberling, A. D. Fox, O. Gilg, D. S. Hik, T. T. Høye, R. A. Ims, E. Jeppesen, D. R. Klein, J. Madsen,

A. D. Mcguire, S. Rysgaard, D. E. Schindler, I. Stirling, M. P. Tamstorf, N. J. C. Tyler, R. Van Der Wal, J. Welker, P. A. Wookey, N. M. Schmidt, and P. Aastrup. 2009. Ecological dynamics across the Arctic associated with recent climate change. Science 325:1355–1358.

Postma, L. D., L. P. Dueck, M. P. Heide-Jørgensen, and S. E. Cosens. 2006. Molecular genetic support of a single population of bowhead whales (*Balaena mysticetus*) in Eastern Canadian Arctic and Western Greenland waters. Canadian Science Advisory Secretariat, 2006/051: 23.

Priest, H., and P. J. Usher. 2004. Nunavut wildlife harvest study. Nunavut Wildlife Management Board: 816.

Pritchard, J. K., M. Stephens, and P. Donnelly. 2000. Inference of population structure using multilocus genotype data. Genetics 155:945–959.

Proulx, S. R., D. E. L. Promislow, and P. C. Phillips. 2005. Network thinking in ecology and evolution. Trends in Ecology and Evolution 20:345–353.

Ramsay, M. A., and I. Stirling. 1986. On the mating system of polar bears. Canadian Journal of Zoology 64:2142–2151.

Ray, N., M. Currat, and L. Excoffier. 2003. Intra-deme molecular diversity in spatially expanding populations. Molecular Biology and Evolution 20:76–86.

Reeves, R. R. 1998. Distribution, abundance and biology of the ringed seal (*Phoca hispida*) in the Arctic, pp. 9–45 in Ringed seals (*Phoca hispida*) in the North Atlantic (M. P. Heide-Jørgensen and C. Lydersen, eds.). North Atlantic Marine Mammal Commission (NAMMCO).

Reeves, R. R., and S. E. Cosens. 2003. Historical population characteristics of bowhead whales (*Balaena mysticetus*) in Hudson Bay. Arctic 56:283–292.

Reeves, R. R., E. Mitchell, A. Mansfield, and M. Mclaughlin. 1983. Distribution and migration of the bowhead whale, Balaena mysticetus, in the Eastern North American Arctic. Arctic 36:5–64.

Richard, P. R. 2005. An estimate of the Western Hudson Bay beluga population size in 2004. DFO Canadian Science Advisory Secretariat, Res. Doc. 2005/017.: 33.

Roberts, C. M., and J. P. Hawkins. 1999. Extinction risk in the sea. Trends in Ecology and Evolution 14:241–246.

Robinson, R. A., H. Q. P. Crick, J. A. Learmonth, I. M. D. Maclean, C. D. Thomas, F. Bairlein, M. C. Forchhammer, C. M. Francis, J. A. Gill, B. J. Godley, J. Harwood, G. C. Hays, B. Huntley, A. M. Hutson, G. J. Pierce, M. M. Rehfisch, D. W. Sims, B. M. Santos, T. H. Sparks, D. A. Stroud, and M. E. Visser. 2009. Travelling through a warming world: climate change and migratory species. Endangered Species Research 7:87–99.

Roth, J. D. 2002. Temporal variability in Arctic fox diet as reflected in stable-carbon isotopes; the importance of sea ice. Oecologia 133:70–77.

Roth, J. D. 2003. Variability in marine resources affects Arctic fox population dynamics. Journal of Animal Ecology 72:668–676.

Rugh, D. J., G. W. Miller, D. E. Withrow, and W. R. Koski. 1992. Calving intervals of bowhead whales established through photographic identifications. Journal of Mammalogy 73: 487–490.

Simmonds, M. P., and S. J. Isaac. 2007. The impacts of climate change on marine mammals: early signs of significant problems. Oryx 41:19–26.

Smith, T. G., and M. O. Hammill. 1981. Ecology of the ringed seal, *Phoca hispida*, in its fast ice breeding habitat. Canadian Journal of Zoology 59:966–981.

Smith, T. G., and I. Stirling. 1975. The breeding habitat of the ringed seal (*Phoca hispida*). The birth lair and associated structures. Canadian Journal of Zoology 53:1297–1305.

Stanley, H. F., S. Casey, J. M. Carnahan, S. Goodman, J. Harwood, and R. K. Wayne. 1996. Worldwide patterns of mitochondrial DNA differentiation in the harbor seal (*Phoca vitulina*). Molecular Biology and Evolution 13:368–382.

Stewart, R. E. A. 2002. Review of Atlantic walrus (*Odobenus rosmarus rosmarus*) in Canada. Canadian Science Advisory Secretariat, 2002/091: 21.

Stewart, R. E. A. 2008. Redefining walrus stocks in Canada. Arctic 61:292–308.

Stewart, B. E., and R. E. A. Stewart. 1989. *Delphinapterus leucas*. Mammalian Species 336:1–8.

Stewart, R. E. A., S. E. Campana, C. M. Jones, and B. E. Stewart. 2006. Bomb radiocarbon dating calibrates beluga (*Delphinapterus leucas*) age estimates. Canadian Journal of Zoology 84:1840–1852.

Stirling, I. 1983. The social evolution of mating systems in pinnipeds, in Advances in the study of mammalian behavior. American Society of Mammalogists, Special Publication 7: 489–527 (J. F. Eisenberg and D. G. Kleiman, eds.).

Stirling, I. 1997. The importance of polynyas, ice edges, and leads to marine mammals and birds. Journal of Marine Systems 10:9–21.

Stirling, I., and A. E. Derocher. 1993. Possible impacts of climatic warming on polar bears. Arctic 46:240–245.

Stirling, I., and H. P. L. Kiliaan. 1980. Population ecology studies of the polar bear in northern Labrador. Canadian Wildlife Service, number 42: 19 pp.

Stirling, I., and C. L. Parkinson. 2006. Possible effects of climate warming on selected populations of polar bears (*Ursus maritimus*) in the Canadian Arctic. Arctic 59:261–275.

Stirling, I., and T. G. Smith. 2004. Implications of warm temperatures and an unusual rain event for the survival of ringed seals on the coast of southeastern Baffin Island. Arctic 57:59–67.

Stirling, I., and J. A. Thomas. 2003. Relationships between underwater vocalizations and mating systems in phocid seals. Aquatic Mammals 29.2:227–246.

Stirling, I., N. J. Lunn, and J. Iacozza. 1999. Long-term trends in population ecology of polar bears in western Hudson Bay in relation to climate change. Arctic 52:294–306.

Stirling, I., N. J. Lunn, J. Iacozza, C. Elliott, and M. Obbard. 2004. Polar bear distribution and abundance on the southwestern Hudson Bay Coast during open water season, in relation to population trends and annual ice patterns. Arctic 57:15–26.

Storfer, A., M. A. Murphy, J. S. Evans, C. S. Goldberg, S. Robinson, S. F. Spear, R. Dezzani, E. Delmelle, L. Vierling, and L. P. Waits. 2007. Putting the 'landscape' in landscape genetics. Heredity 98:128–142.

Taylor, M. K., S. Akeeagok, D. Andriashek, W. Barbour, E. W. Born, W. Calvert, H. D. Cluff, S. Ferguson, J. Laake, A. Rosing-Asvid, I. Stirling, and F. Messier. 2001. Delineating Canadian and Greenland polar bear (*Ursus maritimus*) populations by cluster analysis of movements. Canadian Journal of Zoology 79:690–709.

Turgeon, J., P. Duchesne, L. D. Postma, and M. O. Hammill. 2009. Spatiotemporal distribution of beluga stocks (*Delphinapterus leucas*) in and around Hudson Bay: Genetic mixture analysis based on mtDNA haplotypes. DFO Canadian Science Advisory Secretariat 2009/011: 14.

Tynan, C. T., and D. P. Demaster. 1997. Observations and predictions for Arctic climate change: potential effects on marine mammals. Arctic 50:308–322.

Wang, J., L. A. Mysak, and R. G. Ingram. 1994. Interannual variability of sea-ice cover in Hudson Bay, Baffin Bay, and the Labrador Sea. Atmosphere-Ocean 32:421–447.

Wang, Y.-H., K.-C. Yang, C. L. Bridgman, and L.-K. Lin. 2008. Habitat suitability modelling to correlate gene flow with landscape connectivity. Landscape Ecology 23:989–1000.

Whitehead, H. 2008. Precision and power in the analysis of social structure using associations. Animal Behaviour 75:1093–1099.

Woodworth, L. M., M. E. Montgomery, D. A. Briscoe, and R. Frankham. 2002. Rapid genetic deterioration in captive populations: causes and conservation implications. Conservation Genetics 3:277–288.

Woollett, J. M., A. S. Henshaw, and C. P. Wake. 2000. Palaeoecological implications of archaeo-logical seal bone assemblages: case studies from Labrador and Baffin Island. Arctic 53:395–413.

Wright, S. 1943. Isolation by distance. Genetics 28:114–138.

Wrigley, R. E., and D. R. Hatch. 1976. Arctic fox migrations in Manitoba. Arctic 29:147–158.

Understanding and Managing Wildlife in Hudson Bay Under a Changing Climate: Some Recent Contributions From Inuit and Cree Ecological Knowledge

D. Henri, H.G. Gilchrist, and E. Peacock

Abstract In the last few decades, traditional and local ecological knowledge (TEK/LEK) have contributed information to understanding and managing wildlife. In the Hudson Bay region of the Canadian Arctic, there have been numerous initiatives to document Inuit and Cree knowledge regarding animal ecology, and this information has occasionally complemented ongoing scientific research. This chapter presents an overview of the existing TEK/LEK literature concerning two animal species of cultural significance in the Hudson Bay marine region, namely the common eider (*Somateria mollissima sedentaria* and *Somateria mollissima borealis*) and the polar bear (*Ursus maritimus*). For each of these species, some key insights offered by TEK/LEK are reviewed. Examples include population size and trends, animal health and behaviour, as well as the perceived effects of changing climate and ice conditions on animal populations. In addition, the following discussion compares and contrasts available TEK/LEK information on common eiders and polar bears with existing scientific knowledge in the same geographic region. In doing so, it identifies some of the challenges and opportunities generated by applying both local knowledge and western scientific information in wildlife management. Finally, potential areas of convergence between scientific expertise and TEK/LEK for understanding climate change and its impacts on marine mammals and marine birds in the Hudson Bay region are discussed.

Keywords Traditional/local ecological knowledge • Inuit • Cree • Common eider • Polar bear • Climate change • Hudson Bay

D. Henri (✉)
School of Geography and the Environment, Oxford University Centre for the Environment, South Parks Road, Oxford, OX1 3QY, UK
e-mail: dominique.henri@ouce.ox.ac.uk

S.H. Ferguson et al. (eds.), *A Little Less Arctic: Top Predators in the World's Largest Northern Inland Sea, Hudson Bay*, DOI 10.1007/978-90-481-9121-5_13,
© Springer Science+Business Media B.V. 2010

Introduction

Over the last few decades, traditional and local ecological knowledge (TEK/LEK) have contributed information to understanding animal ecology, as well as gained importance in wildlife management. Because of their acknowledged connection with environmental sustainability and the empowerment of local communities, traditional and local ecological knowledge have been incorporated into several environmental decision-making processes worldwide (Mailhot 1993). Indeed, the proponents of TEK/LEK have argued that their inclusion can benefit scientific research and resource management by providing regional information as well as the meaningful involvement of resource users in making decisions that affect them (Berkes and Henley 1997).

In northern Canada, there has been a growing interest in aboriginal TEK/LEK, much of it related to their use in land claim processes (White 2006, 2008), wildlife co-management (Nadasdy 1999, 2003; Stevenson 2004), environmental impact assessments (Stevenson 1996), natural resource management and climate change research (Duerden and Kuhn 1998; Riedlinger and Berkes 2001). Historically, northern aboriginal peoples have developed a strong knowledge base relating to animal ecology and weather, and snow and ice conditions. Arguably, this is linked to a long-standing, yet constantly evolving, reliance on their environment for various aspects of their livelihoods. Although collaborative attempts that apply both western science and aboriginal TEK/LEK have increasingly occurred, the acceptance of their combined use remains controversial (Bielawski 1995; Dowsley 2007, 2009; Fraser et al. 2006). We offer the view that the complexity of the ecological questions and environmental issues faced today require that the best knowledge available be sought out.

Here, we critically examine the contribution that TEK/LEK can offer to contemporary wildlife management and research grounded in western scientific traditions through a review of some of the key insights offered by the TEK/LEK literature on two animal species of cultural significance in the Hudson Bay: the common eider (*Somateria mollissima sedentaria* and *Somateria mollissima borealis*) and the polar bear (*Ursus maritimus*). In addition, some potential areas of convergence between scientific expertise and aboriginal TEK/LEK for understanding the effects of climate change on marine mammals and birds in Hudson Bay are discussed. In doing so, we explore mutually-affirming ways in which scientists and TEK/LEK holders can engage in issues of joint concern.

Defining Traditional and Local Ecological Knowledge

In the mid-1980s, a new field of inquiry emerged called *traditional ecological knowledge* (TEK) that arose from two different approaches: ethnoscience and human ecology. Although the term *traditional ecological knowledge* came into widespread use only a few decades ago, its practice appears to be as old as ancient hunter-gatherer cultures. Many definitions have been proposed for TEK (Johnson 1992; Pierotti and Wildcat 2000), which can be most thoroughly defined as:

> A cumulative body of knowledge, practice, and belief, evolving by adaptive processes and handed down through generations by cultural transmission, about the relationship of living beings, including humans, with one another and with their environment. (Berkes 1999: 8)

According to this operational definition, traditional ecological knowledge is both cumulative and dynamic, building on experiences and responding to changes. It is an attribute of societies with historical continuity in resource use in a particular land, such as the Cree and Inuit groups inhabiting the Hudson Bay area today. The above mentioned definition refers to a practice-knowledge-belief system. It implies a component of local knowledge of species and other environmental phenomena, a component of practice in the way people carry out their livelihood activities, and a further component of belief in peoples' perception of their role within ecosystems, as ecological aspects of tradition cannot be divorced from the social and spiritual. Through the process of documenting local ecological perspectives on the Hudson Bay region, some researchers and scholars have preferred to use the term local ecological knowledge (LEK) rather than traditional ecological knowledge to refer minimally to "fact-based detailed knowledge of a place or a practice as it is carried out in a place [about integrated living systems]", and generally to "all the [ecological] knowledge attached to a place, regardless of who bears it, uses it, or the means of transmission to others" (Bielawski 2005: 950). This conceptual tool has been employed as the type of ecological knowledge often gathered through interviews with Hudson Bay residents for the purpose of incorporating their views to wildlife management and scientific research has referred specifically to current ecological knowledge acquired more recently over the lifetime of individuals (Gilchrist et al. 2005). It should be stated that both TEK and LEK in general do not represent a single unified body of knowledge, catalogued and accepted by everyone as universal truth. As Dowsley and Wenzel (2008: 183) argue, "there is much variation in life experience, analysis of observations, and ability to integrate various pieces of information" among individuals that hold TEK/LEK.

The Hudson Bay region is currently home to about 20,000 Cree and 14,000 Inuit living in 31 communities (Statistics Canada Census 2006). The ecological knowledge held by these groups originates from interaction with the flora and fauna living in the marine waters, islands and lands surrounding the camps and communities of Hudson Bay, James Bay, Hudson Strait and Foxe Basin (Fig. 1). While the ancestors of the Inuit arrived in the central Canadian Arctic around AD 1200, the Cree began occupying the Hudson Bay area around 5,000–6,000 BC after the retreat of the Laurentide Ice Sheet (Irving 1986; McGhee 1990; Taylor 1968; Wright 1968). Inuit and Cree groups have thus relied on the resources of the Hudson Bay region for centuries. Throughout this time, a collective body of knowledge has been gathered and transmitted through generations from their observations and experiences, to which this chapter now turns.

Fig. 1 Contemporary Inuit and Cree communities of the Hudson Bay region (Map: J. Akearok)

Combining Local Knowledge and Scientific Expertise About the Common Eider: Some Examples of Meaningful Collaboration

The common eider (*Somateria mollissima*) is a well-studied sea-duck, mainly because of its large numbers, its economic importance, and its significance in subsistence and sport harvests (Reed and Erskine 1986; Fig. 2). The common eider inhabits Arctic and subarctic coastal marine habitats and has a circumpolar distribution. Most populations migrate in winter to avoid freeze-up and starvation (Abraham and Finney 1986). In the Hudson Bay region, two out of the six or seven recognized subspecies of common eider can be encountered: the northern common eider (*S. m. borealis*), which nests primarily in Ungava Bay and Hudson Strait and migrates to

Fig. 2 Female Hudson Bay eider swimming on a polynya in winter among the Belcher Islands (Photo: H.G. Gilchrist)

winter along the coast of the Gulf of St. Lawrence, Newfoundland and southwest Greenland (Reed and Erskine 1986); and the Hudson Bay eider (*S. m. sedentaria*), which breeds and overwinters in polynyas (*i.e.*, permanent areas of open water surrounded by ice) and along ice edges near the Sleeper and Belcher Islands (Gilchrist and Robertson 2000). Common eiders typically nest colonially during the summer months at coastal islands, islets, and sometimes narrow, low-lying points of land as well as islands in freshwater lakes and river deltas near marine waters.

The common eider has had a long-standing cultural significance for subsistence hunters in northern Canada. This species, for instance, has been an important source of meat, eggs and feather down for the Inuit (Nakashima 1991). In past centuries, all parts of the eider duck could be used by the Inuit for food, medicinal purposes, clothing and tools. Use varied across geographical regions, mostly depending on the availability of resources for subsistence. The range of tools and clothing crafted from eider products was diverse: eider wings were sewn together and used as whisk brooms; secondary wing feathers were used to clean out gun barrels and as arrow fletching; and skins were used to make slippers, accessories and parkas. Other parts of the eider were also used: down was dipped in old fat from *qulliq* (a fat-burning stone lamp) and saved to make candles; eider feet were used as water containers for short hunting trips or as toys for children; and the leftover bits of neck trimmings and tails served as hand towels. Different parts of the eider were also used for medicinal purposes: tufts of down were used to clean oozing wounds; down covered skin was used as pads and to clean newborn babies. In the past, every part of the eider duck could be eaten, including crushed bones (Oakes 1990).

Among the many uses that Canadian Arctic Inuit have made of the eider, some remain part of the Inuit lifestyle today (Fig. 3). Hunters currently harvest eiders for their meat, eggs and down, and women make down-filled parkas, pants, mittens,

Fig. 3 Inuit uses of the common eider in Cape Dorset and Kimmirut, Nunavut (Photos: D. Henri)

hats and blankets, as well as various accessories from the eiders, including brooms and toys for children. Inuit use of the common eider thus has a long history that has led to the accumulation and transmission of a specialized body of ecological knowledge that remains well alive today (Nakashima 1991; Henri 2007).

Recent advances in the documentation of Inuit knowledge on this species has shed light on ways in which both local and scientific perspectives can be used in a complementary fashion to provide resource users, wildlife biologists and managers with the best possible available information on its ecology and status. For example, in July 2004, hunters from Ivujivik, Nunavik (northern Quebec) observed many hundred common eider ducks dying from disease among nesting colonies in the northern Hudson Bay and western Hudson Strait. Laboratory analyses of eider carcasses later confirmed that they had died from avian cholera. Every summer since, avian cholera has been observed among eider colonies along the northern coasts of Quebec and at the East Bay colony on Southampton Island in Nunavut (Gilchrist et al. 2006b), where research on common eiders has been conducted since 1996. Avian cholera is a naturally occurring disease that affects birds, and especially wild waterfowl (Friend 1999). While significant cholera outbreaks have been reported among more southerly common eider populations (Goudie et al. 2000), to the best of scientific knowledge, it had never been identified in northern populations prior to 2004. Inuit TEK has helped to determine whether mass die-offs have occurred historically in the Hudson Bay region and had been previously missed by western science, or alternatively had recently migrated into the north to infect eiders. In 2007, a survey of Inuit TEK on the frequency and geographical distribution of disease outbreaks among common eiders was conducted in the communities of Kimmirut, Cape Dorset and Coral Harbour, Nunavut. It was concluded that avian cholera has only recently migrated into the western Hudson Strait and northern James Bay area, rather than being a cyclical and recurring phenomenon in that region (Henri 2007). Results from this study also suggest that Inuit TEK on the common eider is based on observations made at specific spatial and temporal scales. Inuit observations of eiders occur within a specific geographical range around communities, follow seasonal patterns of travel and wildlife use, and show variation between individuals (Henri 2007) (Fig. 4).

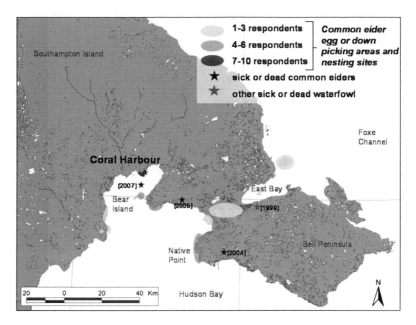

Fig. 4 Inuit use and knowledge of the common eider reported by Coral Harbour residents (Adapted from Henri 2007: 44)

This example illustrates how combining Inuit and scientific observations made at complementary spatio-temporal scales can yield a more comprehensive understanding of common eider ecology. Furthermore, it suggests that TEK/LEK can contribute information required for sound eider duck management by providing historical baseline data that, in this case, western science could not have generated. It also identified areas for further scientific research, as well as contributed to the establishment of effective disease monitoring practices. These contributions are particularly valuable in the Hudson Bay region where standard scientific approaches for monitoring avian disease are often impractical and for which limited scientific data is available.

Knowing and Managing the Polar Bear Under a Changing Climate: Contrasting Views From Science and TEK/LEK

The relationship between the polar bear and aboriginal groups that have lived in the Hudson Bay region is complex and multi-dimensional. For centuries, polar bears have been admired, feared, respected and formed part of the subsistence practices of the Inuit and Cree. Indeed, archaeological records indicate that the species has probably been hunted in the eastern Canadian Arctic for some 6,000 years (McGhee 1990). However, such evidence also highlights the fact that, historically, polar bears have had an overall minor impact on Cree and Inuit material cultures given their peripheral role in their subsistence economy (Lytwyn 2002; Wenzel 2005). In contrast, the species has occupied a dominant role in the collective imagination of these aboriginal groups (Randa 1986a).

In fact, scholars have argued that no animal has as large a symbolic place in Canadian Inuit culture as the polar bear, which figures predominantly in Inuit cosmology and mythology (Saladin d'Anglure 1990). The significance of the symbolic status occupied by polar bear in Inuit culture could be explained by the many parallels, transpositions and analogies which can be drawn between *Homo sapiens* and *Ursus maritimus*. Polar bears, due to their predatory nature, their reliance mainly on marine mammals and fish for sustenance and their ability to build snowhouses and stand on two legs, were seen as spiritually powerful figures whose existence, in many ways, paralleled the human experience (Hallowell 1926; Randa 1986b; Trott 2006). In today's Hudson Bay communities, people still speak of a unique human-bear relationship as polar bears are accorded sentience and intelligence (Tyrell 2006).

A source of food, clothing, tools and money, the polar bear has been abundantly represented in oral tradition. It has been, and continues to be, a significant component of Inuit and, to a lesser extent, Cree cultural ecology. Over the last few decades, a few initiatives have documented the vast amount of orally transmitted knowledge that has resulted from the long-standing experience of Cree and Inuit with polar bears. For instance, testimonies collected in the 1990s from elders and representatives from 27 communities across the Hudson Bay region provided detailed information on the seasonal activities of polar bears (Fig. 5). Numerous qualitative observations about distributional shifts, population trends, as well as changes in body condition and behaviour were also documented. For example, community residents throughout Hudson Bay reported that polar bears have recently lost their fear of humans and dogs and that they are becoming increasingly aggressive and more dependent on foraging dump-sites, camp sites and meat caches (McDonald et al. 1997: 91).

For the greater part of Inuit and Cree history, the use and hunt of polar bears was regulated by sets of taboos, rituals and ethical codes of conduct performed in an attempt to maintain appropriate attitudes to ensure animal capture (Randa 1993). Since the 1960s, however, polar bear management has been framed within the context of wider bureaucratic and institutional structures, as well as international conservation agreements. At the international level, Canada signed the 1973 International Agreement on the Conservation of Polar Bears and their Habitat (Lentfer 1974). Within Canada, polar bear management largely falls under the jurisdiction of provincial and territorial governments. In the Hudson Bay region specifically, current management plans allow for the regulated harvest of polar bears by aboriginal groups in Quebec, Ontario and Nunavut (Aars et al. 2006). In Nunavut, where two-thirds of the world's polar bears occur, approximately 450–500 bears are harvested annually by both Inuit and non-aboriginal hunters following a quota system based on both scientific and traditional ecological knowledge (GN 2005; Fig. 6). However, the process of combining those two sources of knowledge for managing polar bears has recently been a subject of controversy.

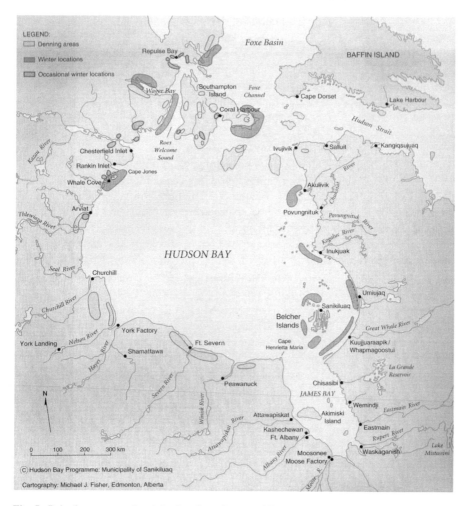

Fig. 5 Polar bear seasonal activity in selected areas of Hudson Bay (McDonald et al. 1997: 90)

Fig. 6 Polar bears are currently harvested by Inuit for their meat and hide (Photos: D. Henri)

Scientific information on boundaries of the subpopulations of polar bears, population size and trend are central to the current management of the harvest of polar bears. The Hudson Bay region supports three recognized polar bear subpopulations, namely the Western Hudson Bay (WH), Southern Hudson Bay (SH), and Foxe Basin (FB) subpopulations (Taylor and Lee 1995; Taylor et al. 2001; Peacock et al., this volume). As one of the southernmost polar bear subpopulations, Western Hudson Bay may be among the first to respond to climatic warming in the Arctic (Derocher et al. 2004; Gough et al. 2004). Indeed, several peer-reviewed articles have related the status of Western Hudson Bay polar bears to climate change and related ice conditions (Stirling and Derocher 1993; Stirling and Parkinson 2006; Stirling et al. 1999, 2004; Regehr et al. 2007). Among the scientific community, the current prevailing view is that the Western Hudson Bay subpopulation is under increasing nutritional stress. A causal link is hypothesised between this stress and a decrease in polar bears' ability to access their primary prey, the ringed seal (*Pusa hispida*) (Fig. 7). Decreased access to ringed seals is, in turn, presumably related to earlier ice break up, which forces bears to arrive on land earlier in the year. This reduces their critical spring hunting season while simultaneously prolonging their seasonal fast. As a result, the condition of adult female polar bears and several demographic parameters (including abundance) in Western Hudson Bay have been shown to have declined significantly over the past 20 years (Regehr et al. 2007; Stirling et al. 1999).

However, not all sources of information regarding the status of polar bears in Western Hudson Bay suggest a struggling subpopulation; Inuit and Cree living

Fig. 7 Polar bears feeding on a ringed seal on sea ice in Hudson Strait (Photo: D. Henri)

along the shores of Hudson Bay have reported seeing more bears in recent years, which has been interpreted as evidence by community members that the subpopulation is increasing (Dowsley and Taylor 2006; McDonald et al. 1997; Tyrell 2006; NTI 2007b). For instance, in the community of Arviat, Nunavut, sightings seem to have increased at certain times of year, even posing concerns for public safety:

> During late October, when the ice sea begins to form, sightings of polar bears close to and in the community become more common. Along the coast, hungry bears awaiting the formation of sea ice are attracted by the smell of beluga whale carcasses left along the shoreline by hunters following the beluga whale migration and harvest in early fall. Bears are frequently seen on the outskirts of the community and at the garbage dump, just a short distance from the edge of town. At this time of year, school is occasionally closed early due to the close proximity of bears. (Tyrell 2006: 194)

While Inuit and Cree residents have attributed the increase in human-bear encounters around their communities to increases in the polar bear population, scientists suggest that receding floe edges and longer ice-free seasons have concentrated polar bears in areas where humans are more likely to encounter polar bears (Stirling and Parkinson 2006), and that the Western Hudson Bay subpopulation is overall in decline due to the effects of earlier ice break up on demographic rates (Regehr et al. 2007).

In Nunavut, where polar bear hunting quotas are established based on both TEK and scientific expertise, such contrasting views have led to controversial management decisions (Dowsley and Wenzel 2008). In 2005, based on Inuit hunters' testimonies suggesting an increase in polar bear abundance (M. Taylor, personal communication 2009), Nunavut hunting quotas were modified for the five Western Hudson Bay communities from a total of 47 to 56 bears per year. This decision was later reversed in 2007 as the quotas were brought down to a total of eight polar bears per year, based on newly-published scientific evidence documenting a population decline in Western Hudson Bay (NWMB 2007; Regehr et al. 2007). In an attempt to resolve dissent between scientific and Inuit perspectives, some scientists have hypothesized that differences in the geographic and temporal scales of the observations made by biologists and TEK/LEK holders could partly explain why such diverging views exist regarding polar bear population trend in Western Hudson Bay. In fact, scientists and wildlife managers largely corroborate the Inuit sense of a dramatic increase in polar bear abundance since the 1960s and 1970s, when quotas were enacted, and readily use this example to demonstrate the effectiveness of harvest management. Yet, some scientists are concerned with more recent and subtle decline associated with climate change. However, this hypothesis has not been verified, and thus highlights the need for further collaboration and information exchange between biologists and TEK/LEK holders to progress on this issue.

Adding yet another layer of complexity to this debate are the contrasting views regarding the actual methods used for researching polar bears, such as the physical capture and chemical immobilization of bears (Fig. 8). In Western Hudson Bay specifically, where 55% of the population is currently physically tagged, polar bears have been intensively studied since the 1960s (Regehr et al. 2007). The intensity of this research has fuelled controversy on both the effects of the methods

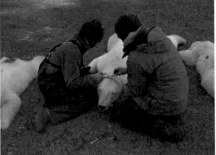

Fig. 8 Satellite collars and ear tags are put on polar bears to better understand their movements (Photos: E. Peacock)

employed on polar bear behaviour and the decline of the subpopulation (Dyck et al. 2007; Stirling et al. 2008; Tyrell 2009). Indeed, in recent years, Canadian Inuit have publicly expressed mounting concerns about the impacts of scientific research on this species of cultural significance:

> The Inuit of Canada have serious and increasing concerns about the intrusive methods by which western science researchers are handling and gathering research data on polar bears that is causing stress and harm to the animals – namely through methods such as the use of tranquilizers, direct handling of the animal, taking bodily samples, collaring, paint marking, recapturing, and using aircraft to track, chase, and seize the animals (...) The Inuit of Canada are deeply concerned that these research methods ignore, do not take into account, or do not respect Inuit knowledge (...) on polar bears. (ITK 2009; see also NTI 2007a)

In response to such views, polar bear biologists and managers have emphasized the importance of scientific research, especially in the context of the increasing uncertainties posed by climate change, and have highlighted their responsibilities and legal requirements to conduct research programs relating to the conservation and management of polar bears in their jurisdictions (PBSG 2009). Many members of the scientific community have stressed the need for the effective monitoring of polar bears based on techniques involving physical capture and satellite-collaring (which have proven critical to understanding polar bear ecology) while simultaneously striving to develop less invasive research protocols. For instance, Peacock et al. (2008) describe recent initiatives in Nunavut to incorporate two dimensions of aboriginal TEK/LEK into polar bear research and management, namely empirical ecological observations, as well as values and beliefs regarding polar bears. The authors cite the development of non-invasive aerial survey techniques as an acknowledgement of Inuit values, reducing the need to capture polar bears for monitoring population size (Stapleton 2009), and the integration of TEK/LEK with satellite telemetry data on habitat use of polar bears to generate predictive habitat models for polar bears in Foxe Basin (Sahanatien 2009; Peacock et al. 2008).

Thus, there continues to be diverging perspectives about how to ethically gather information about polar bears as well as ongoing initiatives to address the issue in the Hudson Bay region. Again, this situation sheds light on the importance of collaboration and ongoing communication between scientists, managers and resource users for the establishment of methods of gathering knowledge on polar bears that are respectful to aboriginal views while satisfying the need for acquiring sound scientific data required for ensuring the presence of healthy, viable polar bear populations well into the future.

The examples discussed above clearly reveal that polar bear conservation under changing climatic conditions raises complex socio-ecological issues. We take the view that concerted research and management initiatives can only take place to address such challenges if stakeholders not only engage in constructive information exchange but also work collaboratively to overcome the mistrust that has characterized recent debates surrounding polar bear management in Western Hudson Bay and elsewhere in Nunavut (George 2009). Scientists, resource users and managers involved in this process are either legally, morally or culturally interested in conserving and protecting polar bear populations, and thus ultimately share similar objectives. If polar bear populations are to be negatively impacted by the loss of sea ice, all three groups should be working towards the goal of determining what might be a socially acceptable and ecologically sound strategy to research and manage polar bears.

Assessing the Ongoing Effects of Climate Change in the Hudson Bay Region

According to the Intergovernmental Panel on Climate Change, climate change is expected to be the most dramatic and rapid in the polar regions, and will cause major physical, ecological, sociological, and economic impacts in the Arctic (IPCC 2007a). Aboriginal communities in the Arctic are considered to be some of the most vulnerable to the environmental changes associated with global warming, given their close relationship with the land and sea, coastal geographic location, reliance on their local environment for aspects of their diet and economy, and the current dynamic state of social, cultural, economic and political change (ACIA 2005; Ford et al. 2006; Ford and Furgal 2009; Huntington and Fox 2005; IPCC 2007b). However, in spite of much scientific research, a considerable amount of uncertainty still exists concerning the rate and extent of climate change in the Arctic, and how such phenomena will affect regional climatic processes and the diverse components of northern ecosystems (IPCC 2007a).

In this context, and given the unique source of environmental expertise offered by the Cree and Inuit, one must consider whether an expanded scope of knowledge and inquiry could augment scientific understandings of the effects of climate change in the Hudson Bay region. In fact, in recent years, many communities in the Canadian North have experienced environmental changes that differ from normal variability, suggesting that the effects of a change in climate are already recognizable (Ford et al. 2006, 2007; Furgal 2006). Indeed, changes in the seasonal extent and distribution of sea ice, fish and wildlife abundance and health, permafrost thaw, and patterns of soil erosion have been reported and, in some instances, are considered to be without precedent from human memory (Krupnik and Jolly 2002). Community assessments of environmental change are based on cumulative knowledge of local trends, patterns and processes, and are derived from generations of reliance on terrestrial and marine environments. Can these assessments, based on local observations and traditional ecological knowledge, enrich and expand scientific understandings of Arctic climate change?

In Hudson Bay, environmental change associated with variation in weather and climate has not gone unnoticed by aboriginal communities. In fact, the Hudson Bay Programme was one of the first attempts to understand environmental change through TEK (McDonald et al. 1997). This community-based research documented Cree and Inuit perspectives on environmental change due to large-scale industrial development in the Hudson Bay bioregion, and concomitantly addressed climate-related change. The project demonstrated that Inuit and Cree inhabiting the region were able to distinguish subtle patterns, cycles and changes in ecosystem structure, and that accumulated knowledge of the land and environmental indicators gave Inuit the ability to interpret and understand seasonal-change processes in the Hudson Bay region. However, in the early 1990s, Cree and Inuit participants were finding that some environmental indicators that had been used for generations did not coincide with the existing weather system. People living along Hudson Strait, northwest Hudson Bay and eastern James Bay were less confident in their predictions:

> We cannot make predictions anymore. We don't know if the water is going to freeze or not. We used to know what was going to happen at certain seasons but, with all the changes in the climate and the different qualities of water, we can't make those predictions anymore.
>
> –Helen Atkinson, Chisasibi, Quebec (McDonald et al. 1997: 28–29)

While mounting empirical evidence has thus suggested that TEK/LEK can enhance scientific understandings of climate change in the northern context (Laidler 2006; Oakes and Riewe 2006), how to combine, in practice, information coming from distinct cultural practices and intellectual traditions often poses a challenge. To bridge the gap between aboriginal and western scientific epistemologies, Riedlinger and Berkes (2001) have identified areas of convergence that may facilitate the use of traditional ecological knowledge and western science as distinctive, yet complementary, sources of knowledge for understanding the ecological impacts of climate change in the Canadian North. These areas of convergence are the use of TEK/LEK (1) as local-scale expertise; (2) as a source of climate history and baseline data; (3) in formulating research questions and hypotheses; (4) in long-term, community-based monitoring.

Combining western scientific knowledge with local expertise can translate global processes such as climate change into local-scale understandings of potential impacts. Forecasts generated by climate and other models, while key to predicting change, typically stimulate phenomena at coarse spatial and temporal scales, and thus can be limited in their capacity to explain changes occurring at local or regional scales. Given the expected regional variability associated with climate change and the immense geographic scale of the Hudson Bay region, TEK/LEK could contribute greatly to detecting change. Such observations could 'ground-truth' scientific predictions and provide supporting evidence for coarse models (Riedlinger and Berkes 2001). For example, Stirling and Derocher (1993) predicted that one consequence of climate change on polar bears would be an increase in human-bear interactions; this has indeed occurred on local levels according to both TEK/LEK and Government of Nunavut data, although the causal mechanism of increased interactions is in dispute (see above). Furthermore, Hudson Bay residents have described local ecological changes through a complex of interacting variables including water temperatures, marine currents, ice-thickness, ice colour and phenology, pressure-ridge and lead distribution, animal health and wind patterns. These perceived changes have been identified by comparing the past and present, which could help explain larger-scale phenomena (McDonald et al. 1997). The table below summarizes several examples of local environmental changes, as recently discussed by members of the communities of Puvurnituq, Kangiqsujuaq and Ivujivik, Nunavik, during a series of workshops that investigated the potential impacts of climate change at the community level and strategies for adaptation (Nickels et al. 2006; Table 1).

A second area of convergence for the use of traditional ecological knowledge and scientific approaches concerns climate history, which is central to understand present and future climate change. TEK, through cumulative experience and oral history, has provided insights into past climate variability and fluctuation, as such a knowledge base is embedded into a social history of wildlife populations, travels,

Table 1 Summary of observations of environmental change in the Canadian Arctic (Adapted from Nickels et al. 2006: 67–68)

Community	Puvirnituq	Kangiqsujuaq	Ivujivik
Observations			
Precipitation			
Decrease in rainfall	□	□	□
Force of rain has increased			□
Decrease in snowfall/ snow on the land	□	□	□
Changes in snow type/characteristics	□	□	
Ice			
Ice is thinner now	□	□	□
Earlier break-up of ice	□	□	□
Later freeze-up of ice	□	□	□
Permanent snow-packs/icepacks are melting/ glaciers are melting		□	□
Water			
Freshwater levels are lower	□	□	□

unusual events, harvesting records and migration. This kind of knowledge can complement data gathered from a variety of scientific and non-scientific sources, including meteorological observations, documentary sources and proxy data (*e.g.*, ice-core and lake sediments records) in the process of reconstructing climate history. It has the potential to therefore provide a more comprehensive baseline against which unidirectional or cyclical changes in climate can be established. As Usher (2000) describes, the historical accounts provided by TEK can provide a baseline for expected deviations from 'normal' conditions. Indeed, Cree and Inuit from the Hudson Bay region have been reporting that while cyclical environmental changes have always been observed (such as the timing of seasons, freeze-up, break-up and fluctuations in wildlife migrations), the differences observed in recent years emphasize that some of these changes are beyond the range of expected variability:

> Even if we try to predict what it is going to be like tomorrow (…) the environmental indication isn't that the Elders said it would be. Sometimes, it is still true but sometimes it isn't. In the past, when they said, "it's going to be like this tomorrow," it was. But our weather and environment are changing so our knowledge isn't true all the time now. We're being told [in Hudson Strait] that maybe if we put January, February, or March one month behind, our knowledge of weather would be more accurate because the weather in those months isn't the same anymore.

> –Lucassie Arrangutainaq, Sanikiluaq, Nunavut (McDonald et al. 1997: 28)

A third way in which both scientific and traditional or local perspectives can be combined for a better understanding of climate and environmental change relates to the process of formulating research hypotheses. The method of western science can be characterized as hypothetico-deductive; constructing hypotheses based on observation and testing them empirically (Popper 1959). The formulation of research hypotheses constitutes a crucial component of this method as it strongly determines the research that follows. Scientific researchers tend to formulate hypotheses based on the range of questions of which they are aware and of specific situations that they have personally observed. In this context, TEK/LEK may contribute to expand the range of concepts and possibilities upon which to base scientific research questions and formulate hypotheses through insights into ecological relationships (Berkes and Folke 1998). A good example of this stemming from the Hudson Bay region has been discussed elsewhere in this book (Mallory et al. this volume). Indeed, Inuit observations of major changes in winter currents and regional sea-ice conditions around the Belcher Islands were related to increases in winter kills of common eiders, which were later corroborated in scientific studies (McDonald et al. 1997; Robertson and Gilchrist 1998; Gilchrist et al. 2006a).

One last area of convergence for the use of TEK/LEK and scientific approaches in relation to understanding climate change and its impacts on marine mammals and birds of the Hudson Bay region relates to the potential of TEK/LEK to enhance environmental monitoring. A primary difference between western science and TEK/LEK is that traditional or local knowledge holders are the resource users themselves and are typically present year-round (Berkes 1999). As a consequence, environmental monitoring performed by TEK/LEK holders occurs in the context of seasonal rounds of resource harvesting and land-, sea- or coastal-based activities which are closely tied to travel routes, as well as the times and places of harvesting. This presence is often beyond the potential of western scientific monitoring programs in remote regions. This type of monitoring ensures that ecological relationships are noted, and such monitoring recognizes and uses environmental indicators. As our discussion of Cree and Inuit TEK/LEK on common eiders and polar bears exemplifies, such bodies of knowledge hold great potential to contribute to monitoring environmental change by adding site-specific information, bringing attention to signs or indicators, highlighting relational information, as well as capturing anomalous events that otherwise would go undocumented by western science. Furthermore, community-based monitoring projects in the Hudson Bay region can provide ongoing and cost-effective means of establishing baseline data and monitoring change (Moller et al. 2004).

Although TEK/LEK can play an important role in understanding and monitoring climatic change in the Hudson Bay region, this is not without challenges. Indeed, northern aboriginal groups have experienced substantial cultural changes throughout their recent history. This phenomenon has led to profound changes in the ways in which TEK/LEK are now acquired and transmitted. The shift from dispersed to centralised settlements that has occurred throughout the twentieth century in the Canadian North brought on profound economic, social and cultural changes for the Inuit and the Cree, as they moved from small, all-native hunting-trapping base camps to much larger villages of mixed ethnicity. As a consequence, northern peoples have experienced drastically altered social environments (Damas 2002; Lytwyn 2002). While subsistence activities remain central to Inuit and Cree livelihoods today, substantial changes in the environmental knowledge and related land-base skill sets have been documented in various Cree and Inuit communities. For instance, through their work among western James Bay Cree women, Ohmagari and Berkes (1997) found that TEK tends to be transmitted later in life and incompletely, and that there has been a change in the skill sets and kind of land-based knowledge held and transmitted. An erosion of Inuit knowledge and land-based skills has also been documented among the younger generation of Inuit throughout Nunavut (Rasing 1999; Aporta 2004) and in the Canadian Arctic generally (Condon et al. 1995; Newton 1995; Collings et al. 1998). Such phenomena have been attributed in part to southern educational requirements, which result in decreased time to participate in hunting, increased dependence on wage employment, a general shift in social norms, the advent of exogenous forms of control of the land and natural resources traditionally used by aboriginal peoples, and an ongoing segregation of the young and older generations (Kral 2003; Takano 2004). Far from being static, unchanging or fixed in the past, TEK/LEK is both cumulative and dynamic, building on experiences and responding to environmental and cultural changes (Berkes 1999; Mallory et al. 2006). The changing nature of TEK/LEK and the social histories which underlie such knowledge practices should thus be critically examined and acknowledged in the process of linking the unique knowledge base constituted by Cree and Inuit TEK/LEK with scientific information to understand environmental and climate-related changes (Davis and Wagner 2003; Mallory et al. 2006).

Conclusion

We have reviewed several research initiatives and managerial practices that have sought to combine both TEK/LEK and scientific information for gaining a better understanding of animal ecology and managing wildlife under changing climatic conditions in the Hudson Bay region. As our case studies on common eiders and polar bears illustrate, Inuit and Cree residents of Hudson Bay can bring significant and original contributions to scientific endeavours and environmental decision-making, notably by contributing ecological observations made at precise spatial and temporal scales, providing unique historical baseline data, identifying areas for further scientific research, as well as monitoring environmental change. However, such a collaborative process continues to face significant challenges, particularly when scientific and LEK/TEK findings do not concur. These include: (1) the difficulty of combining both TEK/LEK and scientific knowledge in wildlife management policies due to conflicting perspectives over the interpretation of observations relating to animal ecology; (2) the challenge of finding culturally relevant ways of assessing the validity of knowledge claims rooted in multiple cultural traditions; and (3) the necessity of reconciling diverging ethical views about research practices. Furthermore, in spite of evidence that scientific practices in the Hudson Bay region and elsewhere in northern Canada are becoming more pluralistic and participatory in nature, aboriginal groups still question conventional research processes and increasingly demand more involvement throughout the implementation of research projects that may ultimately affect their communities (Gearheard and Shirley 2007), and quite possibly conflict with their held views. At the same time, scientists and wildlife managers suggest that contributions from TEK/LEK should not be accepted and incorporated into environmental decision-making without undergoing some degree of scrutiny (Davis and Wagner 2003; Ellis 2005; Gilchrist and Mallory 2007). This situation highlights the need for further collaboration among aboriginal communities, scientists and policy makers. Clearly, there is room for more progress in accepting both TEK/LEK and western science as sources of knowledge and understanding, not in the abstract, but in practice, through adequate research, policy initiatives and communication between stakeholders.

More than a decade ago, Agrawal (1995) argued that it was time to bridge the divide between TEK/LEK and scientific knowledge by identifying mutually affirming ways in which scientists and aboriginal communities could use all knowledge that offers information to deal with the issues being faced. We suggest that such thoughts still hold true today and offer the view that neither western science nor traditional or local ecological knowledge is sufficient in isolation for understanding the complexities of the effects of global climate change on animal ecology in the Hudson Bay region and elsewhere. It is therefore imperative to further explore the ways in which such perspectives can enter a constructive dialogue.

References

Aars, J., Lunn, N.J., and Derocher, A.E. 2006. *Polar Bears: Proceedings of the 14th Working Meeting of the IUCN/SCC Polar Bear Specialist Group, 20-24 June 2005, Seattle, Washington, DC, USA*. Gland and Cambridge: IUCN.

Abraham, K.F. and Finney, G.H. 1986. Eiders of the Eastern Canadian Arctic. In: Reed, A. (ed.) *Eider Ducks in Canada*. Report Series No. 47. Ottawa: Canadian Wildlife Service, pp. 55–73.

Agrawal, A. 1995. Dismantling the Divide Between Indigenous and Scientific Knowledge. *Development and Change* 26(3): 413–439.

Aporta, C. 2004. Routes, Trails and Tracks: Trail Breaking Among the Inuit of Igloolik. *Études/ Inuit/Studies* 28(2): 9–38.

Arctic Climate Impact Assessment (ACIA). 2005. Cambridge University Press: Cambridge.

Berkes, F. 1999. *Sacred Ecology: Traditional Ecological Knowledge and Resource Management*. Philadelphia, PA: Taylor & Francis.

Berkes, F. and Folke, C. (eds.). 1998. *Linking Social and Ecological Systems. Management Practices and Social Mechanisms for Building Resilience*. Cambridge: Cambridge University Press.

Berkes, F. and Henley, T. 1997. Usefulness of Traditional Knowledge: Myth or Reality? *Policy Options* 18 (3): 55–56.

Bielawski, E. 1995. Inuit Indigenous Knowledge and Science in the Arctic. In: Johnson, D.R. and Peterson, D. L. (eds.). *Human Ecology and Climate Change: People and Resources in the Far North*. Washington, DC: Taylor & Francis, pp. 219–228.

Bielawski, E. 2005. Indigenous Knowledge. In: Nuttal, M. (ed.). *Encyclopedia of the Arctic*. New York: Routledge, pp. 950–955.

Collings, P., Wenzel, G. and Condon, R. 1998. Modern Food Sharing Networks and Community Integration in the Central Canadian Arctic. *Arctic* 51(4): 301–314.

Condon, R., Collings, P. and Wenzel, G. 1995. The Best Part of Life: Subsistence Hunting, Ethnicity and Economic Adaptation among Young Adult Inuit Males. *Arctic* 48(1): 31–46.

Damas, D. 2002. *Arctic Migrants/Arctic Villagers. The Transformation of Inuit Settlement in the Central Arctic*. Montreal and Kingston: McGill-Queen's University Press, Native and Northern Series 32.

Davis, A. and Wagner, J.R. 2003. Who Knows? On the Importance of Identifying "Experts" When Researching Local Ecological Knowledge. *Human Ecology* 31: 463–489.

Derocher, A., Lunn, N. and Stirling, I. 2004. Polar Bears in a Warming Climate. *Integrative and Comparative Biology* 44: 163–176.

Dowsley, M. 2009. Community Clusters in Wildlife and Environmental Management: Using TEK and Community Involvement to Improve Management in an Era of Rapid Environmental Change. *Polar Research* 28: 43–59.

Dowsley, M. 2007. Inuit Perspectives on Polar Bears (*Ursus maritimus*) and Climate Change in Baffin Bay, Nunavut, Canada. *Research and Practice in the Social Sciences* 2(2): 53–74.

Dowsley, M. and Taylor, M.K. 2006. *Management Consultations for the Western Hudson Bay (WH) Polar Bear Population (01-02 December 2005)*. Final Wildlife Report. Iqaluit: Government of Nunavut, Department of Environment.

Dowsley, M. and Wenzel, G. 2008. "The Time of Most Polar Bears": A Co-management Conflict in Nunavut. *Arctic* 61(2): 177–189.

Duerden, F. and Kuhn R.G. 1998. Scale, Context, and Application of Traditional Knowledge of the Canadian North. *Polar Record* 34: 31–38.

Dyck, M.G., Soon, W., Baydack, R.K., Legates, D.R., Baliunas, S., Ball, T.F., Hancock, L.O. 2007. Polar Bears of Western Hudson Bay and Climate Change: Are Warming Spring Air Temperatures the 'Ultimate' Survival Control Factor? *Ecological Complexity* 4(3): 73–84.

Ellis, S.C. 2005. Meaningful Consideration? A Review of Traditional Knowledge in Environmental Decision Making. *Arctic* 58(1): 66–77.

Ford, J. and Furgal, C. 2009. Climate Change Impacts, Adaptation and Vulnerability in the Arctic. *Polar Research* 28(1): 1–9.

Ford, J., Pearce, T., Smit, B., Wandel, J., Allurut, M., Shappa, K., Ittusujurat, H. and Qrunnut, K. 2007. Reducing Vulnerability to Climate Change in the Arctic: The Case of Nunavut, Canada. *Arctic* 60(2): 150–166.

Ford, J., Smit, B. and Wandel, J. 2006. Vulnerability to climate change in the Arctic: A case study from Arctic Bay, Canada. *Global Environmental Change* 16: 145–160.

Fraser, D.J., Coon, T., Prince, M.R., Dion, R. and Bernatchez, L. 2006. Integrating Traditional and Evolutionary Knowledge in Biodiversity Conservation: A Population Level Case Study. *Ecology and Society* 11(2): 4.

Friend, M. 1999. Avian Cholera. In: Friend, M. and Franson, J.C. (eds.) *Field Manual of Wildlife Diseases: General Field Procedures and Disease of Birds*. Reston: US Geological Survey, Biological Resources Division Information and Technology Report 1999–2001, pp. 75–92.

Furgal, C. 2006. *Ways of Knowing and Understanding: Towards the Convergence of Traditional and Scientific Knowledge of Climate Change in the Canadian North*. Ottawa: Minister of Public Works and Government Services Canada.

Gearheard, S. and Shirley, J. 2007. Challenges in Community-Research Relationships: Learning from Natural Science in Nunavut. *Arctic* 60(1): 62–74.

George, J. 2009. "Baffin hunters threaten revolt over Nunavut government polar bear quotas". *Nunatsiaq News*, September 30, 2009.

Gilchrist, H.G. and Mallory, M.L. 2007. Comparing Expert-Based Science With Local Ecological Knowledge: What Are We Afraid of? *Ecology and Society* 12(1):r1. [online] URL: http://www.ecologyandsociety.org/vol12/iss1/resp1 (consulted on 16/03/10).

Gilchrist, H.G. and Robertson, G.J. 2000. Observations of Marine Birds and Mammals Wintering at Polynyas and Ice Edges in the Belcher Islands, Nunavut, Canada. *Arctic* 53: 61–68.

Gilchrist, H.G., Mallory, M. and Merkel, F. 2005. Can Local Ecological Knowledge Contribute to Wildlife Management? Case Studies of Migratory Birds. *Ecology and Society* 10(1): 20.

Gilchrist, H.G., Heath, J., Arragutainaq, L., Robertson, G., Allard, K., Gilliland, S. and Mallory, M.L. 2006a. Combining Science and Local Knowledge to Study Common Eider Ducks Wintering in Hudson Bay. In: Riewe, R. and Oakes, J. (eds.). *Climate Change: Linking Traditional and Scientific Knowledge*. Winnipeg, MB: Aboriginal Issues Press, pp. 284–303.

Gilchrist, H.G., Robertson, M., Dallaire, A., Gaston, T. and Butler, I. 2006b. Avian Cholera Confirmed Among Northern Common Eider Ducks Nesting in Northern Hudson Bay, Nunavut 2004–2005. Information Leaflet. Ottawa: Canadian Wildlife Service.

Goudie, R.I., Robertson, G.J. and Reed, A. 2000. Common Eider (*Somateria mollissima*). In: Poole, A. and Gill, F. (eds.) *The Birds of North America*. No. 546. Philadelphia, PA: The Birds of North America.

Gough, W.A., Cornwell, A.R., Tsuji, L.J.S. 2004. Trends in Seasonal Sea Ice Duration in Southwestern Hudson Bay. *Arctic* 57: 299–305.

Government of Nunavut. 2005. *Polar Bear Management Memorandum of Understanding Between Arviat HTO, Baker Lake HTO, Aqigiq HTO (Chesterfield Inlet), Aqiggiaq HTO (Rankin Inlet), Issatik HTO (Whale Cove), Kivalliq Wildlife Board and the Department of Environment for the Management of the "Western Hudson" Polar Bear Population*. March 9. Iqaluit: Government of Nunavut.

Hallowell, I. 1926. Bear Ceremonialism in the Northern Hemisphere. *American Anthropologist* 28(1): 1–175.

Henri, D. 2007. *The Integration of Inuit Traditional Ecological Knowledge and Western Science in Wildlife Management in Nunavut, Canada: The Case of Avian Cholera Outbreaks Among Common Eider Ducks in the West Hudson Strait and North James Bay Area*. Masters' thesis. Oxford: University of Oxford.

Huntington, H. and Fox, S. 2005. The Changing Arctic: Indigenous Perspectives. In: *Arctic Climate Impact Assessment*. Cambridge: Cambridge University Press, pp. 61–98.

Inuit Tapiriit Kanatami. 2009. *Inuit Tapiriit Kanatami Resolution: Polar Bear Research Methods*. Resolution passed on June 10th during the Annual General Meeting held in Nain, Nunatsiavut.

International Panel on Climate Change. 2007a. *Climate Change 2007: The Physical Science Basis*. Contribution of Working Group I to the Fourth Assessment Report. Cambridge: Cambridge University Press.

International Panel on Climate Change. 2007b. *Climate Change 2007: Impacts, Adaptation and Vulnerability*. Contribution of Working Group II to the Fourth Assessment Report. Cambridge: Cambridge University Press.

Irving, W.N. 1986. The Barren Grounds. In: C.S. Beals (ed.). *Science, History and Hudson Bay. Volume 1*. Ottawa: Department of Energy, Mines and Resources, pp. 26–54.

Johnson, M. 1992. Research on Traditional Environmental Knowledge: Its Development and its Role. In: Johnson, M. (ed.). *Lore: Capturing Traditional Environmental Knowledge*. Hay River, NWT: Dene Cultural Institute and the International Development Research Centre, pp. 3–22.

Kral, M. 2003. Unikkaartuit: *Meaning of Well-Being, Sadness, Suicide and Change in Two Inuit Communities*. Final Report to the National Health Research and Development Programs. Ottawa: Health Canada Project #6606-6213-002.

Krupnik I. and Jolly, D. (eds.). 2002. *The Earth is Faster Now: Indigenous Observations of Arctic Environmental Change*. Fairbanks, AK: Arctic Research Consortium of the United States in cooperation with the Arctic Studies Center, Smithsonian Institution.

Laidler, G. 2006. Inuit and Scientific Perspectives on the Relationship between Sea Ice and Climatic Change: The Ideal Complement? *Climatic Change* 78(2–4): 407–444.

Lentfer, J. 1974. Agreement on the Conservation of Polar Bears. *Polar Record* 17(108): 327–330.

Lytwyn, V.P. 2002. *Muskekowuck Athinuwick: Original People of the Great Swampy Land*. Winnipeg, MB: University of Manitoba Press.

Mailhot, J. 1993. *Traditional Ecological Knowledge. The Diversity of Knowledge Systems and Their Study*. Great Whale Environmental Assessment, Background Paper No. 4. Montreal: Great Whale Public Review Support Office.

Mallory, M.L., Gilchrist, H.G. and Akearok, J. 2006. Can We Establish Baseline Local Ecological Knowledge on Wildlife Populations? In: Riewe, R. and Oakes, J. (eds.). *Climate Change: Linking Traditional and Scientific Knowledge*. Winnipeg, MB: Aboriginal Issues Press, pp. 24–33.

McDonald, M., Arragutainaq, L. and Novalinga, Z. 1997. *Voices from the Bay: Traditional Ecological Knowledge of Inuit and Cree in the Hudson Bay Bioregion*. Ottawa: Canadian Arctic Resources Committee.

McGhee, R. 1990. *The Peopling of the Arctic Islands*. In: Harington, C. R. (ed.). *Canada's Missing Dimension. Science and history in the Canadian Arctic Islands. Volumes 1–2*. Ottawa: Canadian Museum of Nature, pp. 666–676.

Moller, H., Berkes, F., Liver, P.O. and Kislalioglu, M. 2004. Combining Science and Traditional Ecological Knowledge: Monitoring Populations for Co-management. *Ecology and Society* 9(3): 2–17.

Nadasdy, P. 2003. *Hunters and Bureaucrats. Power, Knowledge and Aboriginal-State Relations in the Southwest Yukon*. Vancouver, BC: UBC Press.

Nadasdy, P. 1999. The Politics of TEK: Power and the Integration of Knowledge. *Arctic Anthropology* 36: 1–18.

Nakashima, D. J. 1991. *The Ecological Knowledge of Belcher Island Inuit*. Ph.D. thesis. Montreal: McGill University.

Newton, J. 1995. An Assessment of Coping with Environmental Hazards in Northern Aboriginal Communities. *The Canadian Geographer* 39(2): 112–120.

Nickels, S., Furgal, C., Buell, M. and Moquin, H. 2006. *Unikkaaqatigiit – Putting the Human Face on Climate Change: Perspectives from Inuit in Canada*. Ottawa: A joint publication of Inuit Tapiriit Kanatami, Nasivvik Centre for Inuit Health and Changing Environments at Université Laval and the Ajunnginiq Centre at the National Aboriginal Health Organization.

Nunavut Tunngavik Incorporated. 2007a. *NTI Demands Intrusive Scientific Wildlife Research be Halted*. NTI News Release, December 5, 2007.

Nunavut Tunngavik Incorporated. 2007b. *West Hudson Bay Polar Bear*. Wildlife Briefing Note Submitted to the Nunavut Wildlife Management Board, April 24–26, 2007.

Nunavut Wildlife Management Board. 2007. *West Hudson Bay Polar Bear Population: Total Allowable Harvest*. Record of Decision, July 2007.

Oakes, J. 1990. *Eider Coats*. Winnipeg, MB: Aboriginal Issues Press.

Oakes, J. and Riewe, R. (eds.). 2006. *Climate Change: Linking Traditional and Scientific Knowledge*. Winnipeg, MB: Aboriginal Issues Press.

Ohmagari, K. and Berkes, F. 1997. Transmission of Indigenous Knowledge and Bush Skills among the Western James Bay Cree Women of Subarctic Canada. *Human Ecology* 25: 197–222.

Peacock, E., Orlando, A., Sahanatien, V., Stapleton, S., Derocher, A.E. and Garshelis, D.L. 2008. *Foxe Basin Polar Bear Project. Interim Report*. Iqaluit: Government of Nunavut, Department of Environment.

Pierotti, R. and Wildcat, D. 2000. Traditional Ecological Knowledge: The Third Alternative. *Ecological Applications* 10(5): 1333–1340.

Polar Bear Specialist Group of the IUCN/SSC. 2009. *Resolution #1-2009: Effects of Global Warming on Polar Bears*. Resolution from the 15th Meeting of the PBSG in Copenhagen, Denmark, June 29–July 3, 2009.

Popper, K. 1959. *The Logic of Scientific Discovery*. New York: Basic Books.

Randa, V. 1993. Des Offrandes au Système de Quotas: Changements de Statut du Gibier chez les Iglulingmiut de l'Arctique Oriental Canadien. Communication at a Conference Titled "Peuples des Grands Nords (Sibérie-Amérique-Europe). Traditions et Transitions", INALCO/Université de Paris III/Sorbonne Nouvelle, 25–27 March 1993. Paris: UNESCO.

Randa, V. 1986a. *L'Ours Polaire et les Inuit*. Paris: SELAF.

Randa, V. 1986b. Au Croisement des Espaces et des Destins: Nanuq, «Marginal Exemplaire». Un Cas de Médiation Animale Dans l'Arctique Central Canadien. *Études/Inuit/Studies* 10(1): 159–169.

Rasing, W. 1999. Hunting for Identity: Thoughts on the Practice of Hunting and its Significance for Iglulingmiut Identity. In: Oosten, J. and Remie, C. (eds.). *Arctic Identities: Continuity and Change in Arctic and Saami Societies*. Leiden, The Netherlands: University of Leiden, pp. 79–108.

Reed, A. and Erskine, A.J. 1986. Populations of the Common Eider in Eastern North America. In: Reed, A. (ed.). *Eider Ducks in Canada*. Report Series No. 47. Ottawa: Canadian Wildlife Service, pp. 156–169.

Regehr, E.V., Lunn, N.J., Amstrup, S.C. and Stirling, I. 2007. Effects of Earlier Sea Ice Breakup On Survival and Population Size of Polar Bears in Western Hudson Bay. *Journal of Wildlife Management* 71: 2673–2683.

Riedlinger, D. and Berkes, F. 2001. Contributions of Traditional Knowledge to Understanding Climate Change in the Canadian Arctic. *Polar Record* 37: 315–328.

Robertson, G.J. and Gilchrist, H.G. 1998. Evidence of Population Declines among Common Eiders Breeding in the Belcher Islands, Northwest Territories. *Arctic* 51: 378–385.

Saladin d'Anglure, B. 1990. Nanook, Super-mâle: The Polar Bear in the Imaginary Space and Social Time of the Inuit of the Canadian Arctic. In: Willis, R. (ed.). *Signifying Animals: Human Meaning in the Natural World*. London: Routledge, pp. 178–195.

Statistics Canada Census 2006. http://www12.statcan.gc.ca/census-recensement/2006/rt-td/ap-pa-eng.cfm (consulted on 28/08/09).

Stevenson, M.G. 2004. Decolonizing Co-management in Northern Canada. *Cultural Survival* Spring: 68.

Stevenson, M.G. 1996. Indigenous Knowledge in Environmental Assessment. *Arctic* 49(3): 278–291.

Stirling, I. and Derocher, A. 1993. Possible Impacts of Climatic Warming on Polar Bears. *Arctic* 46(3): 240–245.

Stirling, I. and Parkinson, C. 2006. Possible Effects of Climate Warming on Selected Populations of Polar Bears (*Ursus maritimus*) in the Canadian Arctic. *Arctic* 59(3): 261–275.

Stirling, I., Derocher, A.E., Gough, W.A. and Rode, K. 2008. Response to Dyck et al. (2007) on Polar Bears and Climate Change in Western Hudson Bay. *Ecological Complexity* 5: 193–201.

Stirling, I., Lunn, N.J., Iacozza, J., Elliott, C. and Obbard, M. 2004. Polar Bear Distribution and Abundance on the Southwestern Hudson Bay Coast During Open Water Season, in Relation to Population Trends and Annual Ice Patterns. *Arctic* 57: 15–26.

Stirling, I., Lunn, N. and Iacozza, J. 1999. Long-term Trends in the Population Ecology of Polar Bears in Western Hudson Bay in Relation to Climatic Change. *Arctic* 52(3): 294–306.

Takano, T. 2004. *Bonding with the Land: Outdoor Environmental Education Programmes and Their Cultural Contexts*. Ph.D. thesis, Department of Anthropology. Edinburgh: University of Edinburgh.

Taylor, M. and Lee, J. 1995. Distribution and Abundance of Canadian Polar Bear Populations: A Management Perspective. *Arctic* 48(2): 147–154.

Taylor, M.K., Aeeaguk, S., Andriashek, D., Barbour, W., Born, E.W., Calvert, W., Cluff, H.D., Ferguson, S., Laake, J., Rosing-Asvid, A., Stirling, I. and Messier, F. 2001. Delineating Canadian and Greenland Polar Bear (*Ursus maritimus*) Populations by Cluster Analysis of Movements. *Canadian Journal of Zoology* 79: 690–709.

Taylor, W.R. 1968. Eskimo Prehistory of the North and East Shores. In: Beals, C.S. (ed.). *Science, History and Hudson Bay. Volume 1*. Ottawa: Department of Energy, Mines and Resources, pp. 1–26.

Trott, C.G. 2006. The Gender of the Bear. *Études/Inuit/Studies* 30(1): 89–109.

Tyrell, M. 2009. West Hudson Bay Polar Bears: The Inuit Perspective. In: Freeman, M.M.R. and Foote, L. (eds.) *Inuit, Polar Bears and Sustainable Use: Local, National and International Perspectives*. Edmonton, AB: CCI Press, pp. 95–110.

Tyrell, M. 2006. More Bears, Less Bears: Inuit and Scientific Perceptions of Polar Bear Populations on the West Coast of Hudson Bay. *Études/Inuit/Studies* 30(2): 191–208.

Usher, P.J. 2000. Traditional Ecological Knowledge in Environmental Assessment and Management. *Arctic* 53(2): 183–193.

Wenzel, G.W. 2005. Nunavut Inuit and Polar Bears: The Cultural Politics of the Sport Hunt. *Senri Ethnological Series* 67: 363–388.

Wenzel, G. and Dowsley, M. 2005. Economic and Cultural Aspects of Polar Bear Sport Hunting in Nunavut, Canada. In: Freeman, M.M.R., Hudson, R.J. and Foote, L. (eds.). *Conservation Hunting: People and Wildlife in Canada's North*. Edmonton, AB: Canadian Circumpolar Institute, pp. 37–45.

White, G. 2008. "Not the Almighty": Evaluating Aboriginal Influence in Northern Land-Claim Boards. *Arctic* 61(1): 71–85.

White, G. 2006. Cultures in Collision: Traditional and Euro-Canadian Governance Processes in Northern Land-Claim Boards. *Arctic* 59(4): 401–414.

Wright, J.V. 1968. The Boreal Forest. In: Beals, C. S. (ed.). *Science, History and Hudson Bay. Volume 1*. Ottawa: Department of Energy, Mines and Resources, pp. 55–68.

The Future of Hudson Bay: New Directions and Research Needs

M.L. Mallory, L.L. Loseto, and S.H. Ferguson

Abstract What does the future hold for Hudson Bay? The consequences of Arctic warming include widespread melting of sea ice, a longer open water season, and warming of coastal permafrost. These and other alterations associated with climate warming will have an impact on Hudson Bay and the ecosystem it supports. The high trophic level marine mammals and birds have already experienced the cascade of ecosystem changes and are revealing those effects to the global community. The variability and magnitude of effects on these top predators need to be adequately assessed to begin mitigating and adapting to predicted changes. This book summarizes our current scientific understanding of the key oceanographic drivers and wildlife of Hudson Bay. We also have identified gaps in our knowledge and highlight future research needs. We are in the early stages of fully understanding: (a) the impacts of climate change on the Bay; (b) species-specific interactions and requirements of the physical and biological environment; and (c) the degree to which a warming climate will influence marine and coastal habitats, and consequently the Hudson Bay ecosystem and the iconic Arctic wildlife that call it home.

Keywords Anthropogenic stressors • Change • Climate change • Community-based monitoring • Ecotourism • Oil and gas exploration • Hydroelectricity • Sea ice • Shipping • Wildlife management

Introduction

The Hudson Bay region (HBR) encompasses over 1.2 million square kilometers (Stewart and Barber this volume), a huge area approximately the size of the state of Alaska, larger than the Canadian province of Ontario, or the combined size of Finland plus Norway plus Sweden. Two of the islands in this region (Southampton

M.L. Mallory (✉)
Canadian Wildlife Service, P.O. Box 1714, Iqaluit, NU X0A 0H0, Canada
e-mail: mark.mallory@ec.gc.ca

S.H. Ferguson et al. (eds.), *A Little Less Arctic: Top Predators in the World's Largest Northern Inland Sea, Hudson Bay*, DOI 10.1007/978-90-481-9121-5_14,
© Springer Science+Business Media B.V. 2010

and Prince Charles islands) are larger than the Canadian province of Prince Edward Island. It is a massive, remote area. The HBR has been occupied for centuries by different groups of aboriginal peoples (notably Cree, Thule and Inuit; Hoover this volume), and has been visited by European and "western" explorers for over four centuries, starting with Martin Frobisher at the eastern entrance of Hudson Strait in 1578, and of course Henry Hudson and his mutinous crew in 1610 (Encyclopaedia Brittanica 2009), from which the region was named (however, if you believe Farley Mowat (2002), pre-Viking Europeans were in Hudson Strait 1,000 years ago). Commercial whaling activities were active in Hudson Bay between the mid-nineteenth century to the early twentieth century (Ross 1979). The HBR is ringed by communities (Stewart and Barber this volume), with Igloolik, Nunavut at its northern reaches, Churchill Manitoba to the west, Moosonee, Ontario at its southern limits, and Kangiqsualujjuaq, Quebec (Nunavik) in the east. Thus, this vast marine zone borders on four Canadian provinces or territories but is typified by having a very small number of people that use it. In fact, the largest coastal community, Rankin Inlet, had fewer than 3,000 people in the 2006 census (Statistics Canada 2009), and only four of the communities around the Region had populations larger than 2,000 residents (Arviat, Kuujjuaq, Moosonee, Rankin Inlet). Collectively, fewer than 40,000 people inhabit the entire HBR, which remains dominated by First Nation and Inuit peoples of Canada (Atlas of Canada 2009).

With so few people in this area, it is perhaps not surprising that we still know so little about the wildlife of this region. In fact, Prince Charles and Air Force islands were the last major land masses to be charted in North America in 1948 (Johnston and Pepper 2009), testifying to the remote and unknown character of this region. If the research in this book shows one thing, it is that we are just now, at the start of the twenty-first century, starting to unlock many of the marine wildlife mysteries of this vast area. Much like the seminal local knowledge work on the HBR collated by McDonald et al. (1997), this book has highlighted some of the novel information found just in the last decade or so on how some animals move around the HBR (Ferguson et al. this volume; Westdal et al. this volume), and by backcasting, how some populations may have changed (Higdon and Ferguson this volume). In fact, several chapters in the book (Ferguson et al. this volume; Higdon and Ferguson this volume; Henri et al. this volume) have shown the strong benefits of using both aboriginal traditional knowledge and western empirical research to better understand the ecology of top predators in this remote region.

If there is one term that can define the future of Hudson Bay, that term is "change". Change is underway for the habitats, peoples and wildlife of this region on many fronts, some of which are directly related to climate change, and others which are not. Hopefully in going through the chapters of this book, the reader has been exposed to research on some of these changes. However, below we provide a brief review and a few of our expectations for various types of change that we believe are coming for Hudson Bay in the future. It is our hope that these forecasts spur additional physical, biological and socio-economic research in this region.

Physical Change

At a fundamental level, the physical environment of Hudson Bay is changing, irrespective of climate change. A main driver of this effect is postglacial emergence or isostatic rebound, that is, the land area continues to rise following the retreat of glaciers from the last glacial period (Stewart and Lockhart 2005). In effect, much of Hudson Bay is getting shallower, and island archipelagoes like the Belcher Islands are increasing in terrestrial surface area. At the rate of emergence, presumably losses of some coastal habitat due to succession of vegetation are offset by gains as new coast rises from the water. Presumably this alone would not lead to substantial alterations to coastal biological functions in food webs, but over time it will change the nature and distribution of key habitats for wildlife in the region.

As reviewed by Hocheim et al. (this volume), the altered sea ice regime is perhaps the most noticeable, blatant physical change in the Hudson Bay region. For the entire Arctic, the melt season length has increased by about 20 days over the last 30 years with the largest trends of over 10 days per decade seen for Hudson Bay (Markus et al. 2009). Recent calculations show a reduction in the sea ice concentration between 14% and 19% per decade from 1980 to 2005 (Hocheim et al. this volume). Although there are spatial and temporal variations in patterns within the entire region, in general sea ice is forming later and is breaking up earlier in Hudson Bay (Gagnon and Gough 2005; Comiso et al. 2008). Many of the chapters in this book have identified processes by which this physical change will manifest itself in expected changes for wildlife and communities. For Inuit hunters, changes in sea ice conditions are reducing the reliability of local or traditional knowledge for predicting traditional travel routes, weather and areas where wildlife will be at different times of the year (McDonald et al. 1997; Hoover this volume; Henri et al. this volume). In short, conditions are far less predictable than they once were. Reductions in sea ice extent or duration of ice cover are also having direct and indirect effects on marine wildlife in the region. If sea ice formed a barrier to some species for accessing the site, less sea ice (or its earlier disappearance) may be removing that barrier; this seems to be the case for killer whales (*Orcinus orca*; Ferguson et al. this volume). Common sense tells us that species that rely on the ice as a hunting platform (e.g. polar bears, *Ursus maritimus*) or to build breeding sites (e.g. ringed seals, *Phoca hispida*) should be negatively affected by reductions in sea ice, and there is empirical evidence to support this expectation (e.g., Stirling et al. 1999; Ferguson et al. 2005; Peacock et al. this volume; Chambellant this volume). Moreover, changes in the phenology of sea ice cover affect the productivity of the marine food web (Welch et al. 1992; Hoover this volume), which will have long-term consequences for food supplies of species (e.g. thick-billed murres *Uria lomvia*; Kelley et al. this volume; Mallory et al. this volume).

Collectively, the evidence to date suggests that the rapid changes in sea ice cover in Hudson Bay are having a variety of effects on marine food webs of the region, but we are just beginning to understand the complex interactions and breadth of these changes (Hoover, this volume). Of particular concern are the possible rapid food web changes that could shift the sea ice subarctic ecosystem to an open water temperate environment (Picture 1).

Picture 1 Walrus resting on an ice floe near Akpatok Island (Photograph by Manon Simard)

Top predators in the Hudson Bay marine ecosystem will also be affected by warming temperatures and perhaps altered precipitation regimes on the terrestrial environment surrounding the Bay, and the consequent effects on permafrost melting and freshwater discharge into the Bay (e.g., Prinsenberg 1988; Gough and Leung 2002; Stewart and Lockhart 2005). Freshwater fluxes play a major role in oceanographic processes in the Bay, by indirectly increasing nutrient supply and productivity to the marine food web via upwelling of deep water (Kuzyk et al. 2009). Changes to permafrost will affect the suitability of habitats for many of the migratory birds that rely on the HBR as breeding and moulting sites (Mallory and Fontaine 2004), and for species like polar bears that rely on permafrost features for breeding sites (Peacock et al. this volume).

Anthropogenic Change

Although the dominant change affecting the HBR is global warming due to anthropogenic inputs to the atmosphere, there are other human-based activities that are changing the environment of the HBR for the wildlife that occur there. Most of these represent real, or potential, additional stressors on top of changes stemming from climate change.

Reduced sea ice cover is likely to lead to increased shipping activity in the HBR, for several reasons. First, more open water (i.e. less need for ice-breaking) will make it more attractive for sending materials in and out ports like Churchill, or for resupplying Arctic communities (Arctic Council 2009). Second, ship-based ecotourism of Arctic communities and attractions (e.g. seabird colonies) will likely continue to increase in intensity (Hall and Johnston 1995). Third, large development projects for extracting resources in the Arctic are increasing, and will become more feasible with less ice issues (e.g., Baffinland Iron Mines Corporation 2008; Arctic Council 2009). Collectively, increased shipping comes with higher environmental risks, in the form of: (1) greater chances of ships striking marine mammals (Jensen and Silber 2003); (2) increased risks of oil spills (e.g., Dickins et al. 1990); (3) altered ice regimes due to ice-breaking activity (Baffinland Iron Mines Corporation 2008); and (4) deleterious effects of increased noise on wildlife (Richardson 2006; Nowacek et al. 2007; Weilgart 2007).

Future harvest by hunters must also be considered an additional stressor for marine mammals and birds in the HBR. Rights of harvest for wildlife ("country food") are protected for beneficiaries of land claim agreements in the HBR (e.g., Indian and Northern Affairs Canada 1993), although certain species are regulated by a quota system (Hovelsrud et al. 2008). This is a particularly important issue, because communities in the HBR are growing rapidly in population size (Atlas of Canada 2009), which could mean more demand for country foods in the long-term. Furthermore, hunting for these species is currently undertaken principally by motorized vehicles (snowmobiles or boats with outboard motors); increased hunting effort, or for that matter even increased recreational travel by community members, may also translate to increased noise or physical disturbance to marine wildlife.

Another anthropogenic stressor that must be considered especially in the case of top predators is contaminants. Due to the propensity for contaminants to bioaccumulate over time and biomagnify up food webs, higher trophic level species that live long lives are at risk for toxic injury. Contaminants in the HBR typically arrive via long range transport in air and water from southern latitudes, and these accumulate to levels well above "background" in Hudson Bay marine mammals (Gaden and Stern this volume) and birds. Although non-local sources are the main concern, the increased coastal erosion in HBR will result in a release of previously unavailable contaminants such as mercury in addition to release by new land use practices in coastal areas. There is currently little evidence to suggest that contaminants are playing a significant, deleterious role on the health of wildlife populations in the HBR (Braune et al. 2005), yet together with other environmental stressors, the cumulative impacts may put some populations to tipping points. Despite international treaties to eliminate and reduce the use of chemicals that undergo long range transport, their long half lives have resulted in their environmental persistence and, disturbingly, some contaminants continue to increase in Arctic wildlife (e.g., Braune 2007).

In addition to these factors, future anthropogenic changes in the HBR are difficult to predict, as some will be linked to technological advancements and economic cycles, but will likely include stressors new to this region. Oil and gas exploration and possible development is already on the horizon (e.g., Nicolas and Lavoie 2009), as well as future hydrological and oceanographic changes in the HBR due to hydroelectric development (McDonald et al. 1997). There is international concern about the effects of ocean acidification linked to climate warming (Fabry et al. 2008), and in Canada, about acidifying pollutants in the Arctic (AMAP 2006). Much more data are needed to assess how these stressors will affect marine wildlife of the HBR.

Biological Change

Considering that both physical and anthropogenic changes to the HBR will affect biological systems, more research is needed to understand whether systems in flux will be able to adapt. Nonetheless, the focus of this book has been on new knowledge and changes in top predator populations of Hudson Bay in relation to climate change. As outlined in many of the chapters, we are just beginning to understand these changes, and we have much to learn.

Some of the biological changes are quite obvious. For example, certain species appear to be invading or increasing (e.g. razorbill *Alca torda*, Gaston and Woo 2008; killer whale, Higdon and Ferguson 2009), while populations of top predators like polar bears are declining in some parts of the HBR (Stirling et al. 1999; Stirling and Parkinson 2006; Peacock et al. this volume). Another clear change was the substantial shift in the marine food web of northern Hudson Bay detected by monitoring the diet of thick-billed murres (Gaston et al. 2003). This seminal work showed that the ice-associated Arctic cod (*Boreogadus saida*) has declined markedly in the diet of murres at the Coats Island colony, and instead the birds are returning with more subarctic capelin (*Mallotus villosus*; Mallory et al. this volume). Could this change reflect a regime shift, whereby an abrupt alteration occurs and the ecosystem shifts from one persistent state to a quite different persistent state? If so, our ability to adapt to or manage such a regime shift is limited. Adaptation depends upon our understanding of the causes of change and linkages among ecosystem components. Such knowledge is ultimately limited by our observational capabilities. In the examples above, biological changes were documented empirically with strong data, and at present we can be quite confident that "real" change is occurring (Picture 2).

However, other biological changes in the HBR remain anticipated or speculative at this time. These include: (1) the relaxation of winter ice constraints on eiders in the vicinity of the Belcher Islands (Mallory et al. this volume); (2) the effects of increased killer whales and bowhead whales on other parts of the marine food web in the HBR (Ferguson et al. this volume; Higdon and Ferguson this volume); (3) the effects of long-term climate change on ringed seal populations (Chambellant this volume); (4) the importance of sea ice seasonality on habitat suitability for species like narwhal (*Monodon monoceros*) during key periods of the year (Westdal et al. this volume); (5) the movement of novel diseases or parasites into this region, either brought by new species or extending their range into the Arctic due to more suitable climatic conditions, such as avian cholera in eiders (Henri et al. this volume) or phocine distemper from new seal species (Goldstein et al. 2009); and (6) separating effects of unidirectional climate change from long-term climate cycles (Chambellant this volume; Hoover this volume). In all of these situations, scientists have sound reasons for

Picture 2 Research team descending the cliffs at Coats Island to band murres (Photograph by Mark Mallory)

expecting change based on theory or examples from other regions, but we presently lack sufficient data to undertake strong tests of our hypotheses in the HBR.

Ultimately, changes to the ecology of top predators in the HBR, largely in response to either physical or anthropogenic changes in the region, will undoubtedly affect the genetic makeup of the "stocks" we now recognize in this area (Petersen et al. this volume). Highly migratory species may already be largely panmictic, but changes to habitats that reduce population size and gene flow could pose problems and lead to increased risk for parts of the populations. Conversely, certain species (e.g. Hudson Bay common eider, *Somateria mollissima sedentaria,* and ringed and harbour seals) may exhibit local phenotypic and genotypic adaptations which could be lost due to genetic swamping with increased gene flow. The extent to which these stocks can adjust, and how quickly they can accomplish this, remains unknown.

Detecting and Monitoring Change

The key point to remember about monitoring change in the HBR is that there is really no substitute for long term observations for increasing the confidence in our interpretations of "change" based on any available data. This is particularly true for areas like the HBR, which are remote, large, and have small, dispersed human populations. Certainly the work on polar bears (Stirling et al. 1999) or murres (Gaston et al. 2005) highlight the value of annual monitoring at key locations, and

Picture 3 Beluga whales in Churchill River estuary (Photograph by Lisa Loseto)

the work on killer, beluga and bowhead whales (Ferguson et al. this volume; Higdon and Ferguson 2009, this volume; Kelley et al. this volume) underscore the value of mining data from old, natural history observations. Of course, in the Arctic, there is no greater source of long term observations than the knowledge held by Inuit elders and hunters (McDonald et al. 1997), and this will continue to be a critical source of primary information, as well as a valuable dataset against which patterns or hypotheses derived from quantitative, western scientific approaches can be validated. Thus, we consider it self-evident that western science needs to continue and enhance efforts to work hand-in-hand with the indigenous peoples of the HBR, an approach that has worked well in this region, as shown in several chapters of this book. Moreover, for much of the data required in the future, community-based monitoring is a logical mechanism for delivering observational information of change (Picture 3).

Nonetheless, communities and scientists should be open to embracing new techniques and technologies to detect and track changes, as these can yield insights that are difficult to discern with the approaches mentioned above. For example, modern chemical techniques, such as stable isotopes, fatty acids, or genetic analysis, often do not require the killing of an organism to gather a suitable-sized sample. Data gathered from these techniques can provide novel insights into diet variation among social groups or clans (Marcoux et al. 2007; Witteveen et al. 2009), between sexes (Tucker et al. 2007) and species (Gross et al. 2009), and across seasons (de Stephanis et al. 2008; Loseto et al. 2009) and ages (Knoff et al. 2008). This type of information can be critical in understanding seasonal habitat use, for example, which might be relevant for understanding movements or habitat needs in the HBR during changing environmental conditions. Moreover, use of chemical signals in combination with genetic studies can elucidate social networks within a marine mammal population, and assist in providing information necessary to understand how top predators will adapt, or not adapt, to the transformation Hudson Bay is undergoing.

Future Change

The top predators, marine mammals and birds, have already experienced a cascade of ecosystem changes in the HBR, and as outlined in this book, are now revealing those effects to the global community. However, the variability and magnitude of effects on these top predators need to be adequately assessed to begin mitigating and adapting to predicted changes. Clearly we need to learn a lot more about the movements and key habitat requirements of top predators in the HBR, if we are to better protect important sites for them, ensure sustainable harvest levels are maintained, understand new and evolving stressors on their populations, and thus try to manage populations to ensure their long-term health. The technology required to gather this information is available for most species now (Mallory et al. 2006; Westdal et al. this volume). It is unfortunate that there is currently much resistance to using these technologies to track wildlife in the eastern Canadian Arctic (Nunavut Tunngavik Inc. 2007). However, with the examples of successful collaborations employing traditional knowledge and western science, particularly in the HBR, we hope that this impasse can be overcome in the near future.

For some species, it seems inevitable that their recruitment and eventually their population size will decline in the HBR – available data suggests that this will be the case for thick-billed murres (Gaston et al. 2005) and polar bears (Stirling and Parkinson 2006), and probably ringed seals (Ferguson et al. 2005) and narwhal (Hoover this volume). These are all examples of Arctic species at the southern edge of their range, and thus it should be expected that they might feel the effects of global warming the most (Boyd and Madsen 1997; Tynan and DeMaster 1997), resulting in a northward shift in the centre of gravity of their breeding population with climate change.

For other species, our expectations are murkier, forecast through a lens of "what ifs". Less tied to specific sea ice features, other marine mammals and birds may exhibit greater behavioural plasticity in being able to compensate for loss of sea ice than those in the preceding paragraph, assuming that their food supplies and important habitat sites remain similar or increase in abundance. In these cases, the indirect effects of global warming, such as introductions of new predators, new competitors, new diseases, and changes to food quantity or quality, will probably pose a greater threat than simply a reduction of sea ice. Thus, increased knowledge of the life cycle needs of these species in the HBR, and the interactions between species will be critical in developing suitable management plans to sustain their futures in this area.

One point that must not be forgotten in our considerations is that these species have all been through periods of warming before (i.e., changing ice, temperature and

Picture 4 Lucassie Ippak cleaning hunted eider ducks near Sanikiluaq (Photograph by Jeff Higdon)

precipitation conditions), and that their current distributions are influenced largely by biogeography related to the last glacial period (Petersen et al. this volume). In this sense, we know that the species over its entire circumpolar range can adjust to periods of warming and cooling, and the consequent massive spatial adjustments to their habitats brought on by the advance and retreat of glaciation. However, this is the first time in their evolutionary history that each species will have to adjust to habitat changes (in this case, warming and loss of ice) while also undergoing additional anthropogenic stressors such as human harvest (facilitated by fast transport and the ability to access huge areas), disturbance and noise, potential sublethal effects of contaminants, and habitat loss due to human activities. Moreover, the rate at which their environment is changing may be faster than they have ever faced in their evolutionary history (ACIA 2004; Serreze et al. 2007). Thus, from the perspective of the species, the question should really be "how will each respond to climate change at the current rapid rates of warming, and in the face of these novel pressures?" From the perspective of aboriginal peoples in Arctic communities, the question should be "how can we continue traditional activities, preserve those aspects that relate to our culture, and do this in a sustainable manner?" (Picture 4).

Conclusion

Irrespective of the various causes of climate change and other stressors in Hudson Bay, changes are currently manifesting themselves. To begin to address these questions from any perspective, much additional information is required. Scientists and

Arctic community members must work together, using all available information sources that help understand and manage wildlife. This will require using new approaches and technologies, undertaking new research, and digging into the past to draw on old knowledge. There is little time for politics in this situation – action must be taken soon to gather some of the necessary baseline information, before new activities (e.g. industrial development) are initiated and can potentially compound the challenges created by climate change. To this end, a large scale, or given the size of the region, several large scale investigations of the physical and biological environment of the Hudson Bay Region are urgently required to help development suitable and sustainable management plans for marine mammals and birds in this part of the Arctic.

References

ACIA [Arctic Climate Impact Assessment]. 2004. Impacts of a warming Arctic: Arctic Climate Impact Assessment. Cambridge University Press, Cambridge, UK.

AMAP. 2006. AMAP Assessment 2006: acidifying pollutants, Arctic haze, and acidification in the Arctic. Arctic Monitoring and Assessment Programme, Oslo, Norway. 112 pp.

Arctic Council. 2009. Arctic marine shipping assessment 2009 report. April 2009. 187 pp.

Atlas of Canada 2009. The Atlas of Canada. Population. http://atlas.nrcan.gc.ca/site/english/maps/peopleandsociety/population/#abori ginalpopulation (accessed 29 December 2009).

Baffinland Iron Mines Corporation. 2008. Development proposal for the Mary River Project. 89 pp.

Boyd, H., and J. Madsen. 1997. Impacts of global change on arctic-breeding bird populations and migration. Global Change and Arctic Terrestrial Ecosystems (eds. W.C. Oechel, T. Callaghan, T. Gilmanov, J.I. Holten, D. Maxwell, U. Molau and S. Sveinbjornsson), pp. 201–217. Springer, New York.

Braune, B.M. 2007. Temporal trends of organochlorines and mercury in seabird eggs from the Canadian Arctic, 1975 to 2003. Environmental Pollution 148: 599–613.

Braune, B.M., P.M. Outridge, A.T. Fisk, D.C.G., Muir, P.A., Helm, K., Hobbs, P.F., Hoekstra, Z.A., Kuzyk, M., Kwan, R.J., Letcher, W.L., Lockhart, R.J. Norstrom, and G.A. Stern. 2005. Persistent organic pollutants and mercury in marine biota of the Canadian Arctic: An overview of spatial and temporal trends. Science of the Total Environment 351–352: 4–56.

Comiso, J.C., C.L. Parkinson, R. Gersten, and L. Stock. 2008. Accelerated decline in the Arctic sea ice cover. Geophysical Research Letters 35: L01703.

De Stephanis, R., S. Garcia-Tiscar, P. Verborgh, R. Esteban-Pavo, S. Perez, L. Minvielle-Sebastia and C. Guinet. 2008. Diet of the social groups of long-finned pilot whales (*Globicephala melas*) in the Strait of Gibraltar. Marine Biology 154: 603–612.

Dickins, D., K. Bjerkelund, P. Vonk, S. Potter, K. Finley, R. Stephen, C. Holdsworth, D. Reimer, A. Godon, W. Duval, I. Buist, A. Sekerak. 1990. Lancaster Sound Region – a coastal atlas for environmental protection. DF Dickins Associates Ltd., Vancouver, BC.

Encyclopaedia Brittanica 2009. Hudson Strait. http://www.britannica.com/EBchecked/topic/274728/Hudson-Strait (accessed 29 December 2009).

Fabry, V.J., B.A. Seibel, R.A. Feely, and J.C. Orr. 2008. Impacts of ocean acidification on marine fauna and ecosystem processes. ICES Journal of Marine Science, 65: 414–432.

Ferguson, S.H., I. Stirling, and P. McLoughlin. 2005. Climate change and ringed seal (*Phoca hispida*) recruitment in western Hudson Bay. Marine Mammal Science 21: 121–135.

Gagnon, A.S., and W.A. Gough. 2005. Trends in the dates of ice freeze-up and breakup over Hudson Bay, Canada. Arctic 58: 370–382.

Gaston, A.J., and K. Woo. 2008. Razorbills (*Alca torda*) follow subarctic prey into the Canadian Arctic: colonization results from climate change? Auk 125: 939–942.

Gaston, A.J., K. Woo, and J.M. Hipfner. 2003. Trends in forage fish populations in northern Hudson Bay since 1981, as determined from the diet of nestling thick- billed murres *Uria lomvia*. Arctic 56: 227–233.

Gaston, A.J., H.G. Gilchrist, and J.M. Hipfner. 2005. Climate change, ice conditions and reproduction in an Arctic nesting marine bird: Brunnich's guillemot (*Uria lomvia* L.). Journal of Animal Ecology 74: 832–841.

Goldstein, T., J.A.K. Mazet, V.A. Gill, A.M. Doroff, K.A. Burek, and J.A. Hammond. 2009. Phocine distemper virus in northern sea otters in the Pacific Ocean, Alaska, USA. Emerging Infectious Diseases 15: 925–927.

Gough, W.A., and A. Leung. 2002. Nature and fate of Hudson Bay permafrost. Regional Environmental Change 2: 177–184.

Gross, A., J. Kiszka, O. Van Canneyt, P. Richard, V. Ridoux. 2009. A preliminary study of habitat and resource partitioning among co-occurring tropical dolphins around Mayotte, southwest Indian Ocean. Estuarine and Coastal and Shelf Science 84: 367–374.

Hall, C.M., and M.E. Johnston. 1995. Polar Tourism: Tourism in the Arctic and Antarctic Regions. Wiley, New York.

Hovelsrud, G.K., M. McKenna, and H.P. Huntington. 2008. Marine mammal harvests and other interactions with humans. Ecological Applications 18: S135–S147.

Higdon, J.W. and S.H. Ferguson. 2009. Loss of Arctic sea ice causing punctuated change in sightings of killer whales (*Orcinus orca*) over the past century. Ecological Applications 19: 1365–1375.

Indian and Northern Affairs Canada. 1993. Agreement between the Inuit of the Nunavut Settlement Area and Her Majesty the Queen in right of Canada. Indian and Northern Affairs Canada, Ottawa, Ontario. 282 pp.

Jensen, A.S., and G.K. Silber. 2003. Large whale ship strike database. U.S. Department of Commerce, NOAA Technical Memorandum. NMFS-OPR-25, 37 pp.

Johnston, V.J., and S.T. Pepper. 2009. The birds of Prince Charles Island and Air Force Island, Foxe Basin, Nunavut. Canadian Wildlife Service Occasional Paper Series No. 117.

Knoff, A., A. Hohn, and S. Macko. 2008. Ontogenetic diet changes in bottlenose dolphins (*Tursiops truncatus*) reflected through stable isotopes. Marine Mammal Science 24: 128–137.

Kuzyk, Z., A. Macdonald, R.W. Tremblay, and G.A. Stern. 2009. Elemental and stable isotopic constraints on river influence and patterns of nitrogen cycling and biological productivity in Hudson Bay. Continental Shelf Research (2009), DOI:10.1016/j.csr.2009.10.014

Loseto, L.L., G.A. Stern, T.L. Connelly, D. Deibel, B. Gemmill, A. Prokopowicz, L. Fortier, and S.H. Ferguson. 2009. Summer diet of beluga whales inferred by fatty acid analysis of the eastern Beaufort Sea food web. Journal of Experimental Marine Biology and Ecology, 374: 12–18.

McDonald, M., L. Arragutainaq, and Z. Novalinga. 1997. Voices from the Bay. Canadian Arctic Resources Committee, Ottawa. 98 pp.

Mallory, M.L., and A.J. Fontaine. 2004. Key marine habitat sites for migratory birds in Nunavut and the Northwest Territories. Canadian Wildlife Service Occasional Paper No. 109.

Mallory, M.L., J. Akearok, N.R. North, D. Vaughan Weseloh, and S. Lair. 2006. Movements of long-tailed ducks wintering on Lake Ontario to breeding areas in Nunavut, Canada. Wilson Journal of Ornithology 118: 494–501.

Marcoux, M., H. Whitehead and L. Rendell. 2007. Sperm whale feeding variation by location, year, social group and clan: evidence from stable isotopes. Marine Ecology-Progress Series 333: 309–314.

Markus, T., J.C. Stroeve, and J. Miller (2009), Recent changes in Arctic sea ice melt onset, freezeup, and melt season length. Journal of Geophysical Research 114: C12024.

Mowat, F. 2002. The Farfarers. Random House. 528 pp.

Nicolas, M.P.B., and D. Lavoie, 2009: Hudson Bay and Foxe Basins Project: introduction to the Geo-mapping for Energy and Minerals, GEM–Energy initiative, northeastern Manitoba (parts of NTS 54); *in* Report of Activities 2009, Manitoba Innovation, Energy and Mines, Manitoba Geological Survey, pp. 160–164.

Nowacek, D.P., L.H. Thorne, D. Johnston, and P.L. Tyack. 2007. Responses of cetaceans to anthropogenic noise. Mammal Review 37: 81–115.

Nunavut Tunngavik Inc. 2007. NTI demands intrusive scientific research be halted. http://www.tunngavik.com/2007/12/05/nti-demands-intrusive-scientific-wildlife-research-be-halted/ (accessed 30 December 2009).

Prinsenberg, S.J. 1988. Ice-cover and ice-ridge contributions to the freshwater contents of Hudson Bay and Foxe Basin. Arctic 41: 6–11.

Ross, W.G. 1979. The annual catch of Greenland (bowhead) whales in waters north of Canada 1719–1915: a preliminary compilation. Arctic 32: 91–121.

Richardson, W.J. (ed) 2006. Monitoring of industrial sounds, seals, and bowheads whales, near BP's Northstar Oil Development, Alaskan Beaufort Sea, 1999–2004. [Updated Comprehensive Report, April 2006.]

Serreze, M.C., M.M. Holland, and J. Stroeve. 2007. Perspectives on the Arctic's shrinking sea-ice cover. Science 315: 1533–1536.

Statistics Canada. 2009. 2006 Census release topics. http://www12.statcan.gc.ca/census-recensement/2006/rt-td/index-eng.cfm (accessed 29 December 2009).

Stewart, D.B., and W.L. Lockhart. 2005. An overview of the Hudson Bay marine ecosystem. Canadian Technical Report of Fisheries and Aquatic Sciences. 2586: vi + 487 p.

Stirling, I., N.J. Lunn, and J. Iacozza. 1999. Long-term trends in the population ecology of polar bears in western Hudson Bay in relation to climatic change. Arctic 52: 294–306.

Stirling, I., and C.L. Parkinson. 2006. Possible effects of climate warming on selected populations of polar bears (*Ursus maritimus*) in the Canadian Arctic. Arctic 59: 261–275.

Tucker, S., W.D. Bowen, and S.J. Iverson. 2007. Dimensions of diet segregation in grey seals *Halichoerus grypus* revealed through stable isotopes of carbon (delta C-13) and nitrogen (delta N-15). Marine Ecology-Progress Series 339: 271–282.

Tynan, C.T., and D.P. DeMaster. 1997. Observations and predictions for Arctic climate change: potential effects on marine mammals. Arctic 50: 308–322.

Welch, H.E., M.A. Bergmann, T.D. Siferd, K.A. Martin, M.F. Curtis, R.E. Crawford, R.J. Conover, and H. Hop. 1992. Energy flow through the marine ecosystem of the Lancaster Sound region, Arctic Canada. Arctic 45: 343–357.

Weilgart, L.S. 2007. The impacts of anthropogenic ocean noise on cetaceans and implications for management. Canadian Journal of Zoology 85: 1091–1116.

Witteveen, B.H., G.A.J. Worthy, K.M. Wynne, and J.D. Roth. 2009. Population structure of North Pacific humpback whales on their feeding grounds revealed by stable carbon and nitrogen stable isotope ratios. Marine Ecology-Progress Series 379: 299–310.

Index